The Management of Complex Projects

a relationship approach

Edited by

Stephen Pryke
Senior Lecturer in Construction and Project Management
Bartlett School of Graduate Studies
University College London

Hedley Smyth
Senior Lecturer in Construction and Project Management
Bartlett School of Graduate Studies
University College London

Editorial offices:
Blackwell Publishing Ltd, 9600 Garsington Road, Oxford OX4 2DQ, UK
Tel: +44 (0)1865 776868
Blackwell Publishing Inc., 350 Main Street, Malden, MA 02148-5020, USA
Tel: +1 781 388 8250
Blackwell Publishing Asia Pty Ltd, 550 Swanston Street, Carlton, Victoria 3053, Australia
Tel: +61 (0)3 8359 1011

First published 2006 by Blackwell Publishing Ltd

ISBN-13: 978-1-4051-2431-7
ISBN-10: 1-4051-2431-8

Library of Congress Cataloging-in-Publication Data:
The management of major projects : a relationship approach / edited by Stephen Pryke, Hedley Smyth.
p. cm.
Includes bibliographical references and index.
ISBN-13: 978-1-4051-2431-7 (pbk. : alk. paper)
ISBN-10: 1-4051-2431-8 (pbk. : alk. paper)
1. Building–Superintendence. 2. Construction industry–Customer services.
I. Pryke, Stephen. II. Smyth, Hedley.
TH438.M284 2006
690.068'4–dc22
2006008516

A catalogue record for this title is available from the British Library

Set in 10/12 Palatino by SNP Best-set Typesetter Ltd., Hong Kong
Printed and bound in India by Replika Press Pvt Ltd

The publisher's policy is to use permanent paper from mills that operate a sustainable forestry policy, and which has been manufactured from pulp processed using acid-free and elementary chlorine-free practices. Furthermore, the publisher ensures that the text paper and cover board used have met acceptable environmental accreditation standards.

For further information on Blackwell Publishing, visit our website:
www.blackwellpublishing.com/construction

Contents

Preface

Shortly after we joined forces on the staff at the Bartlett School, UCL, we quickly found a common interest in relationships and their importance in managing projects. This led to conversations and debate on the role of relationships in project success. We came to the conclusion that thinking on this had yet to be pulled together in a coherent way, so set our sights on jointly editing a book. We quickly identified a long list of respected individuals who we wanted to approach and from whom we wished to capture ideas.

The result is a book with a wide range of definitions and perspectives on relationships in a project environment. Inevitably there is also a wide range of styles and we have tried not edit out this 'light and shade'. We hope that our book will appeal to other academics, undergraduate and postgraduate students, and the 'thinking practitioner'. While we are happy for our readers to dip into individual chapters that appear most relevant, we also hope many will take the trouble to read the entire book to appreciate the very diverse range of knowledge available and research carried out. The information presented in our book has been located within a broad conceptual framework to assist the understanding of a relationship approach to the management of projects.

The book refers to construction for many of its examples and many of the world's largest and most complex projects involve construction or engineering of some sort. The ideas expounded are, for the most part, also applicable to a wide range of non-construction project environments and we hope that academics and practitioners from other sectors will join the debate.

We thank Blackwell Publishing for their enthusiasm and support; we thank the chapter authors for their valuable contributions and for sharing our enthusiasm. The support of the Bartlett School is also gratefully acknowledged.

S.D.P.
H.S.
London, UK

About the authors

Dr Stephen Pryke is a Senior Lecturer in Construction and Project Management, Bartlett School of Graduate Studies, University College London, and is Director of the Master's course Project and Enterprise Management.

Dr Hedley Smyth is a Senior Lecturer in Construction and Project Management, Bartlett School of Graduate Studies, University College London, and is Director of the Graduate Research Programme.

Professor Peter Morris is Head of the School of Construction and Project Management at the Bartlett School of Graduate Studies, University College London. He is Executive Director of INDECO, a consultancy specialising in the management of projects.

Dr Vanessa Urch Druskat is Associate Professor of Organizational Behavior, Whittemore School of Business and Economics, University of New Hampshire.

Paul Druskat is a Project Management Professional (PMP) who spent 18 years with IBM.

Professor Bernard Cova is Professor of Marketing at Euromed Marseilles Ecole de Management and at Bocconi University in Milan, and is Co-Founder and Animator of the European Network on Project Marketing and System Selling.

Professor Robert Salle is Professor in Business to Business Marketing at EM Lyon.

Dr Suzanne Wilkinson is Senior Lecturer in the Department of Civil and Environmental Engineering, University of Auckland.

Dr Mohan Kumaraswamy teaches and researches in Construction Engineering and Management at the Department of Civil Engineering, University of Hong Kong.

Dr Motiar Rahman is Senior Lecturer in the School of Technology at the University of Glamorgan.

Professor Martin Loosemore is Professor and Associate Dean (Research) in the Faculty of the Built Environment, University of New South Wales.

Professor Stuart Green is Professor of Construction Management, School of Construction Management and Engineering, University of Reading.

Professor Andrew Cox is Professor of Business Strategy and Procurement Management at The Birmingham Business School, the University of Birmingham. He is Chairman of Newpoint Consulting in London.

Dr Paul Ireland is a Research Fellow in Strategy and Procurement at the Birmingham Business School, University of Birmingham.

Graham Ive is a Senior Lecturer in Construction and Project Management, Bartlett School of Graduate Studies, University College London, and is Director of the Master's course Construction Economics and Management.

Dr Kai Rintala was formerly a Postdoctoral Research Fellow at the Department of Construction Economics and Management, University of Cape Town, and is currently consulting and carrying out research for KPMG in London.

Section I

Introduction

Focus

Managing projects is becoming one of the central areas of business management. Projects have been a mainstay of operations for many businesses, including the film industry, defence industries, oil and gas, and construction. These have provided insight and models for other sectors to use. Some industries have adopted projects as part of their core business delivery strategy, especially enterprises whose survival depends on research and development. In addition, projects have been increasingly used by many organisations to deliver activities that are not part of their core operations. This trend has increased over the past 25 years as enterprises have focused upon core activities and outsourced as much as possible.

Some projects are resourced in-house, while other projects have been outsourced: this is the transaction costs question in economics of make or buy. This book primarily focuses on projects that are contracted out for delivery by another organisation. The book has been written with the management of projects as the main focus, drawing on case material from a number of sectors, in particular from the infrastructure and construction industries.

The management of projects is becoming more sophisticated both conceptually and in application. This book focuses on beginning to develop one area that remains underdeveloped in current bodies of knowledge: a relationship approach.

Approaches to project management

Project management has gone through several stages of development, each adding complementary understanding to the bodies of knowledge:

(1) *Traditional* project management approach – techniques and tools for application (for example, Koskela, 1992, 2000; Turner, 1999; cf. Turner & Müller, 2003), which tend to have a production or assembly orientation focused upon efficiency.
(2) *Functional* management approach – strategic, 'front-end' management of projects (Morris, 1994, cf. 2004), for example programme and project strategies; and partnering (Egan, 1998) and supply chain management (Green & May, 2003); and other task-driven agendas that dovetail with the traditional approach, for example the waste elimination application of lean production (Koskela, 1992, 2000).
(3) *Information processing* approach – technocratic input–output model of managing projects (Winch, 2002).

These approaches have embraced many of the key human dimensions for managing projects, such as allocating resources to projects, monitoring performance and motivation. However, the dynamics of relationships from the conception of the project through to the completion of the project or programme of projects has yet to be fully articulated theoretically or practically. This book adopts a relationship approach, and therefore adds a fourth dimension:

(4) *Relationship* approach – project performance and client satisfaction, achieved through an understanding of the way in which a range of relationships between people, between people and firms, and between firms as project actors operate and can be managed.

Aims

The primary aim of this book is to provide frameworks for understanding relationships. We wish to initiate a comprehensive discourse to develop an agenda for managing relations, and hence improving performance on project delivery – adding value that matters to the client, thus satisfying their needs in a way that induces a profit. This aim arises out of exploring a range of theoretical approaches from a number of disciplines. **Chapter 1** will outline the conceptual and practical scope of a relationship approach. Subsequent chapters comprise contributions from a range of authors internationally recognised in their fields. These contributions will provide analysis, insights and case material in some key areas. **Section V** will draw together an agenda for others to develop and explore further in theory and practice.

The client–project coalition represents a nexus of relationships. These relationships, as well as relations with others such as stakeholders and the supply chain, are the substance for this book. Defining the term *relationship* will be explored later in **Chapter 1**, as definition arises out of and is located in context, by which is meant both the nature of the client and the nature of the project. Examination of relationships will help students, academics and practitioners to understand the changing nature of reforms emerging in the wake of successive calls for change by various clients and client groups in the project environment.

Project team networks of relationships are a function of mindsets, behaviour and competencies of individuals. This operates at the level of individuals. The strength of the relationships between organisations is heavily influenced, if not determined, by the internal relations within an organisation (Reichheld, 1996). Various types of relationship exist between individuals and firms; relationships between individuals are frequently embedded within employer and project organisations. As will be developed further in **Chapter 1**, structure and strategy embody relationships, so we take the view in this book that relationships occur at project level as well as between organisations. From this perspective the project team can occupy a position of tension, sometimes paradox. The links and relationships between the teams and their headquarters or senior management can be weak. The temporary or 'virtual' organisation of the project team typically spans organisational boundaries of respective firms. The strength of these intra-organisational relationships is heavily influenced by the relationships internal to the organisations. Therefore both internal and external relationships are for exploration in this book.

There has been an increasing emphasis upon process and competency development in management literature and in commerce generally over recent years. There is no sign of this abating. The efficiency of a range of relationships and the appropriateness of these relationships are significant in the effectiveness of the application of competencies and the ability of these competencies to deliver value to the client and to stakeholders.

Objectives

The objectives of the book are to define the parameters of a relationship approach, conceptually and practically in a framework, and then to develop detailed analysis of the parameters through the contributions in the book. The framework is not designed to provide a unified body of theory, the book embracing different theoretical and conceptual approaches. However, it is proposed to develop a framework of reference in order for a discourse to be developed and continue:

(1) Exploring interpersonal and inter-firm relationships at the project interface: client–project, team–contractor, stakeholders and supply chain relations.
(2) Examining different concepts for understanding and for the inception, development and management of relationships.
(3) Analysing the formation and development of relationships in ways that can aid project delivery for contractors and project success for clients.
(4) Highlighting some of the key issues that require development, theoretically, through applied research and in practice.

The first two objectives will be scoped in the editorial chapter and the introductions to each section of the book. The third objective will be covered primarily through the other individual chapters. The fourth objective will be reviewed in **Section V**. The content of the chapters combines conceptual *and* empirical data from the field.

The primary outcomes will be to:

- Examine the extent to which managing relationships induces successful outcomes for projects, from the stance of both the client and the contractor.
- Provide relevant information and insight for academia and practitioners.
- Set agendas for research and the management of projects for the future.

The objectives in the book are seen as complementary to the traditional, functional and information approaches cited at the start of **Chapter 1**.

The audience for the book is likely to be academic – specifically those engaged in research, those teaching and students learning at Masters level. The practitioner is a further important audience. In particular, those working in enterprises at the forefront of managing projects will derive insight, inspiration and practical solutions to real problems. Indeed, the success will ultimately be measured by the extent to which the body of knowledge develops and enterprises delivering projects derive competitive advantage from adopting the content in this and subsequent work. It is intended that the book will appeal to a multiplicity of project environments by project type, from the local to the global scale.

Delivering added value

Managing projects is about realising value through the delivery of a set of tasks, usually in a tight time frame, in an environment of uncertainty, and hence risk. The management of uncertainty and risk frequently absorbs a great deal of the effort. This management effort is as much about people as it is about the end product or service. People function out of relationships. Function can define relationships, but where relationships govern the function a more effective and usually more efficient outcome is the result.

The project goal is frequently perceived by many involved with the management of projects as 'getting the job done'. Such a focus draws upon a narrow project definition from the time of winning the project to handover. Learning about the client and the project, hence the value to be delivered, actually starts with the first contact with the client and many projects require total care packages well beyond the handover. This is an important statement in what has predominantly been a *task*-orientated business activity and recognises many recent developments in project management, including:

- Projects are created to solve an organisational or business problem, the client being located as the focal point of the supply chain at the point of project conception and inception.
- Strategy for a project is important in realising successful outcomes and hence comes prior to the project management in a traditional sense of applying techniques and tools.
- Where clients have several projects – a programme – as part of a portfolio of investment and activity, the client focus is arguably more important than tasks or project per se.

There are two types of client. Some clients view themselves in a human or social sense as the centre of the project or projects. They are 'hands on'. Other clients are more passive and distant or 'hands off'. Clients that are hands on tend towards the '80:20 rule' where they are far more concerned with the nature of the service quality, while the contractor is far more concerned with completing the project tasks (Leading Edge Management Consultants – see Pratt, 1999), hence the existence of a relationship gap in understanding and expectations. High-quality service experiences are predicated upon high-quality relationships.

When people from project backgrounds are asked to draw a diagram of the relations between the enterprises, they typically draw a diagram of the key players linked by lines representing legal contracts between the parties. This is a very partial view. It gives the impression that there were no prior relationships between the parties and does not reflect the density of the actual contacts in project realisation in terms of frequency of contact and closeness of relationship. In reality, relations start prior to the signing of a contract and these will have been important in establishing the relative position or stance of each organisation to the others. Figure I.I shows the client organisation, the black circles representing the key decision makers within the client organisation.

Figure I.II places the client into the corporate context for the project, that is, with the other principal parties and their key decision makers. For simplicity's sake, other key stakeholders within the supply chains have been excluded, although in practice a complex set of relationships will exist.

In practice the whole organisation is not directly involved in the project, only the key decision makers. The key decision makers in management terms are sometimes referred to collectively as the *decision-making unit* or DMU (see Smyth, 2000). In this context the key decision makers are the effective or executive project team. This is shown in Figure I.III.

Therefore we see a series of actual and potential personal relations that can be configured in a number of ways, which are embedded in their organisations yet are also embedded in a team to form the temporary multi-organisation (Cherns & Bryant, 1983) or project coalition (Winch, 2002). This diagram can be simplified to reflect these two organisational settings for the relationships (see Figure I.IV).

When a timeline is superimposed onto Figure I.IV, the process of developing relationships prior to the commencement of the project in the terms of the traditional description can be plotted. This is shown in Figure I.V.

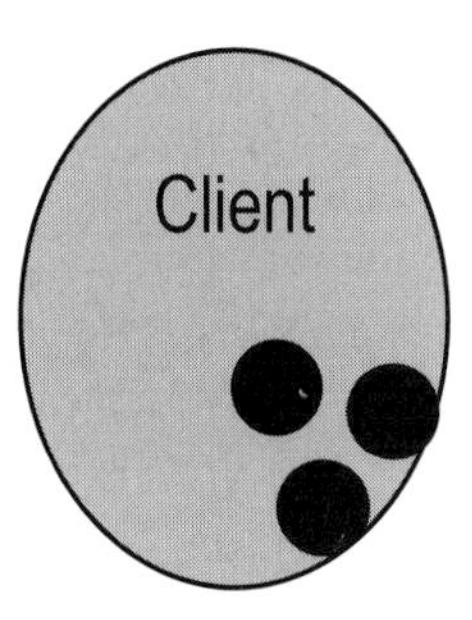

Fig. I.I The client and key project decision makers.

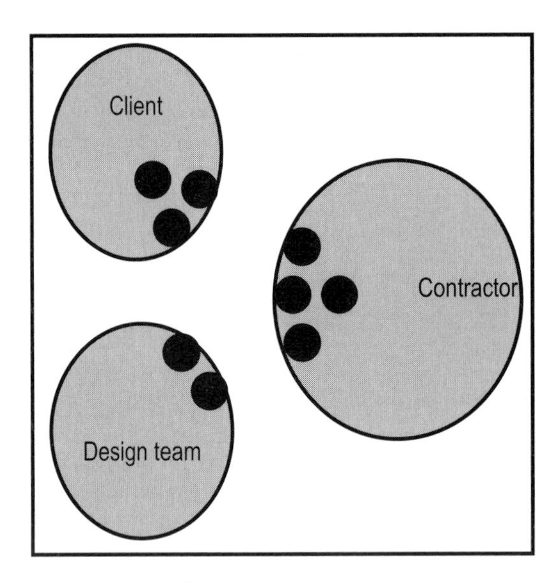

Fig. I.II The corporate environment for projects.

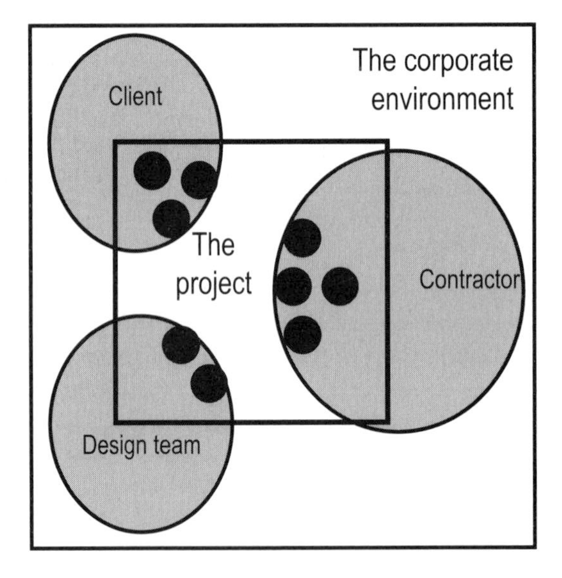

Fig. I.III The project DMU or team.

Figure I.V would be skewed for certain sorts of procurement; for example, a build–own–operate–transfer (BOOT) or design–build–finance–operate (DBFO) contract would have a long tail, where maintaining relations would extend for 25 to 40 years beyond the handover of the physical project component – that is, throughout the facilities management, maintenance and any other operational management period.

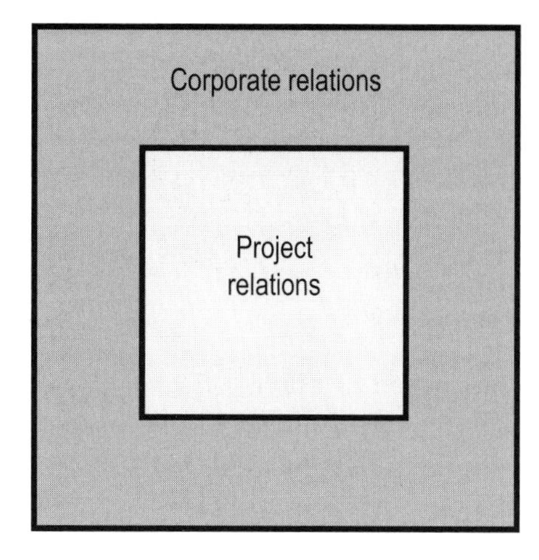

Fig. I.IV Corporate and project relationship context.

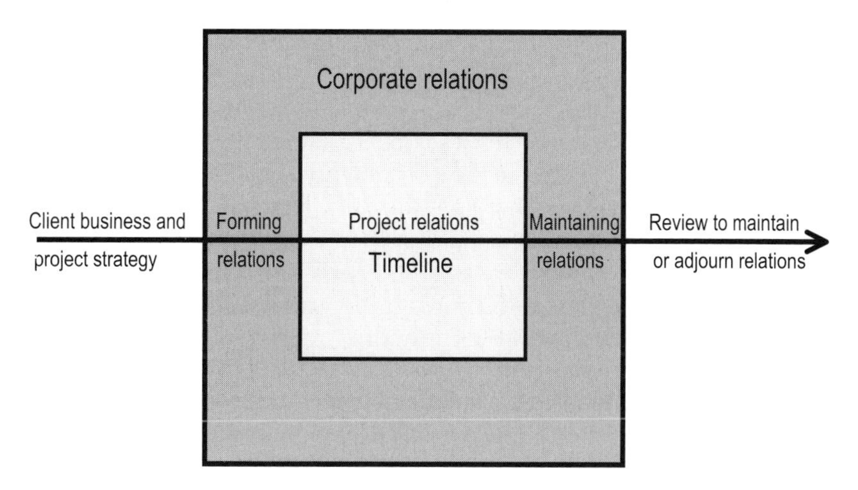

Fig. I.V Superimposed timeline relations.

Defining relationships

What is being asserted is that the people are certainly one key focus because the human relations between them are important in the effective and efficient delivery of a project or projects. The project begins to look different when the 'client' is viewed from a relationship perspective as a primary focus, rather than the project as a technical and functional 'product' or 'service'. The criteria for investment, mobilising project resources, managing the delivery and hence satisfactory completion all shift ground.

There are different definitions of relationships. Gummesson (2001) identified 30 tangible relationships, ranging from the micro-scale or *nano relationships* to the macro-scale of market and *mega relationships*. Wasserman & Faust say:

> *'A collection of ties of a specific kind among members of a group is called a relation. For example . . . the set of formal diplomatic ties maintained by pairs of nations in the world, are ties that define relations.' (1997, p.20)*

In this definition ties define the relation, and the ties are not always human or social. Yet ties are also the product of relations, which in part depend upon the tangibility of and scale at which relationships are formed and develop.

This is qualified in terms of interaction within a network of relationships. The IMP Group states:

> *'Relationship management involves analysis, investment in relationships and a clear view of the wider value that can be gained from each relationship and which extends beyond the straightforward features of the product that is exchanged.' (Ford et al., 2003, p.5)*

The view taken in this book is that relations are context specific. Relationships are negotiated in context, mostly socially but frequently the ties are legally binding too. There are different contexts for relationships, which operate at different levels:

- Business-to-business or organisation-to-organisation
- Organisation-to-individual representing the business: market and other societal relations (see Gummesson, 2001)
- Individual-to-individual: personal and social relations

Personal relations can be characterised as:

- Authority: management and leadership
- Task related: function and role
- Acquaintance: social obligation
- Friendship: social bonding and reciprocation
- Sense of identity: who you are (not what you do), such as inheritance and societal recognition – through ownership in business, for example

Such personal relations can be developed and sustained outside the boundaries of the enterprise, so that when a person changes jobs, the relationships with others from the former role migrate to the new one. This has been a common, even traditional basis of operation for business development managers in construction.

Organisational relations can be characterised as:

- Individual or personal: the individual represents the organisation.
- Systematic or procedural: personal relations have been enshrined into an approach or systematic way of proceeding in order that the essence of a relationship is replicated at a general level in the future through social or legal obligation (cf. Wenger, 1998).
- Strategies and culture help guide the context in which systems operate, guiding the thinking and behaviour of individuals in order that relations through individuals and systems are aligned.
- Structure of an organisation both reflects relations and governs relations through hierarchy, function and proximity.

Structure, strategies and systems are all the product of relations and influence the form relations will take too. Relationships become embedded within organisations. Project relations occupy a somewhat paradoxical position. The ties can be weak between the project teams and their senior management in the core business or at the main office, especially on site specific projects. The temporary organisations or project coalitions formed will typically span organisational boundaries and close ties can build up, especially if the project is long and complex. Separate cultural and relationship norms can be established. This can be positive for efficient project management and is a theme running through both the Latham (1994) and Egan (1998) reports into the UK construction industries. It can also be effective for serving the client. It may not be in the best interests of the separate organisations, especially in the context of adding service value and serving clients consistently across projects. This will be considered in **Chapter 1**.

Developing a relationship approach is rich and rewarding. The scope is enormous as there is no single theoretical approach for managing projects, hence the application of the same relationship tools will be different in each context. And the choice of tools is varied. This diverse palette means there is tremendous fruitfulness to be derived from applying any single relationship approach and this will be expanded in the next chapter.

A relationship approach offers enormous opportunities for enterprises to invest in the means to be more efficient and more effective, and to differentiate their services, thus achieving competitive advantage or a niche position. This therefore feeds back to enterprise strategy and hence investment – managing the market through distinctive service delivery. This is in contrast to a common approach where size is distinctive and competition is addressed through a zero sum game of undifferentiated services. A relationship approach is not an easy option and there are investments and costs incurred, so it is inevitable that different organisations will concentrate their resources in different areas. This in itself is a potential source of competitive advantage in particular market segments.

On the client side, deciding to undertake a project is a strategic decision. For the client the sources of risk and uncertainty are several. Physical or environmental conditions are one source of risk. External factors of people's own making that are beyond the scope of the project are another source. How people behave is the third factor and probably the main source of uncertainty

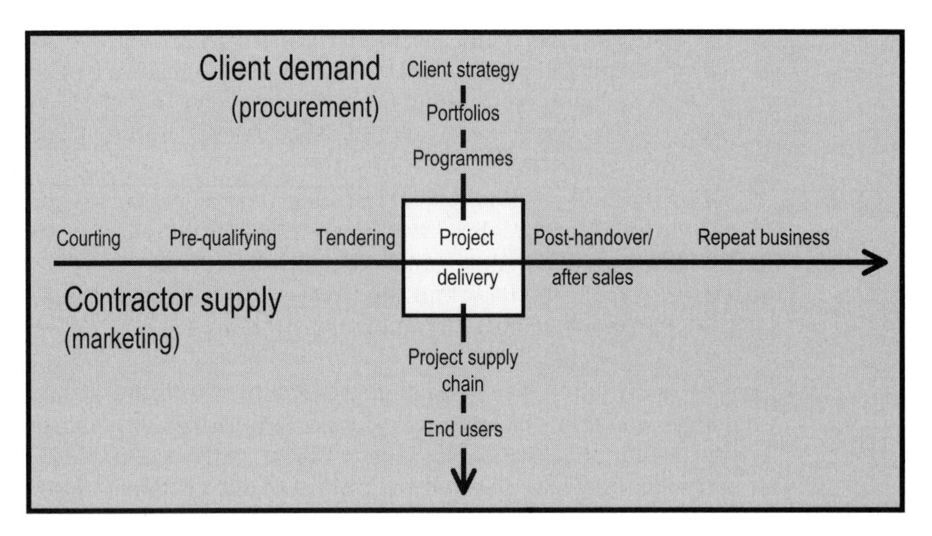

Fig. I.VI Traditional approach to relationships.

– the human risk factors (see **Chapter 1** and Thevendran & Mawdesley, 2004). It is a neglected area in the various bodies of knowledge (BOKs) for project management (PMI, 2000; APM, 2000; IPMA – Caupin *et al.*, 1998). Relations are considered neither systematically, nor always directly. When they are considered they tend to be treated largely in a functional way, which is restrictive and at best rather mechanical.

Project relationships have been divergent, shown as vertical and horizontal lines in Figure I.VI. The vertical relations show the procurement dimension of relationships expressed through client demand. The horizontal relations unpack the relations shown in the timeline (as seen in Figure I.V) in more detail and reflect a marketing dimension to relationships. One of the reasons for divergence is the domination of transaction cost management from the contractor perspective; hence, many market issues are handled through organisational structure (Smyth, 2005a). Another reason is that contractors have emulated the clients, following procurement dominated models by passing the procurement approach along the supply chain rather than responding with its counterpart, marketing (Smyth, 2000, 2005b). Procurement at its simplest is about buying and marketing is about selling; however, procurement in its fullness embodies complex strategic decisions about managing projects, and likewise marketing too embodies strategic decisions that can impact upon projects in subtle and complex ways. The consequence of failing to embody both aspects has been that convergence only occurs at the crossover point – the traditional or historic definition of the project.

What is ideally needed to develop effective relationships is convergence or congruence between these dimensions. It is complex, hence not so easy to depict, particularly because emphasis is context dependent – client and project specific – as well as shifting over the life of the project and longevity of the relationships. The most intensive stage of congruence is during project delivery. Delivering a project is a social activity. People are at the centre of realising the goals, completing the tasks. People add the value. In developing a relationship approach, the primary aim of this book is to develop an agenda for managing relations, hence improving performance on project delivery. A relationship approach cannot be based solely upon theoretical concepts for the richness of understanding will draw upon a full range of models and tools, for example drawn from *relational contracting, relationship management, emotional intelligence* and *social network analysis.* Nor will a relationship approach create a theory for managing projects, a relationship approach complementing and adding value to other approaches.

It is recognised that some theories and concepts are in conflict, even contradict each other. It is also recognised that some theories and concepts have proved and will prove to be more useful in practice and thus an appropriate set of theories, tools and techniques will emerge – the palette from which practitioners will draw. Therefore care is needed in adopting tools and techniques. The outworking of that process begins with analytical rigor and develops through discourse and negotiation. This book sets out to make a contribution and to be an important instrument in this process. The next chapter will develop some of the ideas that have been introduced and will develop a framework from which commentators and practitioners alike may develop their understanding in order to interpret and evaluate what is applicable in any context for the management of projects.

Outline of the chapters

Chapter 1 provides an overview for understanding relationships. The conceptual and practical *scope* of a relationship approach is outlined. Types and different views of relationships are introduced. A relationship framework is set out and then the way in which relationships can be developed within the framework is examined in terms of adding value to the client and securing a profitable return for the contractor and professional service provider.

Each section has an introduction to outline in more detail the focus, which is distinct yet not exclusive of the other sections. **Section II** examines *mindsets, behaviour* and *competencies*. Competencies, arising from a resource-based view in the 1950s (Penrose, 1980), have become a major emphasis in recent years as enterprise strategy focuses more upon process than structure. The introduction and development of competencies is dependent upon the attitudes and behaviours evident in organisations, even though the competencies may seek to change these in positive ways.

In **Chapter 2** Morris examines *learning*. He takes a strategic review of the learning, asking, *How do we learn to manage projects better*? His primary focus is the interface between the bodies of knowledge (BOK) and the firm, where learning has yet to fully take root to influence the way in which the management of projects takes place. Learning is a social process and the BOKs are socially constructed. A secondary focus is between learning in the firm and project performance, as this takes place through social relationships.

This is an important scene-setting chapter, building off the scoping of the relationship approach in **Chapter 1**, in order that developments in this area can be articulated, understood and absorbed and so inform future developments in managing projects.

Chapter 3 by Druskat and Druskat introduces the competency concept *emotional intelligence*. They provide a review of the concept of emotional intelligence that has been pioneered by Goleman (1996). They develop the concept within the field of organisational behaviour and go on to apply it within a project context. Given the historical prevalence of adversarial behaviour in many project environments and current drives to improve relations and performance, emotional intelligence is a key concept in developing responsible and performance-enhancing social relationships within a project environment context.

Trust in teams delivering projects has become a byword for collaborative working, yet it is frequently seen as either an intangible out of reach from a management point of view or as being synonymous with 'open communication'. In **Chapter 4**, *Measuring, developing and managing trust in relationships*, Smyth argues that trust can be identified and measured, developed and managed. To achieve this, trust must be understood in depth. Organisations must also commit resources to actively invest, to develop trust through culture and soft systems.

The argument is supported by the outlining of a framework for understanding trust, covering a range of elements of trust and related concepts and locating these operationally in the organisation of the firm, project team and marketplace. Case study material in private finance initiative (PFI)/public–private partnership (PPP) and commercial development markets is used to support the argument, demonstrating the ability to measure it and to report on the dynamics of trust development or erosion and its fragility under current practices.

Section III has *client, design team* and *contractor relationships* as its primary focus. Relationships are established at first contact with the client. They are developed through the sales or business development stages, develop continuously through bidding and post-tender negotiations, the delivery of the project and after project completion and final account, and potentially continue into the next project if there are repeat business opportunities.

Chapter 5 by Cova and Salle has a marketing and business development approach. The authors contribute to the literature on *Communications and stakeholders*. This is an important chapter, given a growing focus upon the front end of projects – that is, the management of projects rather than project management (Morris, 1994). The emphasis has primarily been placed on the way in which clients develop strategies for projects, locating them in their programmes of expenditure and portfolios of investment. Projects are thus a means to solve a business or organisational problem. The contractor's response to solving the problem concerns their management of projects for the client and the totality of their own activities. Most of the analysis of stakeholders has been located within this context, information and communication flows providing the 'raw materials' and initiative respectively for management (for example, see Winch, 2002, 2004).

Cova and Salle approach stakeholder analysis from the viewpoint of marketing rather than with procurement as a starting point. The marketing approach has the *milieu* as its focus (Cova, 1996), whereby the recognition of a spatial network of relationships is important for contractors to proactively develop their marketing strategy and implement their sales efforts to shape and secure projects. Contractors are therefore locating themselves further up the decision-making process, close to the front end from the client perspective. The chapter therefore places particular emphasis on the nature of the sales effort within the milieu, recognising the formal sales effort conducted through business-to-business relationships and the prior and parallel social relationships that develop, which permit contractors to secure an earlier entry to the market for a project as well as to understand client requirements in greater depth. They provide a rich analysis using conceptual tools of rituals, reciprocity, mutual debt and friendships in their exploration of project stakeholders, drawing upon case study material from the Seine-Maritime *département* in France.

In **Chapter 6** Wilkinson takes a fresh look at *continuity of service*, building on continuity achieved through *account handlers* and *relay teams* (Smyth, 2000). An account handler is one person responsible from cradle to grave for the project and client relations. Others have examined this within relationship marketing and management paradigms, using the term *key account manager* (KAM) as the preferred form. Contractors have traditionally used a *relay team* approach (Smyth, 2000), different teams managing the process from courting during business development, tendering, and frequently delivering the project on site. The problem with the relay team approach is that efforts to minimise transaction costs result in teams failing to pass the baton, or more realistically not carrying one, and there being no real exchange of information, including promises made being transferred between teams. Continuity of service can only be provided where there is a smooth handover with the 'baton' of information and understanding passed at each leg and the client knowing that this is about to happen and being able to watch the effective transfer.

Wilkinson uses this as a springboard to examine an old chestnut: the most appropriate organisation to manage the interests of the client, especially from the viewpoint of service continuity. In other words, taking the client handling models within a firm, she explores how these models apply across organisations. She examines the traditional client's representative, architect or engineer, considers the independent project manager, and considers the contractor too. She applies the account handler and relay team models to these options, examining under what circumstances the models are most suited to client interests.

Applying a relationship approach provides new management insights, drawing upon the results of recently completed research and additional analysis of new research data. Wilkinson develops an analysis of *roles* to be performed at each stage of the project life cycle, which highlights the need for managers in respective organisations handling client interests and service continuity to have or develop particular competencies to improve project performance.

Chapter 7 by Kumaraswamy and Rahman concerns *Applying teamworking models to projects*. The chapter explores the structuring of relationships conceptually, using *teambuilding by objectives* (TBO) and expands upon recently proposed models of *relationally integrated supply chains* (RISCs), *relationally integrated project teams* (RIPTs) and *relationally integrated value networks* (RIVANs). Their aim is to enhance and integrate approaches for better teamworking and improved project performance.

These concepts for integration are then related and applied to case study material involving structuring relationships. The case study concerns a large infrastructure project where partnering has induced favourable outcomes for *relational contracting*. The conceptual work and the application, carried out as a compare and contrast exercise between the two, raises some interesting questions theoretically. While the relationship approach of Kumaraswamy and Rahman is anchored in *relational contracting*, which has its origins in the transactional economics of Williamson (1985), it is also the case that the way they have developed teamworking goes beyond that, dovetailing into and overlapping with *relationship management*, a paradigm that arises out of marketing (Gummesson, 2001) and organisational management (Grönroos, 2000) rather than economics.

Having developed their model of teamworking, they then go on to analyse the implications for value. Both consider their teamworking models alongside other tools, such as Belbin analysis and psychometric testing. The relationship of teamworking to value is also addressed in this chapter.

Loosemore in **Chapter 8** takes a fresh look at the human dimensions of managing project risk, arguing that this is a more important focus than reference to 'objective' risks. While the construction industry has developed sound technical skills in dealing with risk, its ability to deal with the human experience and perception of construction risk is less well developed. His analysis uses Multiplex as a case study, which at the time of editing has strategically highlighted this point in relation to the Wembley Stadium project in the UK. Within projects, environmental degradation as well as wastefulness and poor safety performance have led to irrational responses among external project stakeholders, responses which the industry has largely failed to understand, hence has not addressed. Risk management practices in the construction industry must deal with risk in its full social and cultural complexity.

Loosemore develops the analysis, arguing that it is not merely people's perceptions of risk that need to be understood; their response is also conditioned by the extent to which they feel able to control or influence outcomes. This requires companies involved in construction projects to allocate greater resources to community consultation. Different stakeholders have a legitimate role to play in the development process. In order to understand perceptions, resources are used to value the inputs of stakeholders and provide and develop ways to respond, so that people know they have been heard and understood. Perceived needs are addressed where possible and reasonable, and actions taken.

Many issues arising from projects, such as marketing (**Chapter 5** by Cova & Salle) and risk in relation to stakeholders (**Chapter 8** by Loosemore), cannot be contained within organisational boundaries or even within project teams, as **Section IV** specifically develops within the context of *relationships* across *project clusters* and *supply chains*. Transaction cost theory addresses whether organisations should 'make or buy'. Clients clearly tend to buy. So, projects are typically delivered via the 'buy' option as design and cost control, contracting and subcontracting give rise to a plethora of organisations within a project cluster. The decision not to 'make' has the consequence of loss of management control of many decisions and processes. Partnering, supply chain management and other tools provide ways to increase management control somewhat across organisational boundaries. This section analyses some of the ways in which efforts have been made to increase control indirectly and the role of relationships in facilitating and constraining this process.

In **Chapter 9** on *Projects as networks of relationships*, Pryke takes the view that the construction industry is in a state of transition as it tries to deal with the unfamiliar business environment imposed by post-Latham relational contracting strategies. The move away from a contractually focused approach to project governance has forced construction industry project actors to adopt supply chain management, among other strategies, to prevent deteriorating project performance over multiple-project, strategic partnering-related timescales.

The non-linear, complex, iterative and interactive project environment typical of construction projects is difficult to analyse using traditional task dependence, structural analysis and process-mapping-based approaches. Pryke argues that the solution is to examine the contractual relationships, financial incentives and information exchange networks associated with the principal functions of the project coalition. In this way a comprehensive understanding of the systems that comprise the construction project is possible and it becomes feasible to begin to articulate a social network theory of project coalition, activity and effectiveness. A summary of the findings of current research in this area, involving a number of international construction projects, is provided.

In **Chapter 10** Green places relational contracting and the associated applications and systems in a context of management fashions. He argues that the industry has been coerced into adopting fashionable management strategies and describes the manner in which the theory supporting relational contracting is frustrated in its application through the pursuit, by industry actors, of standard customs or recipes. This makes the resultant practices of relatively little value and at odds with the original concepts supporting relational contracting. Green adopts a fairly polemical position in relation to what he refers to as the 'machine metaphor' for construction organisations and posits that the industry is engaged in a drama involving the competitive adoption of new and innovative management practices, including most notably supply chain management.

Green demonstrates the way in which supply chains managed in practice are frequently at odds with the expectations and mental models people hold. He highlights that there is a need to take social relations more seriously if we are to generate improvements in responsible and effective ways.

Cox and Ireland, on the other hand, take a more cautionary position in **Chapter 11**. They examine *Theories and tools in project procurement*, arguing that concepts of *relational contracting* and *relationship management* can only successfully be applied to a relatively small portion of the market: essentially large and complex projects. They are critical of the claims of the literature in other disciplines and the proclamations of many commentators in relation to project working. This is a timely and important message, providing clarity between on the one hand pursuing espoused argument and rhetoric of what managers of projects should and must do, and on the other, what can be done within the limits of buying rather than making, hence in a market context. Precisely where the limits lie has yet to be established, Cox and Ireland erring upon the side of caution.

Chapter 12 by Ive and Rintala picks up the transaction cost analysis theme and finds that the *Economics of relationships* may be restricted in many circumstances yet positive relationships are to be found within BOOT-type, PFI projects in the UK market. They develop their analysis drawing upon theory and teaching hospital case studies. They argue that while the market *per se* carries structural forces that impose constraints, and hence limit options, it is also the case that structuring projects in particular ways has social implications for managing projects and successful performance. While Ive and Rintala draw upon large and complex project cases, their findings may have wider implications for projects of varying size and complexity in the long run.

Section V draws together the findings from the chapters, combining these with insights from the editors, as a means to set out some parameters for further discourse and an agenda for future research for a *relationship approach* to managing projects.

Summary

Section I provides a focus for the book, setting out the context of a *relationship approach* within the field of managing projects. The aims and objectives for the book have been provided and a start has been made on the complex task of defining relationships. This provides the basis for beginning to explore in greater depth whether relationships are a key means to directly and indirectly add value to projects, especially through the service experience. In effect the interest is in who benefits and whether an improved service experience may lead to competitive advantage, and hence, more repeat and referred project business.

References

Association for Project Management (2000) *Project Management Body of Knowledge* (ed. M. Dixon). APM, High Wycombe.

Caupin, G., Knopfel, H., Morris, P. W. G., Motzel, E. & Pannenbacker, O. (1998) *ICB IPMA Competence Baseline*. International Project Management Association, Zurich.

Cherns, A. B. & Bryant, D. T. (1983) Studying the client's role in project management. *Construction Management and Economics*, **1**, 177–84.

Cova, B. (1996) Construction marketing in France: from reaction to anticipation. *Proceedings of the 1st National Construction Marketing Conference*, July, Centre for Construction Marketing in association with CIMCIG, Oxford Brookes University, Oxford.

Egan, J. (1998) *Rethinking Construction*. HMSO, London.

Ford, D., Gadde, L-E., Håkansson, H. & Snehota, I. (2003) *Managing Business Relationships*. Wiley, Chichester.

Goleman, D. (1996) *Emotional Intelligence*. Bloomsbury, London.

Green, S. D. & May, S. C. (2003) Re-engineering construction: going against the grain. *Building Research & Information*, **31** (2), 97–106.

Grönroos, C. (2000) *Service Management and Marketing*. John Wiley and Sons, London.

Gummesson, E. (2001) *Total Relationship Marketing*. Butterworth-Heinemann, Oxford.

Koskela, L. (1992) *Application of the New Production Philosophy to Construction*. Technical Report 72, Center for Integrated Facility Management, Stanford University, Stanford, CA.

Koskela, L. (2000) *An Exploration towards a Production Theory and its Application to Construction*. Report 408, VTT, Espoo.

Latham, M. (1994), *Constructing the Team: Joint Review of Procurement and Contractual Arrangements in the United Kingdom Construction Industry*. HMSO, London.

Morris, P. W. G. (1994) *The Management of Projects*. Thomas Telford, London.

Morris, P. W. G. & Pinto, J. K. (eds) (2004) *The Wiley Guide to Managing Projects*. John Wiley & Sons, New York.

Penrose, E. (1980) *The Theory of the Growth of the Firm*. Blackwell, Oxford.

Pratt, J. (1999) Understanding clients: the Egan imperative. *Proceedings of the 4th National Construction Marketing Conference*, July, Centre for Construction Marketing in association with CIMCIG, Oxford Brookes University, Oxford.

Project Management Institute (2000), *A Guide to the Project Management Body of Knowledge (PMBOK® Guide)*. PMI, Newtown Square, PA.

Reichheld, F. F. (1996) *The Loyalty Effect*. Bain & Co., Harvard Business School Press, Boston, MA.

Smyth, H. J. (2000) *Marketing and Selling Construction Services*. Blackwell Science, Oxford.

Smyth, H. J. (2005a) Competition. In: D. Lowe & P. Fenn (eds) *Commercial Management of Projects: Defining the Discipline*. Blackwell Publishing, Oxford.

Smyth, H. J. (2005b) Procurement push and marketing pull in supply chain management: the conceptual contribution of relationship marketing as a driver in project financial performance. Special issue on Commercial Management of Complex Projects (eds D. Lowe & R. Leiringer). *Journal of Financial Management of Property and Construction*, **10** (1), 33–44.

Thevendran, V. & Mawdesley, M. J. (2004) Perception of human risk factors in construction projects: an exploratory study. *International Journal of Project Management*, **22**, 131–7.

Turner J. R. (1999) *The Handbook of Project-based Management: Improving the Processes for Achieving Strategic Objectives* (2nd edn). McGraw-Hill, Maidenhead.

Turner, J. R. & Müller, R. (2003) On the nature of the project as a temporary organization. *International Journal of Project Management*, **21** (1), 1–8.

Wasserman, S. & Faust, K. (1997) *Social Network Analysis: Methods and Applications.* Cambridge University Press, Cambridge, UK.

Wenger, E. (1998) *Communities of Practice: Learning, Meaning, and Identity.* Cambridge University Press, Cambridge, UK.

Williamson, O. E. (1985) *The Economic Institutions of Capitalism.* Free Press, New York.

Winch, G. M. (2002) *Managing the Construction Project.* Blackwell, Oxford.

Winch, G. M. (2004) Managing project stakeholders. In: P. W. G. Morris & J. K. Pinto (eds) *The Wiley Guide to Managing Projects.* John Wiley & Sons, New York.

1 Scoping a relationship approach to the management of complex projects in theory and practice

Stephen Pryke and Hedley Smyth

This book provides a framework for understanding relationships. The aim is to improve performance on project delivery by adding value that matters to the client, hence satisfying their needs, in a way that induces a profit for the provider. In order to commence the discussion, the conceptual and practical scope of a relationship approach is outlined in this first chapter.

The management of projects continues to develop. As management strategies and techniques develop they become more sophisticated. Each conceptual approach makes its own contribution, yet no single approach has offered a unified theory, the integration of the approaches providing the nearest thing to completeness. The conceptual approaches offering guidance have been summed up as:

(1) *Traditional* project management approach – techniques and tools for application
(2) *Functional* management approach – strategic, 'front end' management of projects, supply chain management and other task-driven agendas
(3) *Information processing* approach – technocratic input–output model of managing of projects
(4) *Relationship* approach – a paradigm of project performance and client satisfaction, achieved through an understanding of the way in which relationships between people, between people and firms, and between firms as project actors can be managed socially

Hence a relationship approach is an emergent social paradigm for adding value – recognising how people work together has, arguably, a significant and increasingly prominent effect on successful project outcomes.

Deciding to undertake a project is a strategic decision for the client organisation that sets a whole cluster of issues in train. Implementing the project then becomes a series of tactical issues for the organisation. These tactical issues are translated into a project strategy by the project team. Implementing the project strategy then becomes the primary task. It is at this level that traditional project management is located. A number of tools and techniques were therefore developed to help manage the project.

These methods of analysis and graphical presentation comprise *task dependency* analysis (for example, Higgin & Jessop, 1965), *structural analysis* (see Masterman, 2001; Davis & Newstrom, 1989) and *process mapping*

(Tavistock Institute 1966; Curtis, 1989). Task dependency analysis focuses on the time-related implications of dealing with an interdependent group of work packages; Critical Path Analysis exemplifies this approach. Structural analysis refers to the representation of formal organisational relationships, typically line management authority relationships, in a hierarchical or tree arrangement. Process mapping is a slightly more eclectic and conceptual approach involving flow charts and cognitive mapping, as well as linear responsibility analysis. The latter is, perhaps, a means of adding some indications of roles and relationships to what otherwise would be a system representation without actor role information. Winch and Carr (2000) proposed a business approach to process protocols, which models anticipated flows of information between actors in the construction coalition; for a detailed appraisal of these methods, refer to Pryke (2004a).

Articulating the 'gaps' between the programmes and software depends on understanding the functional roles of the tools. The way in which these are set up is dependent upon the project strategy. In the last decade of the previous millennium, a new set of tactical tools were developed for production and began to find their way into projects and construction (Koskela, 1992). Tactical issues, particularly around the procurement part of the production process, were being elevated to strategic level. Although having a new angle and containing a strong functional content, this approach is essentially traditionally based, extending the remit of the traditional approach beyond *scientific management* and *organisational management* (OM) techniques. Included in these procurement drivers are just-in-time management, lean and agile production and supply chain management, which are sometimes labelled 're-engineering', a term that is rather telling. Integration of these tools and techniques requires articulation of functional management.

The functional project management approach therefore considers the systems needed for managing the implementation process, which in turn requires a focus on 'front-end strategy'. Morris (1994) has re-established the importance of the strategic input and building robust systems around this.

Systems provide feedback loops and are hence self-regulatory. The strategy ensures the aims, objectives and underlying assumptions are sound. Success is therefore dependent upon the quality of information input. If everything else is working well then the inputs determine the quality of output. Poor information put in and poor management of the information will lead to problems, hence an information processing approach to project management has arisen (Winch, 2000).

The information processing approach of Winch (2000) runs in parallel with the traditional approach, which is production-based in essence but places more emphasis on wider issues, including the front-end, than the traditional approach. While these are alternatives, they are not substitutes, and so should be embraced. Our proposition is that there could or should be a dovetailing of approaches, not that there needs to be a choice between approaches. Therefore each approach needs to update its 'toolkit' to meet emergent conditions. What this book is proposing is that one key has been neglected – a relationship approach, which constitutes an approach in its own right.

People are remarkable in being able to create complex tools and systems and handle complex information. People are wonderfully creative. However, people also make mistakes, as well as behaving in difficult and unpredictable ways. Behaviour is the result of character traits, such as integrity and honesty, and emotions, such as feelings of satisfaction or disappointment, joy or anger. Behaviour in a project setting is seen through interaction – a social activity. Interaction that is more than a brief encounter, or that is long lived, is a *relationship*.

Projects are initiated, designed, managed, constructed, maintained and serviced by networks of people. These people communicate in pursuit of their project team roles and through these communications establish a sense of mutual understanding about terminology, values and priorities. These networks provide the connections through which 'communities of practice' (Wenger, 1998) are enabled and supported. These project-related interactions also enable the formation of friendships and social ties. The project network is, of necessity, continually reconfiguring to deal with the multitude of complex activities involved in executing a project. These networks enable special interest clusters to form quickly, perform specific multi-disciplinary tasks, link to other clusters and disperse for the purpose of allowing other, different clusters to form.

Firms operate within a network of other firms (Nohria & Eccles, 1992) and institutions. Some of the firms within the network constitute a supply chain for other firms that interface with the client organisation. Firms also become collaborators or competitors. Some firms and organisations are stakeholders within a network (see **Chapter 8** by Loosemore). Institutions provide a regulatory framework within which the network functions. Linking these firms are the interpersonal networks associated with information and resource exchange involved with the execution of projects.

Relationships do not, therefore, exist in a vacuum, but in specific contexts. In a corporate and in a project environment there are *human, organisational* and *information systems* in and through which personal relations are embedded and articulated. It is this combination that constitutes the relationships between organisations. As we have seen, there are also different types of relationship, which operate at several levels:

- Business-to-business or organisation-to-organisation
- Organisation-to-individual representing the business: market and other societal relations
- Individual-to-individual: personal and social relations

And so we have two *contextual dimensions* in which relations are conducted:

(1) Structural context according to the type of relationship in its organisational and legal setting
(2) Operational context according to the system (and culture) in which relationships operate

The structural and operational context will affect the way in which relations are conducted, often overriding or dominating the character, behaviours and dynamics between the individuals at the personal level. We have two *social dimensions*, which affect the way in which relations are conducted:

(1) Position in the structural and operational context – power, status, influence
(2) Disposition of the individuals in relation to the context, hence roles they are performing – character and personality, attitudes and behaviour

The contextual and social dimensions combine. How they combine is mediated, negotiated and synthesised through relationships – people working together to engage structure and operations with people, and people working with each other. The dynamics of this outworking depend upon the types of relationship. Insights from another academic context help us develop our thinking about managing projects. Gummesson (2001) refers to 30 different relationship types arranged into four categories (see Table 1.1):

- Classic market relationships
- Special market relationships
- Mega relationships
- Nano relationships

While Gummesson is ultimately focused on the supplier–customer dimension, the focus above also concerns the internal outworking for a relationship approach. While these have a strong external focus – the context – and Gummesson provides an internal focus via nano relationships, it is important to unpack further internal relationships that do not directly impact on the external relations. We can also produce a list, which, like the Gummesson one, does not claim to be inclusive yet poses a challenge (see Table 1.2).

Reichheld (1996) has shown that the strength of the external relations, hence customer satisfaction and loyalty, is only as strong as the strength of the internal relationships. Indeed, Reichheld (2003) claims that this aspect is sufficiently important that the key measure of the growth potential for an enterprise is closely related to what its customers tell others about it today. Customer experience of internal relationships that translates into product quality and service delivery is key and the customer ideally becomes the 'marketing department', which is simply to say that advocacy and referral markets are central to growth. This is particularly important in a project context, where each project is unique and thus repeat business by project type or client is less than in many other markets. Clients that regularly procure buildings are unlikely to constitute more than 25% of the total construction project market and programmes constitute around 10% of the entire industry output (Blismas, 2001; see **Chapter 11** by Cox & Ireland).

It is evident that the relationship permutations are extensive and each situation will be unique, yet generalisations can be made to help under-

Table 1.1 Market relationships (30Rs) (Gummesson, 2001).

Classic market relationships
R1. The dyad (supplier-customer)
R2. The triad (supplier-customer-competitor)
R3. The network
Special market relationships
R4. Relationships via full-time and part-time marketers (everyone is a part-time marketer)
R5. The service encounter
R6. The many-headed customer and many-headed supplier
R7. The relationship to the customer's customer
R8. Close versus distant relations
R9. The dissatisfied customer
R10. The monopoly relationship
R11. The customer as 'member'
R12. E-relationships
R13. Parasocial relationships (relations to brands and objects)
R14. The non-commercial relationship
R15. The green relationship
R16. The law-based relationship
R17. The criminal network
Mega relationships
R18. Personal and social networks
R19. Mega marketing (the real customer is not always in the marketplace)
R20. Alliances that change market mechanisms
R21. The knowledge relationship
R22. Mega alliances that change the market structure
R23. The mass media relationship
Nano relationships
R24. The internal market
R25. The internal customer relationship
R26. Quality and customer orientation
R27. Internal marketing
R28. The matrix organisation
R29. The relationship to external providers of marketing services
R30. The owner and financier relationship.

standing and the management of effective outcomes. This induces and creates the quality of relationships in a project environment.

The quality of relationships is a key element in the success of a project. The quality may be the product of a range of factors and therefore a consequence of a whole series of dynamic issues. A project team is merely the recipient of those relationships and how they develop, both within the project team and with those who are externally feeding into the project. However, relationships are also *managed*. Conduct can be managed to change and influence relations. Both attitudes and behaviour can be changed to improve relationship performance. Relationships can be managed and will in turn affect project performance. The management of relationships is a competence, perhaps a

Table 1.2 Developing internal relationships.

Internal structural relationships
R31. Line management relations
R32. Mentor relations (cf. R28)
R33. Peer relations
R34. System relationships and communication
Internal functional relationships
R35. HR relations
R36. Immediate job function relations
R37. Other departmental relations
R38. Interdepartmental relations
R39. Circumstantial relations (proximity, friendship networks, serendipity contacts)
Strategic relationships
R40. Formal relationships (strategy development and meetings)
R41. Career and organisational politics
R42. Embedded relationships
R43. Cultural relations
R44. Vision and mission relationships
R45. Strategy identification
R44. Strategy implementation-tactical response
Tactical relationships
R45. Heroes, champions and role models
R46. Nurturers (father/mother-son/daughter relations)
R47. Supporters, members and fan clubs
R47. Surrogate relationships (response to implementation gaps and lack of support)
R46. Relationships of dissent and blame
R47. Jezebel relationships (active yet covert rulers through passive authority figures)
R48. Rebels and Luddites
Personal relationships
R49. The affirming-orientated relationships (trust and confidence based)
R50. The serving-orientated relationships (confidence and humility based)
R51. The performance-orientated relationships
R50. The appearance-orientated relationships
R51. The blame-orientated relationships
R52. The shame-orientated relationships (victim based)

core competence. This is what gives rise to the relationship approach to project management.

The allocation of roles to project actors and the way in which these role-holders are linked have been fundamental features of project management systems commencing with the basic application of organisational structures and delegation of duties, through to some of the more sophisticated means of analysis and decision-making. Indeed the difficulties arising from a lack of clarity in roles have been identified by many analysts of construction systems over half a century or more, from the Simon Report (Ministry of Works, 1944), through Higgin and Jessop (1965) and more recently Latham (1994).

Project teams comprise actors that are allocated specific project function roles, linked by a range of relationships. Clearly the definition and maintenance of these roles and the organisation, monitoring and maintenance of the linkages between these role-holding actors are fundamentally important to the management of network-based organisations or coalitions. This book will explore a number of quite different views and ideas about the nature of the task, in terms of managing the networks and the best ways to achieve such management.

Each project is unique – a mix of different ingredients. Relationships are an ingredient that has often been overlooked. Therefore, relationships are context specific, yet so are project characteristics, a reflexive relationship between the two having the effect of dynamically influencing and changing both.

Uncertainty and risk are two common project factors, and relationships are both a negative source of risk and a positive means to mitigate risk. The sources of risk and uncertainty are several. External factors, such as physical or environmental conditions, are one source of risk, for example conditions in outer space for space travel or ground conditions in bridge building. Other external factors are those risks that are beyond the scope of the project yet are the result of other people's activities, for example malfunctioning technologies or changing political circumstances. Relationships between individuals and between the firms are internal factors. How people behave within and immediately impinging on the project is, therefore, a third factor and probably the main source of uncertainty – human risk factors (Thevendran & Mawdesley, 2004). It is a neglected area in the various bodies of knowledge (BOKs) for project management (see Table 1.3). Human relationships are not considered directly. Leadership and teamwork are, but these elements of BOKs are better understood in terms of conduct in relation to 'professionalism' rather than the content of behaviour, underlying attitudes and emotions.

These aspects of relationships are directly evident as interpersonal behaviours, yet are also embedded in the interrelationship between enterprises. While behaviours can be addressed as they arise, embedded interrelationships are often less obvious, yet their impact and effect is longer term, sometimes carrying on across projects, for example in strategic partnering. Behaviour, whether interpersonal or embedded, has a positive and negative

Table 1.3 Bodies of knowledge and their perspective on relations (PMI, 2000; Caupin *et al.*, 1998; APM, 2000).

PMI	IMPA	APM
Human resource management	Teamwork (19)	Teamwork (71)
Personnel development	Leadership (20)	Leadership (72)
Communications management	Conflict management (21)	Conflict management (73)
	Personnel development (22)	Negotiation (74)
		Personnel management (75)

impact on the 'social account', as opposed to the financial ones. The social account embodies intellectual and human capital.

Businesses frequently state that people are their main asset. Intellectual capital is valued, comprising:

- *Human capital* – expertise, skills and professional competence, which depreciates without investment
- *Structural capital* – strategy, systems, procedures, institutional support and recognition, plus group and social network recognition
- *Social capital* – relationships (including trust and goodwill) brand and market reputation, and social networks (Dawson, 2000)

Relationships are constituted in all these, as well as being a key component of social capital. The question is, do we really believe that people are the main asset of an enterprise? Perhaps such capital can be dismissed as an intellectual concept of no practical worth, in which case people do not add value. Or perhaps it is believed, yet seldom acted upon. If people are the primary asset, why is it that little time and effort is put into developing them beyond their experience on the job? Employees are frequently perceived as an input cost (usually in financial rather than marketing or HR terms), hence a liability rather than an asset. In the UK, the All-Party Parliamentary Group on Management has begun to work towards mandatory human capital reporting by medium-sized and large companies. The UK is perceived to be behind some other leading OECD countries (Task Force for Human Capital Management, 2003), and therefore this issue will have to be taken more seriously over the next years. In the project context, an opportunity is presented to embrace a positive approach to human capital. Relationships provide a prime vehicle for investment in human capital.

Investment in relationships should therefore be expected to yield a return. A relationship approach is needed to develop this neglected yet specific and key dimension to managing projects. This adds to our understanding of managing projects. It should also add value in practice to projects through improved performance. Therefore the relationship approach is complementary rather than an alternative to the existing areas of knowledge.

However, many organisations do not give themselves a focus that creates the option for a relationship approach. The human capital problem identified in industry worldwide (PricewaterhouseCoopers, 2003) is not only echoed but amplified in the management of projects. Projects are primarily managed as a set of activities or tasks to be completed (Handy, 1992; Pryke, 2004; Smyth, 2000). This *task orientation* in projects is equivalent in manufacturing to a *production orientation*. A production orientation is appropriate in some circumstances, for example pharmaceuticals where R&D is the key driver. Many other industries have increasingly adopted a *customer* or *client orientation*, for example aero-engines, where production profit margins are negligible or zero and maintenance and total care services are the source of profit. These examples are extremes, yet the point is that industries have choices. Most project strategies have yet to embrace this choice, although a contracting strategy of securing BOOT or DBFO type projects as a means to generate facilities management (FM) income streams has elements of this. However, there is an absence of widespread evidence for a *client-orientated* approach at many levels. A cursory consideration of the principal bodies of knowledge demonstrates the point. The main bodies of knowledge do consider relationships, yet these tend to be subsumed under task or functional process. The main areas associated with relationships are shown in Table 1.3 and contribute towards a supporting role at most rather than one of the lead roles in the bodies of knowledge (BOKs).

A relationship framework

A relationship approach needs to be recognised as a basis of strategy or vehicle for managing projects. A framework of understanding is needed to set out the primary dimensions. A discourse and negotiation can proceed from the framework. This chapter aims to outline the scope that a framework must, at least initially, cover. Discourse and negotiation will then commence in the rest of the chapters of the book and continue through subsequent debate and contributions. Through this a more definitive framework will emerge, which should then be subsumed into the BOKs in the future.

Projects transform our environment. They transform the natural environment where raw materials and other resources are brought together and fashioned into something usable and valuable. That alone would give a production orientation (see Figure 1.1(a)). Projects are also a social activity. The transformation of the raw materials and resources occurs in the social environment, as production is ultimately a social activity. In the case of projects, services are frequently part of the process (see Figure 1.1(b)) and the management of projects is a service (see Figure 1.1(c)).

Therefore the horizontal arrow representing production in Figure 1.1(a) indicates how raw materials are transformed through work into artefacts. What was in its natural state has been transformed, whether it be a piece of land into a field with hedges, steel or other components into a car, or concrete and glass into a building. This transformation through production is also a process and therefore the arrow also represents a timeline. In a market economy most production is undertaken to produce tradable products. Services, on the other hand, originate in the social environment, transforming the way objects and subjects are used and looked after (see the circular arrow in Figure 1.1(b)).

In the project environment, production and services meet, the management of projects being the service for a diverse and often complex set of activities that are brought together to complete a task or produce an entity. Services are social in nature. The delivery of a service implies human relations. Most services require a range of skills and expertise, and thus cannot be delivered purely by individuals. Therefore the delivery of services is also a social activity involving relationships.

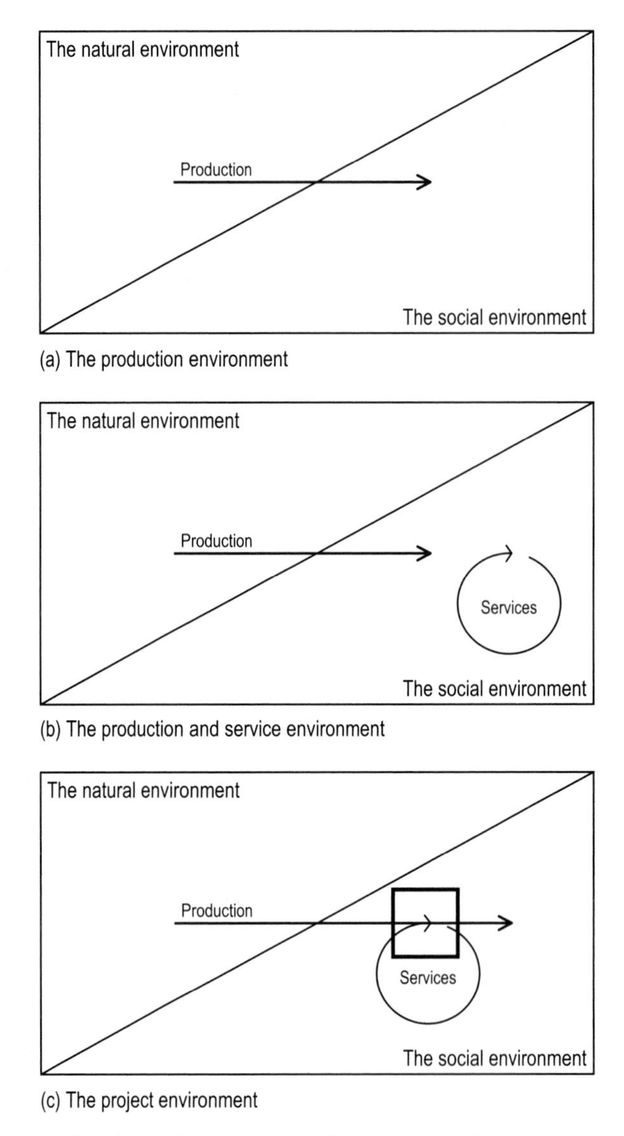

Fig. 1.1 The natural and social environments of production and services.

Thus projects are inherently social. This is not to say that technology, information systems and management tools – and indeed all other complementary approaches to managing projects – are inadequate foci. It is simply to say that social relationships are just as important (see Figure 1.1(c)). Social relationships are therefore central to project outcomes, hence deserve focus in improving project management and being drawn into the project management BOKs, which themselves are a social product derived from analysis and experience. Projects are therefore located in the social environment and typically involve a number of parties (see Figure 1.2).

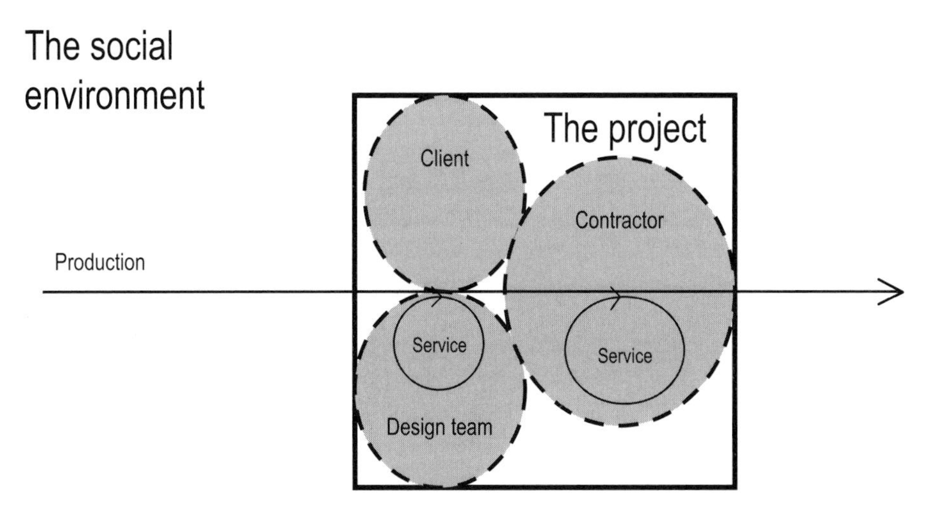

Fig. 1.2 The project environment: internal social context and structure.

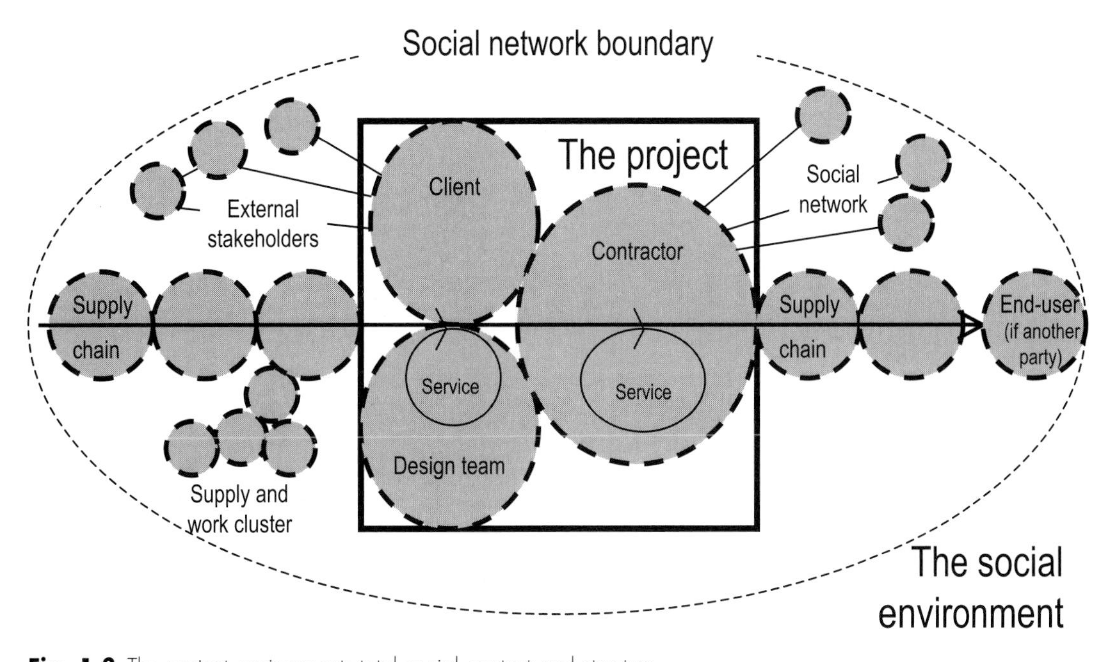

Fig. 1.3 The project environment: total social context and structure.

These parties do not operate in a hermetically sealed environment, the boundaries between the parties and indeed the project being permeable in practice. A network of relationships, including stakeholders (see **Chapter 8** by Loosemore and **Chapter 5** by Cova & Salle) and those involved in supply, are all connected (see **Chapter 4** by Smyth, **Chapter 9** by Pryke and **Chapter 10** by Green). The success of projects is relationship dependent, not only within the project environment but also in its wider social context (see Figure 1.3).

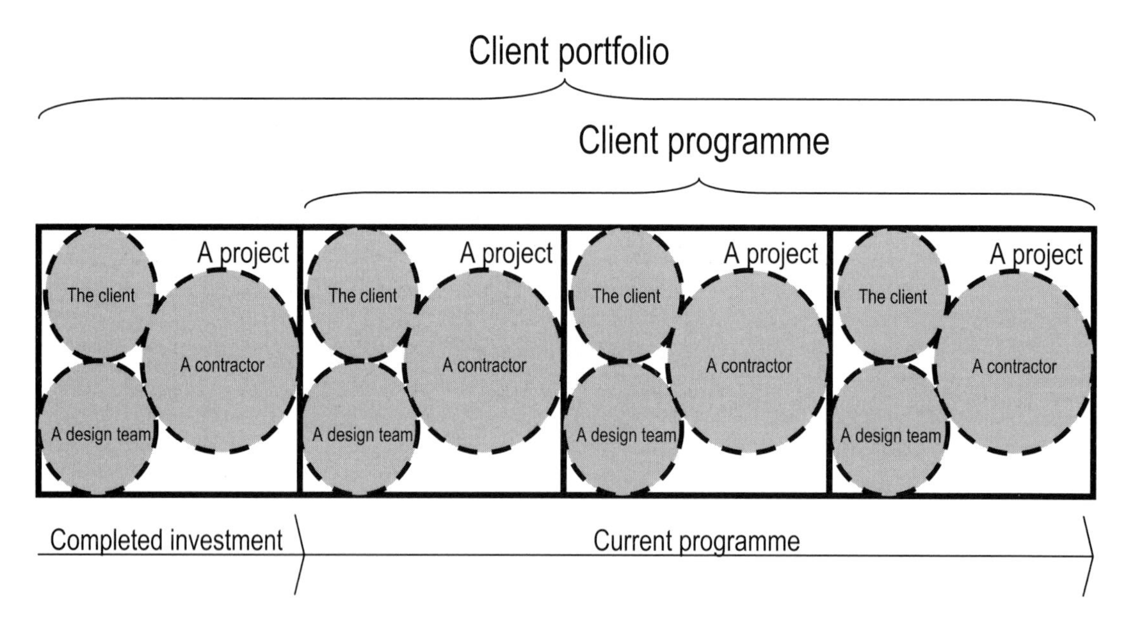

Fig. 1.4 The project environment: internal social context and structure.

Within the project environment, the project will be a solution to a social, often an organisational, problem. Hence the project relates to a policy or strategy. The project may be one piece in a jigsaw, and frequently is part of a series of projects, some of which constitute a current programme and others of which are completed and constitute part of a portfolio (see Figure 1.4).

The project is therefore part of:

- A wider range of strategic activities
- A set of relationships within the project
- A set of social relationships across a network

Yet the mindset has frequently been to see the project apart from its social environment. Management has been conceived this way historically. Projects can be tasks that do not fit neatly into the core operation of an organisation or they can be geographically located away from the centre of operation. Yet this does not make a project less social nor less relationship-based and in many cases may reinforce the need for support, for example an increase in resources.

Such separation is socially artificial. While it is becoming increasingly historic, such separation, sometimes isolation, is still too common. Figure 1.5 (a) shows that this frequently starts at the business development stage (the solid box representing the 'old definition' of the project boundary). Here selling the project services is dislocated from delivery along the project timeline. A common consequence of this separation is that there is a lack of relationship continuity, whereby client expectations – needs and desires – are not adequately transferred to the contract stages.

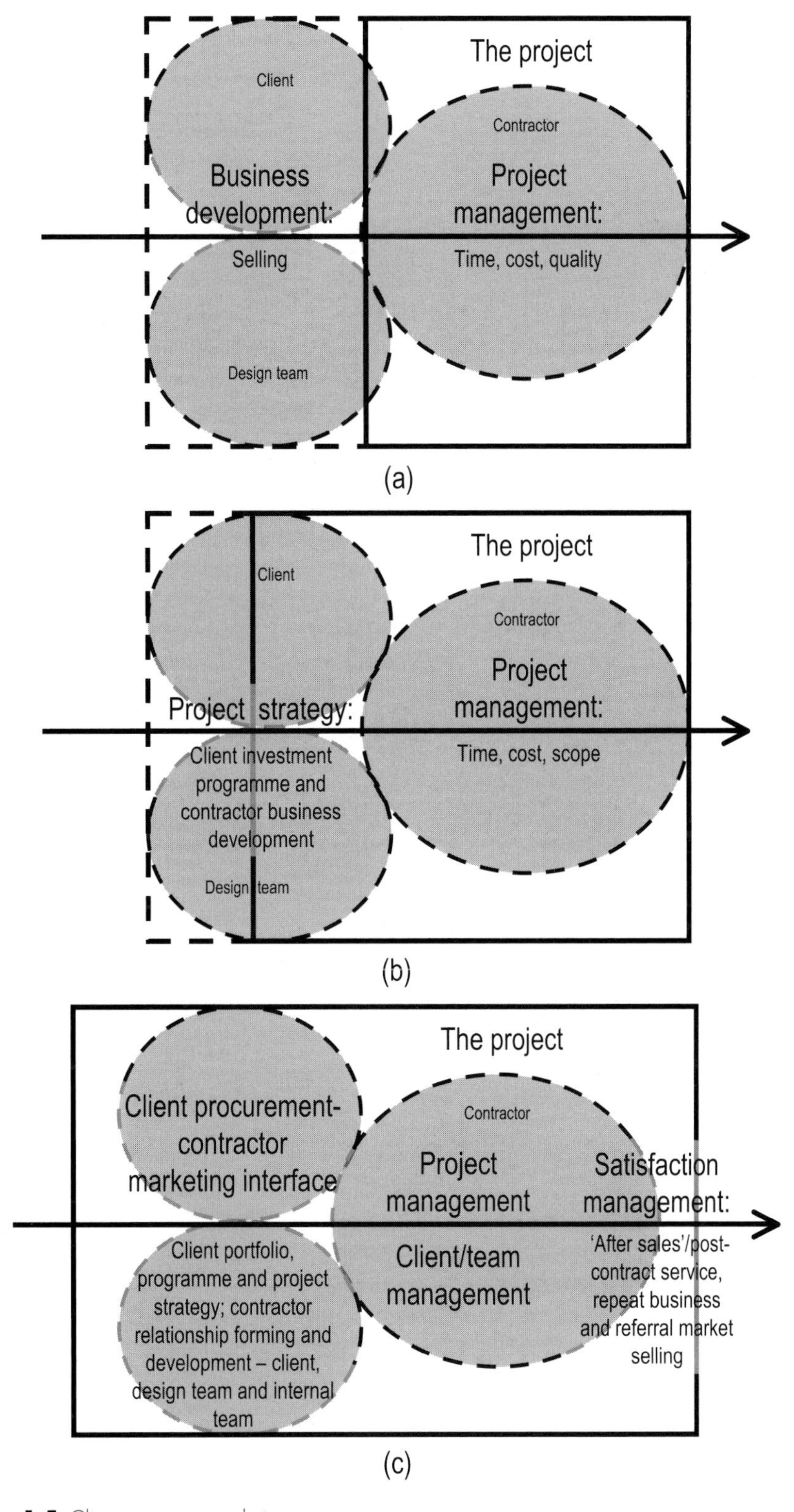

Fig. 1.5 Client–contractor relations.

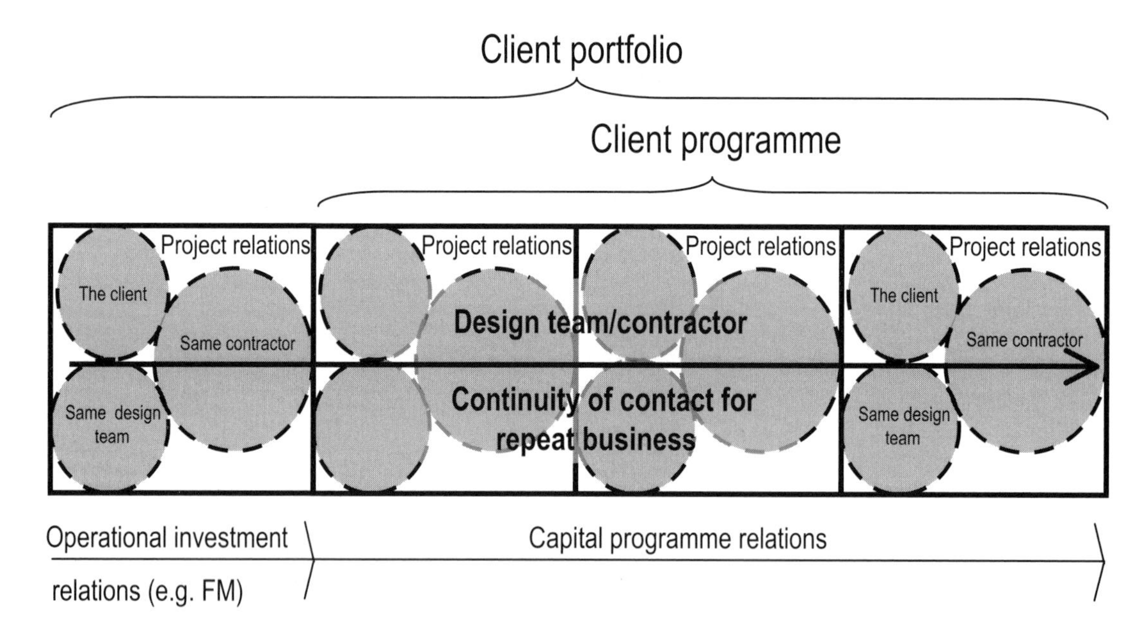

Fig. 1.6 Client relations: client is the 'project'.

Similarly, there is frequently artificial separation from a relationship perspective between internal functions, often formalised as departments such as estimation and procurement, where professional norms and expertise rather than the client requirements guide decisions. In other words, this operates in a dysfunctional way unless a relationship dimension is overtly drawn into the functional approach (see Figure 1.5 (b)).

Figure 1.5 (c) shows the relationship dimension. The project is seen in its broader context – all those directly involved at the client–contractor interface. Once the relationship dimension is drawn in, the traditional perception of the project boundary is removed, and indeed, where there is a series of projects, the contractor or supplier would begin to view the 'client as the project', for the relations with the client span across the application of traditional project management tools and techniques and across the usual functional project tasks. Only people can maintain the continuity for the organisations (see Figure 1.6). Relations and 'soft systems' are the primary links.

Developing the relationship framework

How projects are delivered through effective establishment and maintenance of project team and task cluster relationships is vital. In order to understand the worth of a relationship to both client and contractor, then there must be understanding of:

- How relationships can add value to a service for the client
- How the value derived from a service is translated into profit for the contractor

In other words, a relationship approach needs to be recognised as the basis of a strategy for managing projects, for it is through relationships that continuous improvement is achieved, and specifically relationships add service value. In order to explore this further, consideration of the ways in which value is added to projects is needed.

Historically, contracting has been predominantly based on price-related competition, frequently at the expense of value. Price-based competition leads to what economists term *value added*, that is the value created through the production process and service delivery. *Added value*, on the other hand, is the value that exceeds value added. Added value is what would be expected to arise where competition is addressed through a differentiated strategy. As Porter (1985) has pointed out, the market approach can be low cost or differentiated. Differentiation of services tends to create a market focus in the form of a series of segments (based on service value rather than procurement route), which in turn leads to less intensive competition.

We have argued the importance of relationships as human and social capital – a source of adding value in services – yet there lingers a strong element of mechanistic- and production-orientated thinking in projects today, for example in procurement and supply chain management. It is important to trace this. Taking that example as an illustration, perhaps the trend comes from associated influences. In the example, the *value chain* concept developed by Porter (1985) to analyse the way in which value is added to a product or service is rather mechanical and automated. Similarly, the supply chain for a project may be influenced by this thinking – a sequential series of linked value chains for delivering a project. However, the model presented by Porter (1985) is not easily applied to the project context. Whereas most goods are produced ahead of sale, in the project environment delivery comes after the sale – production and service delivery to contract. How the value chain concept needs adapting to the project environment has been expounded elsewhere (Smyth, 2005a). In summary, this not only reverses the bottom line, placing marketing and sales right at the front end, but also changes the functions.

Supply chains can be conceived as a series of value chains, the reconfiguration for projects changing the interfaces between organisations, and hence the relationships. In addition, the supply chain and supply clusters for projects are far more fluid in structure (see Figure 1.3). Applying to this a relationship approach embodies both demand and response elements – a task-driven procurement approach and a client-orientated value-adding service (Smyth, 2005a). As Green (see **Chapter 10**) demonstrates, the procurement-driven approach to supply chain management is machine-like, yet lacks the fabric to achieve stated goals. Whilst social relations are marginally changed, they cannot be substantially engineered through rhetoric, bullying and fashion. Cultural and systems changes of substance are necessary beyond

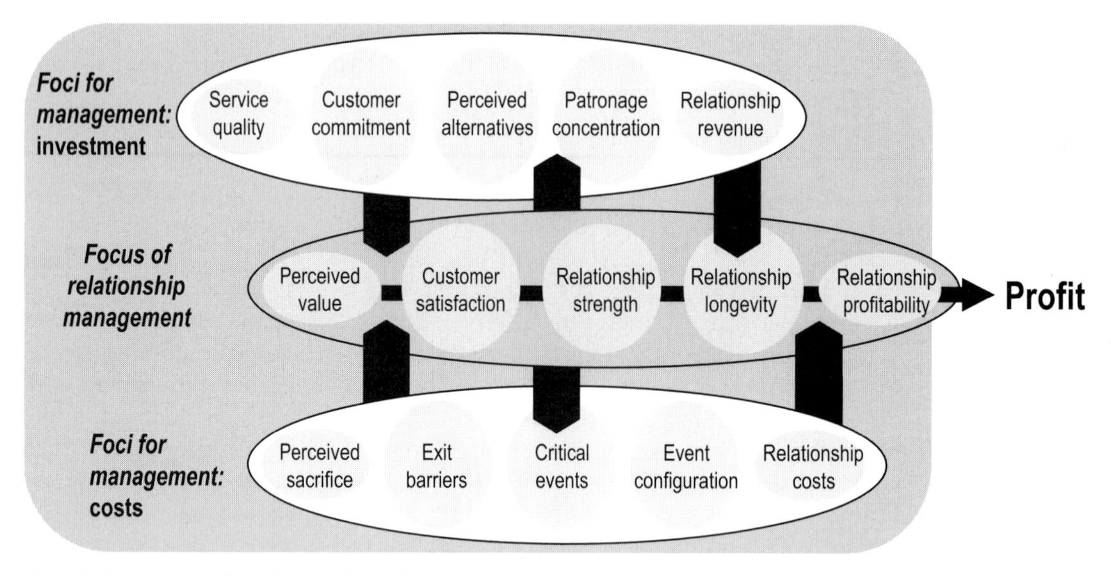

Fig. 1.7 A simplified model of relationship management.

a procurement-driven model to achieve the stated goal of continuous improvement. The selection of any tools or models, including relationship approaches, must be thought through carefully.

There are several relationship management approaches, for example the IMP approach (Ford *et al.*, 2003) and the Nordic School approach (see Grönroos, 2000; Gummesson, 2001). The Nordic model of relationship management will be used in this chapter. The relationship management model has been developed (Gummesson, 2001; Storbacka *et al.*, 1994) and applied in construction in Smyth (2000, 2004, 2005a). A simplified version is summarised in Figure 1.7.

The centre line is the main focus of relationship management. The aim is to create profitable relationships. This starts with the *perceived value* delivered to clients, perceived value being a prior assessment by the client of what the value chain will produce. Perceived value is a prior assessment and therefore based on expectations, derived as follows:

Perceived Value = Added Value to meet Client Expectations

Client Expectations = Client Needs + Desires, which attitudinally can also be expressed as Faith + Hope

Needs + Desires = Explicit Brief + Tacit Expectations to be made Explicit + Unconscious Requirements that surface during the relationship iteratively

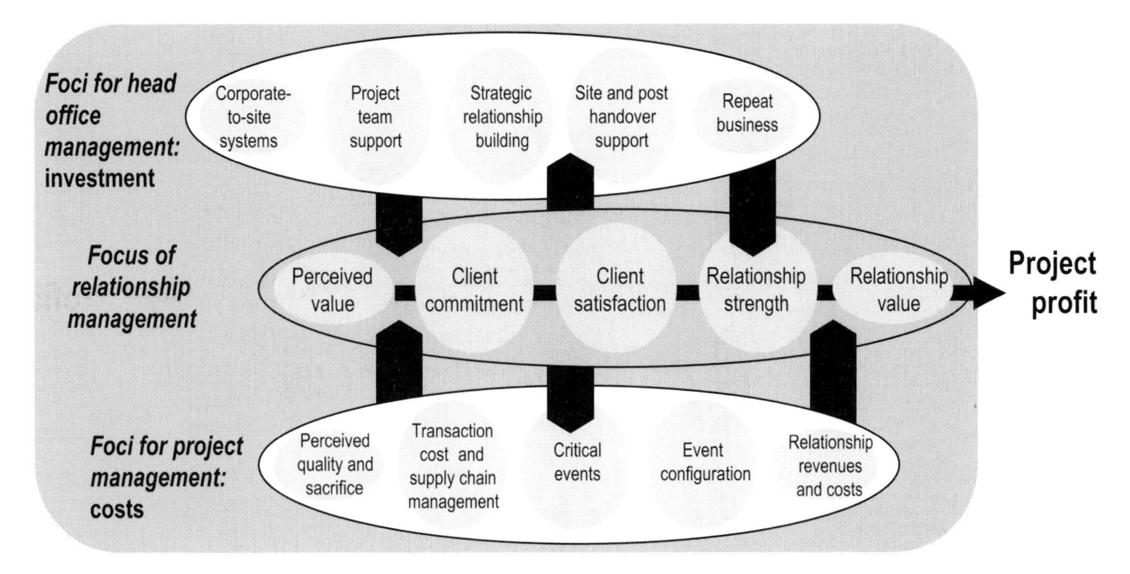

Fig. 1.8 A simplified model of relationship project management.

To meet these expectations through a relationship, the contractor has to invest in *service quality*. In tandem there is the cost incurred of *perceived sacrifice*, going the extra mile. Therefore, in order to manage relationships from the period of developing the relationship where perceived value is established to a profitable outcome, both investments (the top line of Figure 1.7) and costs (the bottom line of Figure 1.7) are incurred. The net relationship value, that is the relationship profitability to the contractor and beneficial value delivered to the client, does not flow automatically, but is the product of careful management of the relationships.

This model applies to the organisation as a whole, in this case the contracting organisations, whether they are main or sub-contractors. Like Porter's value chain model, it does not so easily translate into the project environment. A relationship project management model has been provided (Smyth, 2004), which shows how relationship management principles can be applied at the project level (Figure 1.8). This model is not a substitute for the model at the enterprise or organisational level but sits beneath it, effectiveness from the two working in tandem.

Figure 1.9 clearly indicates that projects cannot operate in a semi-autonomous way from the head office functions. This and the project costs in maintaining the relationships for maximising project efficiency from this perspective give rise to the *relationship revenue*, the net revenue after overheads are discounted, giving rise to the *relationship profitability* shown in Figure 1.7. Investment and continuing support from the head or main office are crucial to success, yet also render the contractor less secure in vulnerable market conditions. It is this inherent risk that will build barriers to entry and increase switching costs through the strategy of differentiation.

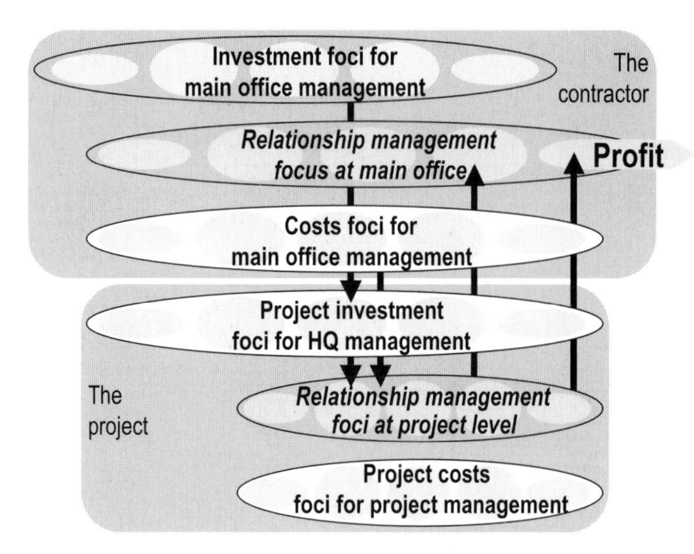

Fig. 1.9 Interrelationship between relationship management for projects and the main office.

Figure 1.10 shows that interrelationship between investment and delivering value through relationships in a more detailed way, drawing together some other common elements on investment and costs in the corporate environment (cf. Smyth, 2004). It is a truism that in the long term a profit can only be yielded from an investment. There are a number of organisational issues that are central to effective relationship management in the development of the framework. First, relationship management requires systems within the organisation – 'soft systems'. The need for soft systems is crucial to address situations where a *personality culture* (Smyth, 2000) dominates. This is so regarding, for example, construction projects, and hence the experience of clients differs from project to project when using the same contractor because the experience depends upon the management style of the team. The direct and extreme opposite – McDonald's – is so systematised that the service can be franchised. While this would be far too rigid for project environments, there is a considerable middle ground that remains virgin territory in many project environments, for example construction. Systems are required to provide consistency of service for clients from the main office in support of every project.

Second, service consistency depends upon behaviour. People's behaviour in delivery needs to be aligned with the service requirements (see **Chapter 7** by Kumaraswamy & Rahman) and therefore behaviour performance needs to be maximised. Around 20% of people's behaviour is informed by what they have learnt and 80% by their emotions, thus behaviour is largely contingent on people's emotions. This is the issue of emotional intelligence (Goleman, 1996, 1998), which Druskat and Druskat address in a project environment (**Chapter 3**). This requires investment in behavioural performance, which is shaped to plug into the soft human systems and vice versa. In some project

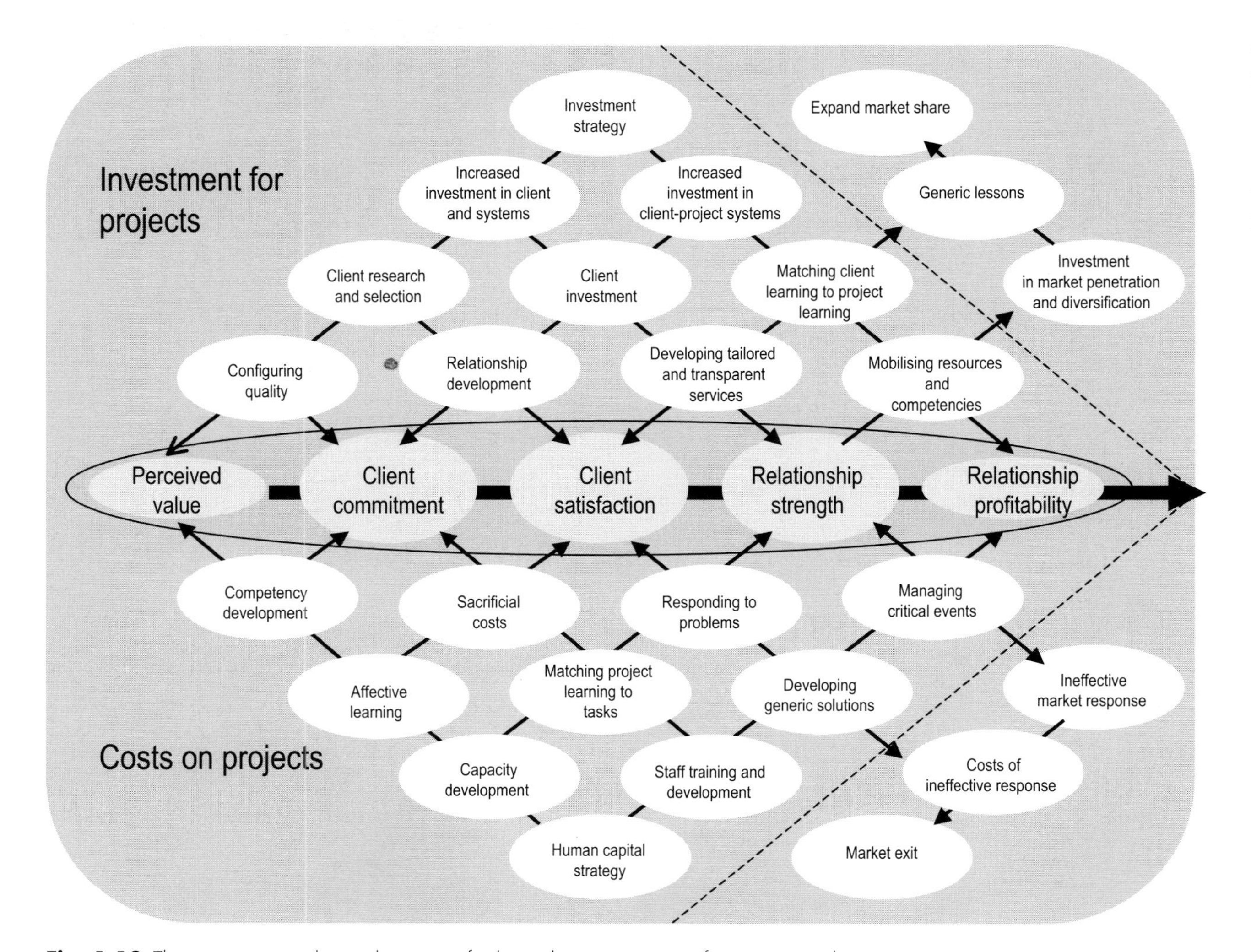

Fig. 1.10 The investment and cost dynamics of relationship management for project working.

environments, especially where uncertainty is high and so transaction costs are high, opportunism prevails (cf. **Chapters 11** by Cox & Ireland and **Chapter 12** by Ive & Rintala). Therefore to overcome opportunism, the relationship value must outweigh the costs. This may not always occur in the short run but must prevail in the longer term, in order to provide reasons to increase barriers to entry and increase switching costs (**Chapter 4** by Smyth; Smyth, 1999; Smyth & Thompson, 1999).

Third, service consistency also depends upon who manages the client and how they are managed. Therefore a client handling model is needed. Two primary models have been identified in construction: *account handler* and *relay team* (Smyth, 2000, 2004). The account handler model essentially provides the single point of contact, which is also referred to in the relationship management literature as the key account manager (KAM) (Kempeners, 1995; McDonald *et al.*, 1997; Millman & Wilson, 1999). Account handler is borrowed from the world of advertising agencies and is the person responsible for the whole account, and therefore all the projects for the client. The role involves both strategic and tactical management, being responsible for developing and nurturing the relationship over its lifetime. The account handler can be a contracts director, business development manager or member of senior management. This achieves continuity of service.

The relay team is the model where the contractor does not keep the same continuity of staff. Changes between business development managers and the contract staff, off-site managers and so on are all acceptable: continuity of service is provided by passing the 'relay baton' on to each successive person with prime responsibility for the client and project. In this way comprehensive and detailed understanding is transferred from the understanding of expectation plus the mapping and profiling of the client organisation (Smyth, 2000), to the information that is project specific in terms of delivery of the technical side of the service. Many projects, as for example in a considerable number of large construction projects, use a relay team approach, but without a baton. The contrast in experience has been summarised in terms of a relay race where school children have no baton. The person on the first leg has to touch the hand of the person running the second leg, except that they run out of steam and the second leg starts off anyway. This is what occurs on many projects. In the mature relay team the transfer is represented by the baton. If you drop the baton, you are disqualified (Smyth, 2004).

For continuity to be achieved using the relay team model, it is important that the client knows in advance what personnel changes are going to be made and when those changes are going to be made. There may be service reasons for making such changes, although generally the relay team model without a baton is motivated by keeping transaction costs to a minimum. In both the relay team (with the baton being passed) and account handler models, continuity is managed and maintained.

In terms of managing projects, different models are better suited to different circumstances, the account handler model being most suited to the relationship management approach. This is the most client-orientated model. It has the potential to optimise service delivery, especially delivery tailored to

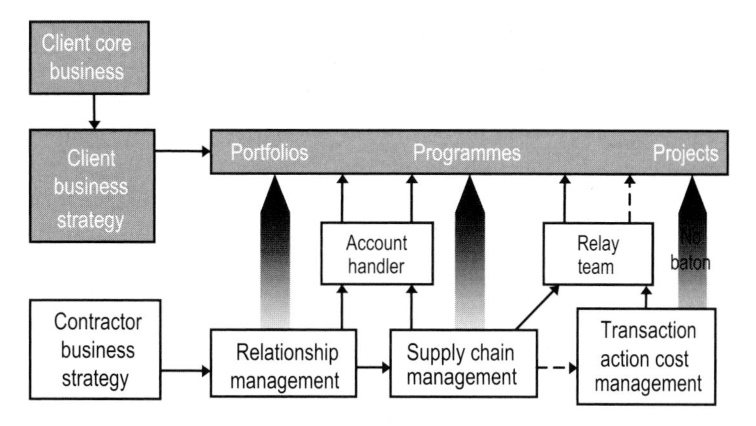

Fig. 1.11 Client handling models and strategies.

client needs. Not all clients want the same service, for expectations differ. Not all added value necessarily is value the client wants. The account handler is best placed to search out the needs and desires of the client and align the services to meet as realistic a package of expectations as possible. The account handler is best placed to respond to emergent needs too. Importantly, the account handler is also able to ensure the client is made fully aware of the value package they are receiving.

The relay team *without* a baton is most suited to the pure transaction cost approach of traditional contracting. The relay team *with* a baton is suited to the middle ground, perhaps where supply chain management is being driven from the client organisation (**Chapter 9** by Pryke and **Chapter 10** by Green), in other words where contractors wish to begin to develop and mobilise competencies to add value and competitive advantage to their service, for example knowledge management (cf. **Chapter 2** by Morris), emotional intelligence (**Chapter 3** by Druskat & Druskat) or relationship management (cf. **Chapter 4** by Smyth, **Chapter 6** by Wilkinson, **Chapter 7** by Kumaraswamy & Rahman and **Chapter 11** by Cox & Ireland). The relative positions for the client handling models are shown in Figure 1.11.

Fourth, teamworking models can be mobilised to enhance continuity and consistency of service at several levels. Initially, each unit of operation needs to work consistently within their own function, for example procurement. There must be alignment across the functions. Reichheld (1996) has repeatedly drawn attention to the fact that the strength of a relationship between customer and supplier can only be as strong as the strength of the internal relationships – 'all singing off the same hymn sheet'. Druskat and Druskat (**Chapter 3**) demonstrate this in relation to emotional intelligence and Smyth in relation to trust (**Chapter 4**). Wilkinson takes the account handler and relay team models, applying them across organisations to explore some of the relationship issues (see **Chapter 6**). One consequence of her analysis is to show how relationships and continuity of service issues have implications for procurement strategy. Developing the need for delivering a seamless service,

Kumaraswamy and Rahman (**Chapter 7**) demonstrate how incentivisation can be used to align teams.

Another level is across project teams, crossing organisational boundaries. Temporary multi-organisational teams (Cherns & Bryant 1983; Pryke, 2005) are often brought together as 'virtual organisations' at the project level, especially for design and on construction sites. Teamworking methods and tools can be used to enhance teamwork performance, for example using Belbin analysis (**Chapter 7** by Kumaraswamy & Rahman). A further level is applying techniques down the supply chain where integration and collaboration have been advocated. These tools and techniques help plug gaps between rhetoric, promises made to the client and practice (cf. **Chapter 9** by Pryke and **Chapter 10** by Green).

Other management concepts, paradigms and tools can be grafted into the framework according to circumstance. Indeed, there will be a need to develop some concepts into more central positions as the discourse and agenda for a relationship approach develops. Sociological issues of power relations, leadership, decision-making and negotiation are all important. Psychological issues that inform behaviour are important too (see, for example, Smyth, 2004). Structure and strategy are built both around and through relationships. In some sense both enshrine the character of relations, the decision makers usually wishing to perpetuate actions beyond the scope of the relationships – structures embody organisational principles that render certain sorts of actions and relationships more sustainable than others, and strategies guide those actions and inform relationships in a more directive way within a given structure. Therefore not only do relationships help form structures and strategy, but these also act back to inform relations.

Although the focus of this book is primarily about process, investment in and commitment to processes require the appropriate structure and strategy. Thus, changes in processes frequently require structural change. Many project organisations have temporary structures, brought together and fashioned around delivery. Where the organisation is responsible for a series of projects, as contractors are, restructuring may be necessary to accommodate any emphasis on relationship development, especially the operational and 'soft systems'. While there is no single way to implement such structural and systems changes, the way being dependent upon what concepts are being developed and on strategic decision-making, the final choice is also a potential source of competitive advantage for the organisation as it forms part of the service differentiation.

The emphasis in this book on process overall is a reflection of the recent trends in management literature, and indeed management practice, which has shifted away from structure towards process. There is an emphasis on resource-based analysis (see, for example, Grant, 1991) and competencies in particular (Prahalad & Hamel, 1990, 1994). Relationships are enhanced by competency development – human skills and expertise rather than technical skills.

However, many project organisations have placed more emphasis on structure. The construction industry offers an extreme example. When the market has changed because of the development of a new procurement route, such as design and build or DBFO contracts, contractors tend to set up a new division to service the procurement route – an operational rather than a client focus (Smyth, 2005b). This approach minimises transaction costs but leaves the initiative with the client to search out the best route, perhaps through their agent or representative. In relationship terms this gives clients the wrong message about their services.

Yet it is important to address whether the potential for added value derived from relationships is important in project delivery. Taking construction as an extreme example, most contractors no longer build anything in many developed countries. All work on site is subcontracted. They are primarily managing transaction costs (Gruneberg & Ive, 2000). They either add value by professionalisation of their services or use construction as an entry point to another market. They use construction to enter an asset-based market via DBFO, BOOT and property development types of activities or they use construction to enter a revenue stream via facilities and maintenance management. While it is still the case that construction is a source of profit – better in terms of returns on capital employed (ROCE) than profit margin – the sector is restructuring; for example, in the UK it could be argued that more effort has gone into restructuring than into improving construction performance. Those firms restructuring to secure subsequent profitable income streams, for example FM, may find this is the sole source of future profit, project delivery simply being the key to the door to the income stream. For others, contracting will only remain a prime source of profit through service differentiation, hence adding value through competencies, especially where relationship management is both a competency and a vehicle for the delivery of other competencies. It is people who add value, and people working together are more effective. People and teams are assets, their value realised through relationships at many levels of operation.

Therefore a relationship approach is important and is likely to become increasingly important, hence the development of a relationship approach to managing projects is timely.

References

Association for Project Management (2000) *Project Management Body of Knowledge* (ed. M. Dixon). APM, High Wycombe.

Blismas, N. G. (2001) *Multi-project environments of construction clients*, unpublished PhD Thesis, Loughborough University, Loughborough.

Caupin, G., Knopfel, H., Morris, P. W. G., Motzel, E. & Pannenbacker, O. (1998) *ICB IPMA Competence Baseline*. International Project Management Association, Zurich.

Cherns, A. B. & Bryant, D. T. (1983) Studying the client's role in project management. *Construction Management and Economics*, **1**, 177–84.

Curtis, C. *et al.* (1991) *Roles, Responsibilities and Risks in Management Contracting*. Construction Industry Research and Information Association, London.

Davis, K. & Newstrom, J. W. (1989) *Human Behaviour at Work*. McGraw-Hill, Singapore.

Dawson, R. (2000) *Developing Knowledge-Based Client Relationships: the Future of Professional Services*. Butterworth-Heinemann, Oxford.

Ford, D., Gadde, L-E., Håkansson, H. & Snehota, I. (2003) *Managing Business Relationships*. Wiley, London.

Goleman, D. (1996) *Emotional Intelligence*. Bloomsbury, London.

Goleman, D. (1998) *Working with Emotional Intelligence*. Bloomsbury, London.

Grant, R. M. (1991) The resource-based theory of competitive advantage: implications for strategy formulation. *California Management Review*, Spring, 114–35.

Grönroos, C. (2000) *Service Management and Marketing*. John Wiley and Sons, London.

Gruneberg, S. L. & Ive, G. J. (2000) *The Economics of the Modern Construction Firm*. Macmillan, Basingstoke.

Gummesson, E. (2001) *Total Relationship Marketing*. Butterworth-Heinemann, Oxford.

Hamel, G. & Prahalad, C. (1994) *Competing for the Future*. Harvard Business School Press, Boston.

Handy, C. B. (1992) *Understanding Organizations*. Penguin, Harmondsworth.

Higgin, G. & Jessop, N. (1965) *Communications in the Building Industry: The Report of a Pilot Study*. Tavistock Publications, London.

Kempeners, M. (1995) Relationship quality in business-to-business relationships. *Proceedings of the 11th IMP Conference*, 7–9 September, Manchester.

Koskela, L. (1992) *Application of the New Production Philosophy to Construction*. Technical Report 72, Center for Integrated Facility Management, Stanford University, Stanford.

Latham, M. (1994) *Constructing the Team: Joint Review of Procurement and Contractual Arrangements in the United Kingdom Construction Industry*. HMSO, London.

Masterman, J. W. E. (2001) *An Introduction to Building Procurement Systems*. E & FN Spon, London.

McDonald, M., Millman, T. & Rogers, B. (1997) Key account management: theory, practice and challenges. *Journal of Marketing Management*, **13**, 737–57.

Millman, T. & Wilson, K. (1999) Processual issues in key account management: underpinning the customer-facing organisation. *Journal of Business and Industrial Marketing*, **14** (4), 328–37.

Ministry of Works (1944) *The Placing and Management of Building Contracts: Report of the Central Council for Works and Buildings (known as the Simon Report)*. HMSO, London.

Morris, P. W. G. (1994) *The Management of Projects*. Thomas Telford, London.

Nohria, N. & Eccles, R. G. (eds) (1992) *Networks and Organisations*. Harvard Business School Press, Boston, MA.

Porter, M. E. (1985) *Competitive Advantage*. Free Press, New York.

Prahalad, C. & Hamel, G. (1990) The core competencies of the organization. *Harvard Business Review*, **63** (3), 79–91.

PricewaterhouseCoopers (2003) *Human Capital Management Practices Outside the UK*, http://www.accountingforpeople.gov.uk

Project Management Institute (2000) *A Guide to the Project Management Body of Knowledge*. Project Management Institute, Newton Square, PA.

Pryke, S. D. (2004a) *Twenty-first Century Procurement Strategies*. RICS Research Paper Series, RICS, London.

Pryke, S. D. (2004b) Analytical methods in construction procurement and management: a critical review. *Journal of Construction Procurement*, **10** (1), 49–67.

Pryke, S. D. (2005) Analysing construction project coalitions: exploring the application of social network analysis. *Construction Management and Economics*, **22** (8), 787–97.

Reichheld, F. F. (1996) *The Loyalty Effect*. Bain & Co., Harvard Business School Press, Boston, MA.

Reichheld, F. F. (2003) The one number you need to grow. *Harvard Business Review*, December, 46–54.

Smyth, H. J. (1999) Partnering: practical problems and conceptual limits to relationship marketing. *International Journal for Construction Marketing*, **1** (2) http://www.brookes.ac.uk/conmark/IJCM/

Smyth, H. J. (2000) *Marketing and Selling Construction Services*. Blackwell Science, Oxford.

Smyth, H. J. (2004) Competencies for improving construction performance: theories and practice for developing capacity. *Journal of International Construction Management*, April, 41–56.

Smyth, H. J. (2005a) Procurement push and marketing pull in supply chain management: the conceptual contribution of relationship marketing as a driver in project financial performance. Special Issue on Commercial Management of Complex Projects (eds D. Lowe & R. Leiringer). *Journal of Financial Management of Property and Construction*, **10** (1), 33–44.

Smyth, H. J. (2005b) Competition. In: D. Lowe & R. Leiringer (eds) *Commercial Management of Projects: Defining the Discipline*. Blackwell Publishing, Oxford.

Smyth, H. J. & Thompson, N. (1999) Partnering and trust. *Proceedings of the CIB Symposium on Customer Satisfaction*, September, Cape Town.

Storbacka, K., Strandvik, T. & Grönroos, C. (1994) Managing customer relationships for profit: the dynamics of relationship quality. *International Journal of Service Industry Management*, **5** (5), 21–38.

Task Force for Human Capital Management (2003) *Accounting for People*, DTI, www.accountingforpeople.gov.uk

Tavistock Institute (1966) *Interdependence and Uncertainty: A Study of the Building Industry*. Tavistock Publications, London.

Thevendran, V. & Mawdesley, M. J. (2004) Perception of human risk factors in construction projects: an exploratory study. *International Journal of Project Management*, **22**, 131–7.

Wenger, E. (1998) *Communities of Practice: Learning, Meaning, and Identity*. Cambridge University Press, Cambridge.

Winch, G. M. (2002) *Managing the Construction Project*. Blackwell Publishing, Oxford.

Winch, G. M. & Carr, B. (2000) Processes, maps and protocols: understanding the shape of the construction process. *Construction Management and Economics*, **19**, 519–31.

Section II

Mindsets, Behaviour and Competencies

Context

Albert Einstein said: 'Common sense is that layer of prejudice laid down before the age of 18.' About 75–80% of what we learn arises out of experience – *affective learning*. In this context learning is from two main sources: the formative years of our life and our career experience. The formative years induce the 'emotional gifts' we have for others and the 'emotional baggage' we carry around with us. The 'gifts' are positive and give us an outward focus from which we all benefit, forming the emotional intelligence we apply to everyday life (see **Chapter 3** by Druskat & Druskat). On the other hand, although when travelling we like to check our baggage in at the airport as soon as we arrive, most of us seem determined to carry our negative 'emotional baggage' around with us into our adult lives, our marriages and of course our jobs.

Our formative years give rise to our beliefs, values and attitudes – what we believe, how we feel and hence how we are predisposed to act. These are our *mindsets*. These mindsets are reinforced by the second main source of learning, our career experience. We tend to select careers that match our positive and negative emotional profile and hence the experience in our careers tends to reinforce the mindset we have. Our *behaviour* patterns are reinforced and we go on to influence the *behaviour* of others by example.

The problem with *mindsets* is that they are 'set', that is to say they are hard to change. *Behaviours* flow from our mindset and the experience of project working tends to remain the same. Indeed, we feel secure with the familiar: 'We've always done it like that!' – perhaps what Spender referred to as standard 'industry recipes' (Spender, 1989). So when new trends come along demanding *competencies* for partnering, supply chain management, knowledge management and the like, we talk the rhetoric yet are reluctant to change our basic *behaviours* and our *mindset*. In other words, emotionally we are reluctant to change our *mindset*. As psychological folklore has it, 80% of our behaviour is informed by our emotions, that is from *affective learning*, and only 20% from what we are taught and instructed, or *cognitive learning*.

If we believe 'You can't teach a dog new tricks', then we need to acknowledge that our *mindsets* and resultant *behaviours* provide limits to adopting new practices. We can adopt all the rhetoric, indeed invest in all the necessary systems, yet the way our *behaviours* map out in our relationships puts a constraint upon improved performance in any dimension and especially those requiring soft *competencies*.

We believe that the project organisations do not employ 'dogs' and that new tricks can be taught. Simply having a greater awareness of the way relationships contribute to project outcomes is a positive step, but more of a willingness to explore relationships as a means for improving management will induce change. The most successful firms in the future will be the early adopters in this area. If firms have people that refuse to explore new ways the result will be that senior management hire people that will conform to the new demands. This change process may be slow, but it is inevitable. Contractors and indeed professional firms can only achieve marginal and temporary advantage from the technical and professional knowledge base and skill-based *competencies*, and so must look to develop non-cognate *competencies* that require management and personal *behaviours* that flow from different and more flexible ways of thinking – *mindflexes* rather than *mindsets*.

Successful implementation of any strategy or tactic requires alignment of our behaviour and thinking to fully achieve successful outcomes – alignment of personal behaviour, interaction with others; alignment of responses to challenges and stressful crises; commitment to objectives in thinking, attitude and action. Every day we face choices as to whether our thinking and behaviour are an emotional investment, or are a withdrawal from the 'bank account' where we feel less aligned or we cause others to come out of line with the firm's objectives. All withdrawals have an adverse impact on effectiveness and hence on project efficiencies – performance experiences 'emotional hold up'! Perhaps *emotional hold up* is more constraining than the *information hold up* problem. We believe it may be, and are confident that it is significant enough to demand attention.

Social identity theory states that people define themselves and others according to their personality and their membership of groups (Greenberg & Baron, 2003). As people also wish to fit in, they will tend to conform to the norms of groups, so if the aim is to improve the managing of projects, then the way in which people work together has to be changed in order to change project functions and tasks successfully. Following Tuckman and Jensen (1977), the *norms* of organisational behaviour for the group have to be changed before the group *performs* in a new way. How relationships are managed is a key part of this and has been largely overlooked in Egan and post-Egan agendas of continuous improvement. Fig. II.I depicts the dynamic of norms in relation to performance for project teams and their parent organisations.

What can be drawn from this is that changing the norms helps to change people's experiences and opens up opportunities for new affective learning. However, many mindset and behavioural issues become embedded in organisations, themselves erecting barriers to change. Threats to organisational change include:

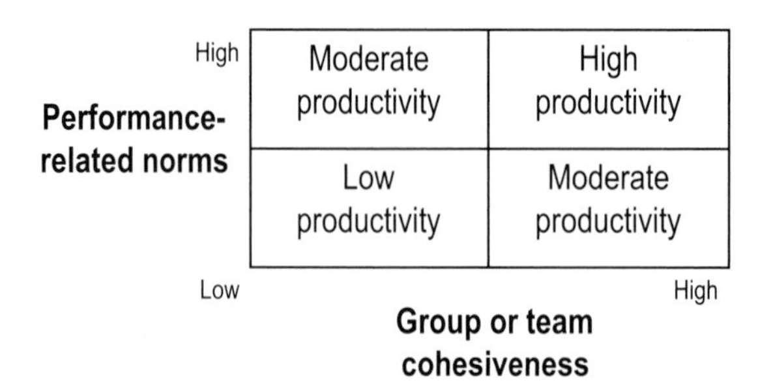

Fig II.I Organisational behaviour and performance. Developed and adapted from Robbins (2003).

- Threat to existing resource allocation
- Structural inertia
- Threats to existing power relationships
- Limited focus of change – as one thing affects another
- Threat to expertise
- Group inertia

In conclusion, mere understanding of the significance of relationships, as affected by *mindsets* and *behaviour*, may help induce marginal change. More extensive changes may prove more challenging. The leading performers will include many who meet and overcome the challenges.

What has been the prevailing culture? Handy (1992) stated that projects are best suited to a *task culture*. This task orientation is equivalent to a *production orientation* in other industries (see **Chapter 1**). A task orientation has a focus upon things – objectives, hence tasks to get the physical built – and therefore has at best a secondary regard for people and relationships. Task-orientated behaviour thus has primary regard for things. Failure to meet objectives, to get the physical built, results in blame being apportioned. This starts internally, building up an internal *blame culture*, which in projects frequently flows from the *task culture* (Smyth, 2000, 2004). Internal blame tends to flow into the marketplace, leading to the adversarial tradition of project environments. This has traditionally been supported in the marketplace by the features of transaction cost management in projects, characterised by high levels of uncertainty, opportunism, asset specificity and infrequency of transaction (Gruneberg & Ive, 2000; cf. Winch, 2002).

The result of the blame culture within the firm is that people become defensive and risk averse. Such behaviour can lead to an internal 'emotional hold up' problem, which mirrors the marketplace 'information hold up' problem (Milgrom & Roberts, 1992) that is so prevalent in projects. Defensive behaviour tends to result in people working as individuals. Despite the call for teamwork, teamworkers tend to be few among those in construction if analysis of those taking university courses that take the Belbin test is anything to go by. Transaction costs tend to reinforce this as senior management keep internal administrative costs, hence related investment expenditure, to a minimum. For example, project managers and site managers receive little corporate support in terms of systems for management. Hence, another way of looking at the culture is that it is a *personality culture* rather than a corporate one (Smyth, 2000, 2004).

It has been argued that professional skills and technical competencies are insufficient to gain competitive advantage. Likewise, in buying in services, advantage cannot be achieved through managing transaction costs in the long run, hence the prevalence of the project *winner's curse* (Winch, 2002) – the winning bid is the bid based on the least accurate evaluation of project risks, together with simple arithmetic errors. Advantage in the emerging market comes from *competencies* that are broader than this, many being closely related to managing relationships either directly or indirectly through human systems, such as conceptual competencies of *learning and knowledge management*, *emotional intelligence* and *relationship management*.

Learning, emotional intelligence and trust within a relationship management context provide the raw materials for the chapters in this section. An overview of the contribution made by each chapter is given below.

Contributions of chapters in this section

Chapter 2 by Morris asks *How do we learn to manage projects better?* He considers institutional and professionally derived learning through the bodies of knowledge (BOKs). He explores these as socially constructed raw materials for firms and the managers of projects, which are both negotiable and to be applied, and hence adapted in socially specific ways. He considers a largely tacit-to-tacit approach, but also the *socialisation* approach where there is a sharing of knowledge that is externalised, hence explicit. The externalisation of knowledge is a social process. This emphasises the significance of relationships. As Morris contends, this lines up with the 'people-centred approach' in projects. However, this is two edged, depending upon the functionality of relationships, which has been outlined above in terms of the personality culture with firms largely determining that they will not be systematic in management to the extent found in many other sectors. Morris sums it up in this way:

> *'In general, culture, strategic intent and organisational constraints were found to inhibit the application of well perceived knowledge management and organisational learning best practices.'*

Indeed, where systems were in place using manuals rather than relationships, manuals tended not to be used. This shows existing constraints, yet also shows the potential for a vigorous relationship approach to management.

Applying emotional intelligence in project working is the subject for **Chapter 3** by Druskat and Druskat. They outline the conceptual competency that concerns the ability of people to monitor their own emotions and the emotions of others, to discriminate among emotions, and to use emotional information to guide thinking and actions. This is particularly important in a project environment where people come together and have to perform to high levels quickly. They examine the dimensions of *emotional intelligence* – self-awareness, self-management, social awareness and relationship management – and then go on to apply it within groups and project teams specifically. They conclude by examining how firms can develop *emotional intelligence*.

Chapter 4 by Smyth examines *Measuring, developing and managing trust in relationships*. Trust is a competence that advocates of new approaches to managing projects have put great store by; however, understanding and managing this competency has been rather overlooked. This chapter defines *trust* and provides a framework of understanding. Smyth then reviews evidence for trust within project coalitions for both a large speculative development project in London and a consortium or special purpose vehicle (SPV) for a series of PFI/PPP projects. The findings give pointers as to what the respective organisations may do to actively manage the development and maintenance of trust.

Locating the chapters within a relationship framework

Each chapter is located within the framework outlined in **Section I**. Once again it is stated that the framework is provided to anchor the work as it is developed. However, debate will help to develop understanding; hence the framework itself becomes amended or perhaps redundant.

In terms of the *objectives* of the book as set out in **Section I**, conceptual development of a relationship approach is provided by Morris through the examination of *learning*. His contribution to improving project success arises from an examination of how we acquire the tools and organise these into bodies of knowledge (BOKs). This contribution helps solve organisational and business problems, aiding programme and project strategy development at the meta-level or external from the firm (Mintzberg *et al.*, 2003), hence the specific context is organisation-to-organisation and indirectly concerns authority relations, informing management and leadership within organisations.

Druskat and Druskat make their contribution through application of the conceptual competency *emotional intelligence* and Smyth develops the competency of trust from a *relationship management* perspective, although emotional intelligence offers an equally appropriate backdrop. These two chapters focus upon interpersonal and inter-firm relationships at the project interface, having a bearing on project delivery and project success. Implementation of both these competencies requires strategic decisions of investment at the level of the firm. The specific context is organisation-to-organisation, operating particularly in the relations between the organisation and individuals both personally and as teams. Emotional intelligence requires systematised and procedural aspects to underpin relations. Trust does not do so directly, but the management and development of trust does. It is here that the contribution of Morris concerning learning links to the other two chapters, the way in which we learn informing how in practice issues of emotional intelligence and relationship management are addressed: ignored or adopted, implemented and developed.

Therefore the contribution of Morris concerns the function and structure of projects within their sectoral context and for the firm, whereas both Druskat and Druskat, and Smyth concern the operational context, particularly the disposition of individuals in relation to the operational context, and hence the roles they perform that are affected by character and personality, attitudes and behaviour. In terms of Gummesson's classification of relationships,

Table II.I Contributions within market relationships. Developed from Gummesson (2001).

Relationships	Chapter 2 Morris	Chapter 3 Druskat and Druskat	Chapter 4 Smyth
Classic market relationships			
R1. The dyad (supplier-customer)	Very indirect	Indirect	Direct
R2. The triad (supplier-customer-competitor)			Direct
R3. The network			Direct
Special market relationships			
R4. Relationships via full-time and part-time marketers			Direct
R5. The service encounter			Direct
R6. The many-headed customer and many-headed supplier			Direct
R7. The relationship to the customer's customer			Direct
R8. Close versus distant relations			Direct
R9. The dissatisfied customer			Direct
R10. The monopoly relationship			Direct
R11. The customer as 'member'			Direct
Mega relationships			
R18. Personal and social networks			Direct
R19. Mega marketing			Direct
R20. Alliances that change market mechanisms		Very indirect	Indirect
R21. The knowledge relationship	Direct		
R22. Mega alliances that change the market structure		Very indirect	
Nano relationships			
R24. The internal market		Direct	Direct
R25. The internal customer relationship	Indirect	Direct	Direct
R26. Quality and customer orientation	Direct	Direct	Direct
R27. Internal marketing	Very indirect	Indirect	Indirect

Chapter 2 by Morris particularly aligns with *mega relationships*, while **Chapter 3** by Druskat and Druskat primarily has a focus on nano relationships and how these affect *classic market* and *special market relationships*. **Chapter 4** by Smyth concerns *nano*, *classic* and *special market relationships* (Gummesson, 2001). Table II.I addresses the most relevant of the relationships (Rs) in more detail. To an extent the allocation of the degree of importance is subjective; however, it provides some guidance. The allocation tries to take into account the content of each chapter, and therefore the spread may look different from another chapter covering the same issues from a different angle.

Gummesson confined his relationships to those that affect marketing. **Chapter 1** recognised a wider range of *nano* or internal relationships that are relevant (see Table 1.2). Table II.II addresses internal relations for this section. This can then be located within the context of portfolios, programmes and the social environment for projects (Figure II.II).

Figure II.II shows the players immediately involved with project delivery – client, design team and contractor. However, in reality there is a distinction

Table II.II Contributions for internal relationships.

Relationships	Chapter 2 Morris	Chapter 3 Druskat and Druskat	Chapter 4 Smyth
Internal structural relationships			
R31. Line management relations		Direct	Direct
R32. Mentor relations (cf. R28)		Direct	Indirect
R33. Peer relations		Direct	Direct
R34. System relationships and communication	Direct	Direct	Direct
Internal functional relationships			
R35. HR relations		Direct	Direct
R36. Immediate job function relations	Indirect		Indirect
R37. Other departmental relations	Indirect		Indirect
R38. Interdepartmental relations	Direct		Direct
R39. Circumstantial relations		Indirect	Direct
Strategic relationships			
R40. Formal relationships	Direct	Direct	Direct
R41. Career and organisational politics	Indirect	Indirect	Indirect
R42. Embedded relationships	Direct	Direct	Direct
R43. Cultural relations	Indirect	Direct	Indirect
R44. Vision and mission relationships	Direct	Direct	
R45. Strategy identification	Direct	Direct	Direct
R44. Strategy implementation-tactical response	Direct	Direct	Direct
Tactical relationships			
R45. Heroes, champions and role models	Indirect		
R46. Nurturers		Direct	Direct
R47. Supporters, members and fan clubs	Indirect	Direct	Direct
R47. Surrogate relationships	Indirect		Direct
R46. Relationships of dissent and blame			Direct
R47. Jezebel relationships		Indirect	Direct
R48. Rebels and Luddites		Indirect	Direct
Personal relationships			
R49. The affirming-orientated relationships		Direct	Direct
R50. The serving-orientated relationships	Indirect	Direct	Direct
R51. The performance-orientated relationship	Direct	Indirect	Indirect
R50. The appearance-orientated relationships	Indirect	Indirect	

between the contracting organisation as a firm and its involvement in any project as part of its own range of projects, and the management of the firm rather than projects. This range of projects has been called a 'portfolio' in Figure II.II in order to distinguish it from a client programme that the contractor is undertaking singly or with competitors, which would also be a programme for the contractor. Figure II.III endeavours to capture connections between the portfolio management by the firm and the projects, applying the relationship management paradigm derived from Figure 1.10.

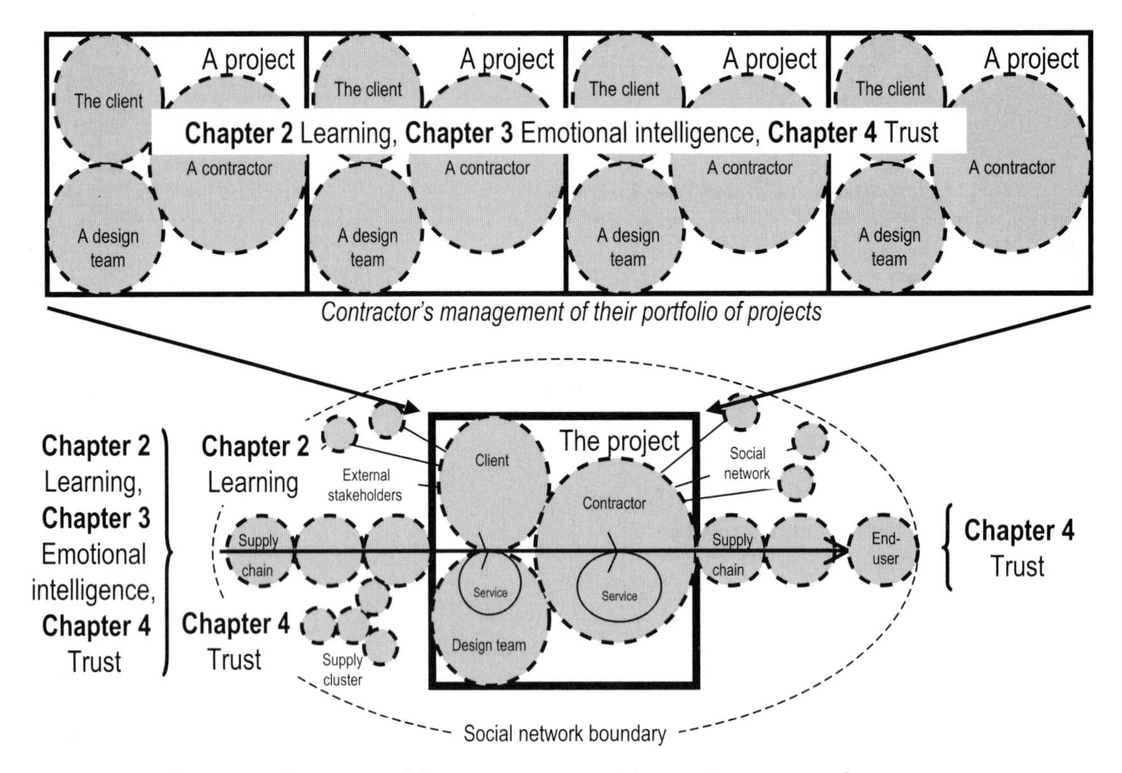

Fig II.II Contributions in relation to portfolios, programmes and the social environment for projects.

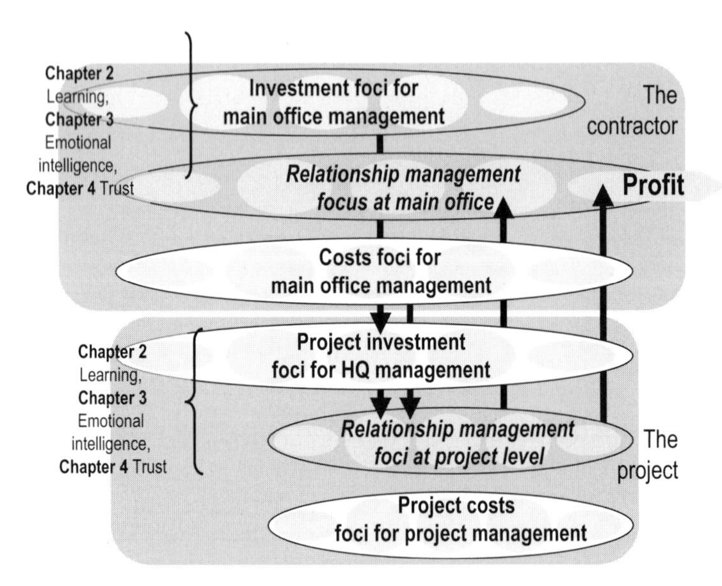

Fig II.III Contributions to the interrelationship between relationship management for projects and the main office.

This introduction to **Section II** has focused on the context, contributions and framework of the three chapters to a relationship approach with particular emphasis on mindsets, behaviour and competencies.

References

Greenberg, J. & Baron, R. A. (2003) *Behaviour in Organizations*. Prentice Hall, Englewood Cliffs, NJ.

Gruneberg, S. L. & Ive, G. J. (2000) *The Economics of the Modern Construction Firm*. Macmillan, Basingstoke.

Gummesson, E. (2001) *Total Relationship Marketing*. Butterworth-Heinemann, Oxford.

Handy, C. B. (1992) *Understanding Organizations*. Penguin, Harmondsworth.

Milgrom, P. & Roberts, J. (1992) *Economics, Organization and Management*. Prentice-Hall, Englewood Cliffs, NJ.

Mintzberg, H., Quinn, J. & Ghoshal, S. (2003) *The Strategy Process: Concepts, Contexts and Cases* (4th edn). Pearson Education, Harlow.

Robbins, S. P. (2003) *Organizational Behaviour*, Prentice Hall, Englewood Cliffs, NJ.

Smyth, H. J. (2000) *Marketing and Selling Construction Services*. Blackwell Science, Oxford.

Smyth, H. J. (2004) Competencies for improving construction performance: theories and practice for developing capacity. *Journal of International Construction Management*, April, 41–56.

Spender, J-C. (1989) *Industry Recipes: An Enquiry into the Nature and Sources of Managerial Judgement*. Basil Blackwell, Oxford.

Tuckman, B. W. & Jensen, M. A. (1977) Stages of small group development revisited. *Group and Organization Studies*, **2**, 419–27.

Winch, G. M. (2002) *Managing the Construction Project*. Blackwell Science, Oxford.

2 How do we learn to manage projects better?

Peter W. G. Morris

Recent years have seen a wealth of initiatives aimed at improving the delivery of projects – in all sectors, including construction. Most, fortunately, share consistent messages, although there are still divergences of view about what is needed to manage projects effectively. Most end up promoting the need to embed best practices through a programme of learning.

This chapter asks what we really know about the management of projects and how we can best go about learning how to assimilate and disseminate this knowledge. It emphasises the situational and contextual nature of this knowledge and the propensity towards a social basis of learning about it.

Building our knowledge about managing projects

Projects are central to the effectiveness of service delivery in many industry sectors, yet too often they are not managed well and our knowledge of how to manage them better is imperfect and poorly communicated.

The IT sector, for example, is regularly excoriated for its frequent failures in delivering projects on time, in budget or without the benefits intended or requirements met (NAO, 2004a; The Standish Group, 1994; Wateridge, 1998; Yardley, 2002; Yeo, 2000). Delays, cost overruns and poor performance are regularly commented upon in defence projects (NAO, 2004b). Large construction and engineering projects have been the subject of critical enquiry (Flyvberg *et al.*, 2002; Miller & Lessard, 2000). There is now a burgeoning literature on project success and failure (Cooke-Davies, 2004).

It is easy to get carried away by such data, however. Often fundamental issues are masked. Might it not have been in the sponsor's interests to allow budgets to be exceeded if business benefit were to be improved, for example? Indeed, what exactly are the appropriate measures of success (Cooke-Davies, 2004)? What happens if these do not align and we have conflicting measures?

Being sensitive to context

In drawing inferences from such data, it is important to be clear on the perspective being used. Which are the relevant or important stakeholders, for example, and why? Needs as well as practices vary by project type and circumstance (Crawford *et al.*, 2005; Shenhar & Dvir, 2004), and not least by industry sector (Morris, 2003b). We should be wary of too coarse a descriptor of industry sectors, however. ICT (information and communications technology) is often treated as a sector but it is not homogeneous: IT-enabled organisational change, for example, is completely different from software development – embedded or new.

Construction is particularly vulnerable to misleading labelling and inaccurate aggregation. It is bigger than just building and civil engineering, the sectors frequently described synonymously with it. Indeed these can be broken down into significant groupings: housing, commercial, and so on. Many in the oil and gas, chemicals, power, paper and pulp, food and similar process engineering sectors would naturally refer to themselves as being in construction. Rail and some water projects would see themselves as part process engineering, part civil engineering. We would really do better referring to the construction industr*ies*:

- *Building:* housing, commercial, social infrastructure (such as hospitals and schools)
- *Civil engineering:* roads, bridges, harbours, rail, water
- *Process engineering:* power, oil and gas, chemicals, paper and pulp

Each has to some degree its own practices, cultures, and needs (Morris, 2004a). And then even within these sectors the pattern is not homogeneous. The UK building and civil construction industry consists of about 200 000 companies. The top 95 represent about 0.05% of the total yet they generate 21% of the industry output. Small firms account for 93% of all firms yet they only generate 28% of the output. The types of projects, contractual relationships, clients, social relationship patterns, learning and development opportunities, and so on, vary considerably between the bigger and the smaller companies.

If we do not recognise the real nature of these sectoral differences we may be in danger of making inappropriate generalisations and of framing management knowledge inaccurately. As this chapter will argue, management knowledge is contextual; in shaping and communicating our management knowledge we need to be alive to the nuances of context – the project or task, business unit or enterprise, industry sector, national and professional standards, and so on.

This tension between the general inference or recommendation and the more precise is a challenge that arises for any student of management and is one that runs through all attempts to articulate guidance on how best to manage projects. In construction, however, the journey is still at a relatively early stage – curiously given its projects' bias – at least in terms of any formal body of knowledge.

Recommendations from the construction industries on managing projects

With its history of overruns and contractual disputes, the UK building and civil engineering sector has been the subject of many official and semi-official enquiries over the last 50 years (Murray & Langford, 2003). Generally the critiques have addressed the institutional and contractual practices of the industry. Only since the Latham report (Latham, 1994) have the recommendations begun to be framed in terms of some formal expertise in the management of projects. In fact, at the time of writing, despite a few notable academic efforts (for example, Walker, 2002; Winch, 2002), the building and civil engineering industries are only beginning to be explicit in formulating their knowledge about managing projects and in positioning such knowledge as important.[1] It is true there is a guide to construction project management published by the Chartered Institute of Building (2002), but it is quite inadequate in its treatment of the *front-end* – business, developmental and design areas of construction projects – and has a distinctly site-orientated production tone. Latham and the two successor Egan reports are pretty thin in terms of specific guidance on how to manage construction projects better, though in fairness this was not their remit.

The current UK building and civil engineering management scene is heavily orientated around the call to arms represented by the Latham report, *Constructing the Team* (Latham, 1994) and the two subsequent Egan reports, *Rethinking Construction* (Egan, 1998) and *Accelerating Change* (Strategic Forum for Construction, 2002). All three have general recommendations about what must be done to improve UK construction performance. Latham concentrated largely on contractual and procurement matters, though notably, and to the report's credit, much attention was also given to briefing. The project management profession was given recognition in the framing of the report and *The New Engineering Contract* (Thomas Telford, 2005), which is explicitly modelled on project management practice, is recommended as a preferred contract form. This said, it would be difficult to infer guidance on how to manage projects from *Constructing the Team*.

Although *Rethinking Construction* is the more explicit in recommendations that have implications for the management of projects, Egan in fact gave less formal recognition to the discipline than Latham: project management is not

---

1. The process engineering industries are ahead of building and civil engineering here. They were about a decade earlier in recognising the need for improvement with the founding of the Construction Industry Institute (CII) in the USA, following concern at falling productivity in the US construction industry. The CII continues as an active force in the broad US construction sector (www.construction-industry institute.org). Its European process industry equivalent is the European Construction Institute (ECI) (www.eci.org). ECI incorporates many of the lessons learned from the two earlier process engineering initiatives – the offshore CRINE (Cost Reduction In the New Era) of the 1980s and onshore ACTIVE (Achieving Cost Reduction Through Innovation and Value Engineering) of the 1990s (National Economic Development Office, 1991).

even mentioned in either of the Egan reports[2] (and in fact one cannot help suspecting that the authors never seriously considered that there really is such a discipline, at least not one that is of any relevance to construction). The report sets out a number of targets that should be achieved on an annual basis and identifies the following as key to the industry's improvement:

- *Five key drivers* – committed leadership, customer focus, integrated teams, quality, and a commitment to people
- *Four key processes* – partnering the supply chain, product improvement, user-focused integration, and waste elimination in the construction process

Detail behind how to implement these recommendations was generally lacking, however, and other groups were assigned the task of working out how they were to be achieved. *Accelerating Change* follows a similar style, though it is if anything even less specific.

By 2005 *Constructing Excellence in the Built Environment* (www.constructingexcellence.org), an industry–government-funded 'best practice' ginger organisation, had become the principal body for realising the post-Egan agenda within the UK building and civil engineering industries. Its advice on how to manage projects is still quite diffuse and difficult to pin down, as a word-search in its web area on 'project management' or 'projects' would, at the time of writing, quickly demonstrate and as a wider search through its 'best practice knowledge' area, or elsewhere, would further substantiate.

Though looking at data on delivery performance and difficulties is a good way of seeing where management attention may be needed, neither *Rethinking Construction* nor *Accelerating Change* actually furnish such data and hence fail to pinpoint what actually should be addressed. The UK Treasury's *Review of Large Public Procurement in the UK* (www.hm-treasury.gov.uk, 2002) is more helpful. Though its focus is on optimism bias in estimating (cf. Flyvberg et al., 2002), it contains evidence of risk areas impacting upon the management of projects (see p.8 and Appendix E), these being, in descending order:

- Inadequate business case
- Environmental impact
- Disputes and claims
- Economic factors
- Late contractor input into design
- Contractual complexity
- Legislation
- Inadequate innovation

2. 'Project implementation' is referred to, however, being positioned as project execution, following 'project development'. The execution orientation too often associated exclusively with the discipline is a point we shall pick up later with reference to the *PMBOK Guide®*.

- Poor contractor capability
- Inadequate project management team

It then makes recommendations on project management actions that should be taken to address these risk areas:

- Optioneering – benefits management, strategy
- Risk management
- Change management
- Stakeholder management
- Communications
- Purchasing

Though general and high level, and clearly biased towards the *front-end*, the recommendations are a useful counterbalance to the more downstream CIOB document on construction project management (Chartered Institute of Building, 2002).

Benchmarking data in the process engineering industries reinforces the importance of the *front-end*. It clearly demonstrates that greater emphasis and more time spent on front-end issues are correlated with better outturn performance (www.cii-benchmarking.org). Indeed, it is this greater attention to front-end definition (Morris, 2005) and alignment with business benefits (Morris & Jamieson, 2004) that begins to mark out the broader perspective that I have called 'the management of projects' (Morris, 1994; Morris & Pinto, 2004).

The case for a broader management of projects view of the knowledge required to manage projects begins to be further supported by the two post-Egan National Audit Office (NAO, 2001, 2005; cf. NAO, 2004a, 2004b for IT and defence) reports on construction. The first, *Modernising Construction*, showed 73% of public sector construction being delivered over budget and 70% being delivered late. Major barriers to improving construction performance were shown to be:

- Inadequate appreciation of the sponsor's role
- Poor need identification, poor briefing and requirements definition and lack of focus on the business case
- Use of prescriptive specifications, weaknesses in design, separate early appointment of designers, late variations, too little use of prefabrication and standardisation
- Little understanding of value management and insufficient appreciation of value engineering or whole life costing
- Limited project management skills, unclear roles and responsibilities, inadequate quality assurance, and a tendency to transfer risk inappropriately through standard contract clauses
- Selecting contractors on the basis of price rather than quality and insufficient incentives for contractors to propose innovative solutions
- Lack of post-project appraisals and learning

An Office of Government Commerce (OGC, 2003) report on progress in implementing Egan's recommendations, *Building on Success* (www.ogc.gov.uk), claimed to show substantial attainment of key recommendations – fully empowered sponsors, training, '93% [of government agencies audited] apply 100% best practice project management', and so on – though interestingly, there is no mention of the impact on outturn performance, and the figures themselves seem questionable (for example, is 100% really plausible?). The slight sense of incredulity is underlined by a later NAO review, *Improving Public Services through better construction* (NAO, 2005). NAO found an improvement in performance – 55% of projects were being delivered on budget and 63% on time, both poor figures – and saw opportunities for improving value-for-money in:

- Improved productivity through streamlined planning and procurement, starting sooner on site, faster building, and larger work programmes
- Collaborative working through integrated teams, earlier contractor involvement, use of non-adversarial forms of contract
- Savings in whole life costs through better design

Specific recommendations included:

- Greater emphasis on 'programme management' – inter-project scheduling and better allocation of resources; greater attention to value and whole life issues
- Organisational support for building enterprise-wide project management capability, better informed sponsoring management boards
- Spending more time in 'front-end' planning
- More bespoke procurement solutions and more intelligent tender management
- More attention to building integrated teams across the whole supply chain
- Better performance evaluation and embedding of project learning

It is interesting how much of the advice coming out of Latham, Egan and the NAO is about people and organisations – leadership, teams, respect for people, roles, supply chain partnering – and how little about the tools and techniques of project management. And interesting, too, how terms like programme management, enterprise-wide project management support, and maturity are present – very much vogue words of the time, all quite heavily promoted by the OGC; the government department specifically tasked with improving public procurement and delivery performance.

NAO in its 2005 report is quite explicit in its recommendations to the public sector to go to OGC for advice on how to manage projects effectively. OGC is much closer to mainstream project management-speak than the tenor of the construction advice offered above; much more about practices and tools and less about people and learning – less about relationships but more about technology – whereas other construction recommendations have been emphasising the people/social dimensions to improving performance and barely mentioning the formal tools. The reason for this probably lies at least partly in OGC's provenance, which we shall turn to next. Whether it over-errs towards tools and away from people, relationships and the social basis of learning is something this chapter seeks to question, however.

More generic views on managing projects

OGC has its origins, both organisationally and intellectually, in the ICT sector. The government's Central Communications and Telecommunications Agency (CCTA) is an important root of OGC. The OGC has been remarkably influential in promoting project management as a professional discipline within government (and beyond), doing this through a series of actions – training, mandatory practices such as gateway reviews for major projects, and guidance documents and methodologies, most notably PRINCE and *Managing Successful Programmes*.

The CCTA developed PRINCE (**Pr**ojects **in** a **C**ontrolled **E**nvironment) in 1989 from an information systems project management method known as PROMPTII developed in 1975 (CCTA, 1990). (PROMPTII was adopted by the CCTA in 1979 as the standard to be used for all government information system projects.) Subsequently PRINCE2 was launched in 1996 (OGC, 2002). Both PRINCE and PRINCE2 are methodologies for managing 'single projects' initially, with PRINCE clearly orientated towards the IT environment but PRINCE2 more ambitiously targeted towards any project 'in a controlled environment'. In reality the orientation is still very biased towards IT-enabled organisational change projects, as can be seen, for example, in the emphasis on Configuration Management, which would seem quite strange to construction executives. OGC also issued guidance on multiple projects related to achieving strategic change programme management in 1999 with its *Managing Successful Programmes* (OGC, 1999).

The promotion of tools and methods by OGC arose very largely in response to the high failure rate of ICT projects without, so far as I am aware, any evidence that the methodologies used necessarily correlate with improved outturn performance, a point we shall return to at the end of this chapter with the comments on 'systemic knowledge assets'. With the government's increasing concern over improving delivery performance in general, OGC was transferred in 2000 to the Treasury and in 2005 was given a remit for support overview for all government departments, including defence, for delivery efficiency.

The point of this slight historical detour is to explain why public sector construction departments, when offered advice on how best to manage construction projects, are referred to a unit offering IT-flavoured project management advice, as may be seen, for example, in OGC's *Successful Delivery Toolkit* (www.ogc.gov.uk), which covers both programme management and project management, with the latter including business case management, requirements management, planning and control, project closure and hand-over, project review/evaluation, risk management, quality management, procurement, contract management, teams, change agents, performance management, HSE (health, safety and environment), and regulatory compliance. (Note that the *Successful Delivery Toolkit* website has a separate Construction area called Achieving Excellence.)

OGC has provided a real service to the project management community but its locus is clearly the public sector. The professional project management associations such as the US headquartered, but very international, Project Management Institute (PMI) (www.pmi.org), with over 160 000 members, or the UK-based Association for Project Management (APM) (www.apm.org.uk), with over 14 000 members, are probably the two groups that have done most to try to document more generally our knowledge on how to manage projects effectively. This has been done via project management 'bodies of knowledge' (APM, 2000, 2005; PMI, 2002). There are significant other groups too. The French, Swiss and German professional bodies have versions highly aligned to the APM body of knowledge (BOK). The International Project Management Association (www.ipma.ch), a federation of some 50 national project management associations, similarly has a generic quasi BOK, called the *International Competency Baseline* – the ICB (Caupin *et al.*, 1998) – based on the APM, French and German BOKs. China has a BOK based on the ICB, as does the Czech Republic. The Engineering Advancement Association of Japan (www.enaa.or.op.jp) has an interesting, more radical version, only partly translated into English at the time of writing (PMCC, 2002).

The BOKs represent a vibrant attempt by the project management professional communities for practitioners to produce a socially based articulation of management knowledge (with all the challenges and tensions this inevitably entails), the explicit aim being to improve performance via learning and qualifications (similarly difficult areas). This effort to 'reify' socially based learning within a professional context is now beginning to attract the attention of the research community (Morris *et al.*, forthcoming).

More specifically within the context of this chapter, however, the BOKs are interesting for at least two reasons. One is the scope of what they consider to be the knowledge required to manage projects. A second is the argument raised by some critics against the very attempt to try to define explicitly knowledge in such a domain.

Differing formal models of project management knowledge – the different bodies of knowledge

PMI defines project management in the *PMBOK Guide*®, its body of knowledge, as 'the application of knowledge, skills, tools and techniques to project activities to meet project requirements' (PMI, 2004, p.8). The key point is the last four words, for the *PMBOK Guide*® takes the project requirements as already defined. The emphasis is totally on execution delivery. This is despite the evidence that much of what goes wrong in projects can be traced to issues arising at the project's front-end (Miller & Lessard, 2000; Morris, 1994; Morris & Hough, 1987), and similarly, as we saw above, that the greatest opportunities for optimising the project lie in the front-end stages (Morris, 2005; Morris & Jamieson, 2004).

Hence the *PMBOK Guide®* is silent on strategy and business objectives, technology management, requirements management, value management, governance, stakeholder relationships and leadership as well as a whole range of similar front-end issues. Instead its content focuses on managing integration, scope, time, cost, quality, human resources, communications, risk, and procurement. Perhaps it is this extreme identification of project management as an execution-only activity that has, arguably, rendered the topic uninteresting intellectually to so many people, such as Egan and his committees (Green, 2005; Morris, 2001, 2003a, 2003b). As Aristotle said, defining the problem is half the solution!

A direct consequence of this execution-orientated view of the discipline is that several commentators have criticised it as overly control-orientated and mechanistic (Hodgson & Cicmil, 2006; Koskela & Howell, 2002; Thomas, 2005; Williams, 2004), pointing out the inappropriateness of such a methods-based approach to a management challenge that is inevitably highly contextual, and have called for a broadening out of the subject (Packendorf, 1995). This broader view seems to be developing some resonance (Davies & Hobday, 2005; Söderlund, 2004a, 2004b; Winch, 2002), but an inescapable consequence of these different perspectives is that different formal views of the domain now exist. There are at least two others: 'programme management' and 'the management of projects'.

Programme management, at least in the UK usage of the term (US usage is more variable), is essentially concerned with the management of interrelated projects to achieve a strategic objective (OGC, 1999; Reiss, 1996; Thiry, 2004). Some see programme management as a distinct discipline separate from, and above, project management. The 'management of projects', however, looks at the discipline in a more rounded, holistic manner, seeking to cover all that needs addressing to develop and deliver the project successfully (Morris, 1994; Morris & Pinto, 2004). The Association for Project Management body of knowledge (APM 2000, 2005) is 'management of projects'-based – BS 6079 (BSI, 2002) and other standards such as those of OGC (1999, 2002) or CIOB (2002) are less comprehensive and lie somewhere in between.

The 'management of projects' perspective arises from studies of the factors giving rise to success and failures in projects (Cooke-Davies, 2004; Morris & Hough, 1987). This research emphasised the importance of managing contextual and definitional aspects of projects in addition to the usual execution, planning and control type issues. This perspective on the scope of the discipline, and hence the knowledge needed to manage projects effectively, has strongly informed the framework, and the content, used in the APM BOK, including such things as the importance of:

- Understanding context
- Aligning with business strategy and shaping project and programme strategy

- Value managing the project definition and having a benefits management programme in place
- Balancing opportunity with risk
- Making quality and HSE central
- Having a clear stage gate life cycle, extending into operations and maintenance, with a number of assurance reviews
- Having clear sponsor/project team roles and responsibilities, including briefing and requirements management (including testing)
- Technology and design management

This vastly larger scope is much more aligned to the overall reality of developing and delivering most projects, but none of these topics are in the *PMBOK Guide*®.

Nor, crucially, are people and the behavioural areas, including such items as leadership, teams, negotiation and influencing, conflict management, decision-making, communications, and, not least, learning and development. In fact, most of the bodies of knowledge, and national and international standards, are weak in the people factors (though the APM BOK has more than the other bodies of knowledge). Projects, and programmes, essentially begin and end with people doing things, and their behaviour in implementing practice determines how effectively projects will be managed. Yet the generic BOKs, standards and guides are generally light if not downright deficient at describing people issues – perhaps because of their largely engineering origin and methodology nature, perhaps because formalising this aspect of our knowledge fairly succinctly will always be difficult.

Context and tacit knowledge

While the underemphasis on people almost certainly is a serious omission, having more than one view of what constitutes knowledge for managing projects may not necessarily be wrong. After all, management of knowledge is pre-eminently contextual (Griseri, 2002) and situated (Lave & Wenger, 1991). What would be wrong would be if incomplete or fallacious models were being proposed, particularly as standards. (Here the *PMBOK Guide*®, with its 'execution-only' orientation, is particularly vulnerable as it has been adopted and is being promoted as an ANSI (American National Standards Institute) standard. And BS (British Standards) 6079 is probably vulnerable, too, to the charge of underplaying the importance of people.) Hence Hodgson and Cicmil (2005) are right, surely, to warn against attempts by project management authors to be too rigid in their definitions of the subject, though in fairness the *PMBOK Guide*® is quite explicit in saying that it is up to the project manager to decide which elements of the *Guide* should be applied in any one situation.

We might therefore quite reasonably expect to find different versions of 'the body of knowledge' for different project contexts: one perhaps for building and civil engineering; one for process engineering; one for software development; one for IT-enabled organisational change – or in fact several in each case: one for company A, one for company B. In fact, as has been argued elsewhere (Morris, 2004b), one could conceive of different versions being appropriate for each project or programme within each business unit within each company within each sector. Unfortunately, it might not be cost effective to develop one at the level of individual projects or even business units. The generic bodies of knowledge, standards, OGC guides and suchlike offer a ready-made point of departure from which individual competency support and development initiatives can take off.

Criticisms that it is inappropriate to try to make explicit knowledge that is inevitably to some extent contextual, situated and tacit, such as are to be found in the post-modernist analyses of Thomas, Hodgson and Cicmil, and even Bresnen (in Hodgson & Cicmil, 2006) miss the point. Communicating socially derived knowledge is inevitably incomplete and inadequate, but is often worthwhile or necessary. Why? Because, hopefully, it helps; it provides a framework around which the real learning can occur. It helps us see things better, helps us understand what we have difficulty with, and helps keep us up to date. Even strong proponents of socially based learning such as Wenger see the making explicit of socially derived knowledge ('reification') as essential (Wenger, 1998). But what is important is that:

(a) The right points are being made – and in this sense the *PMBOK Guide*® is, it is contended, seriously deficient
(b) Such knowledge is acknowledged as being grounded in a social reality, both personal and organisational – simply telling people to read manuals, look up lessons-learned registers, or attend two- or three-day training events will not generate real knowledge on how to manage projects

The social basis of learning about managing projects

People learn in many different ways, and there are many theories of the way they do so, including those orientated around the individual (psychological and physiological ones) and those around the organisation. While understanding how the individual learns is clearly an important starting point, much attention has been paid over the last decade or so to the ways that organisations manage their knowledge and learn as entities in their own right. Again, there are many perspectives on how this may be addressed. Easterby-Smith (1997) has suggested six basic frameworks:

- Psychology and organisational development
- Management science

- Sociology and organisation theory
- The strategic perspective
- The production management perspective
- The cultural perspective

Importantly, of course, these theoretical frameworks overlap. Griseri (2002) has argued strongly that, while these differences may make sense theoretically, in practice managers create and contrive a holistic conceptual framework for dealing with the problems they face. Similarly, practitioners of the management of projects will want to adopt elements of such theoretical approaches that best suit their contextual needs. But what all will find is that just as projects largely start and end with people, so social relations underscore most of what they need to learn and apply about the management of projects.

Two of the most influential theories of organisational learning in the last decade, for example, position socialisation at the heart of their theories, though in quite different ways (and taking different usages of the term).

Communities of practice and knowledge creation

Wenger *et al.* (Wenger, 1998; Wenger *et al.*, 2002), in their theories of communities of practice, put social participation at the heart of learning, with knowledge arising directly as the result of participation in the pursuit of 'valued enterprises' and learning arising from doing (practice), becoming (identity), belonging (community), and experiencing (meaning: situated experience). Knowledge and learning is inevitably, for a social science like management, developed from practice, and this practice is generally community based. To develop such knowledge and learning, contact needs to be maintained with other practitioners, people who are expert in their practice area and appropriately alive to its context. Individuals (subject matter experts) and groups of people (communities of practice practitioners) become primary engines of learning support. Learning is often best carried out through social interaction such as joint work activities or games. Learning support should encourage such social interaction – via, for example, facilitation, group work, or IT tools such as portals, which mix explicitly presented information with links to people.

An important dimension of subject matter experts' value is that so much of their expertise is represented as tacit knowledge – that which, following Polanyi (1958), is embedded in a person's experience and often difficult to articulate clearly; explicit knowledge being that which is 'readily available'. This has become the basis of a model by Nonaka and Takeuchi (1995) in which fresh knowledge is created in organisations. Research conducted at UCL (Morris & Loch, 2004) shows that the tacit-to-tacit exchange, termed by Nonaka and Takeuchi 'Socialisation', is the dominant mode in projects, and would seem particularly preferred in construction-based organisations.

Socialisation as the preferred mode of knowledge interaction in projects and construction

The research began by looking at the value of web-based knowledge portals; the specific task was to develop a knowledge system to support a contractor in preparing bids for Liquefied Natural Gas (LNG) projects. The design of the system was based on the framework developed by Dixon on the most appropriate form of 'knowledge transfer' (Dixon, 2000). Dixon proposed that knowledge transfer will be strongly influenced by:

(a) The nature of the task – specifically whether or not it is frequent and/or routine
(b) Whether the type of knowledge is tacit or explicit

The research found that the 'nature of the task' schema proposed by Dixon did not really work – the LNG bidding knowledge transfer did not quite fit any of the categories Dixon had proposed (Morris *et al.*, 2003). The distinction between tacit and explicit knowledge resonated strongly, however. The contractor found that one of the most useful features of the system was its ability to use explicit knowledge to point users to key people (subject matter experts) who could be interrogated for their tacit knowledge on the subject (Ayas & Zeniuk, 2001). It was realised that the interplay between tacit knowledge and explicit knowledge was central to the organisation's ability to learn, and that merely managing knowledge without achieving organisational learning was of limited business benefit. Accordingly, we set up a study to look at the interaction between tacit and explicit knowledge in project-based organisations, comparing particularly practice in construction and ICT organisations.

A decade or so ago, most theories of organisational learning considered that knowledge development alone constituted learning (Senge, 1990). In projects, the practice of systematically collecting 'lessons learned' and putting these into some kind of database for future reference became the standard formula for best practice project-based learning (Brander-Löf *et al.*, 2000). But given the insights from the earlier research we concluded that, to improve performance, learning requires the generation, and application, of *new* knowledge.

Several models have been posed in terms of knowledge acquisition, diffusion, sense-making, action and storage. The well-known 'double loop' feedback model of Argyris and Schön (1978), where more strategic and cognitive adjustment is supposed to happen, can be contrasted with simpler 'single loop' models such as those of Levinthal and March (1993) and Miner (1990). More recently spiral models have been proposed by Nonaka and Takeuchi (1995) and by Boisot (1998). Nonaka's model is the more well-known and developed, though it is not without critics.

The Nonaka model proposes, inter alia, that knowledge creation proceeds in a spiral form as tacit knowledge is shared via 'socialisation' – person-to-person exchange of primarily tacitly held knowledge – then moves from the tacit form to the explicit via 'externalisation', which is then 'combined' with

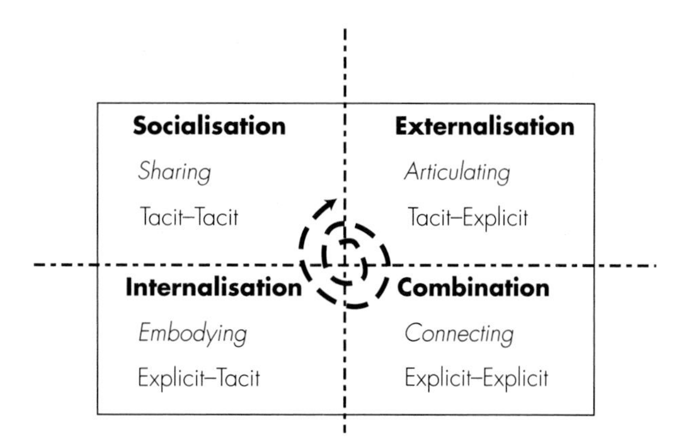

Fig 2.1 The Nonaka and Takeuchi knowledge-creation spiral.

other explicit knowledge into a new explicit form, from whence it is 'internalised' – that is, converted into new tacit knowledge – afresh (Figure 2.1). Many commentators dispute the proposed clarity of what must, after all, be a very complex affair, and note the lack of empirical evidence of such a progression. To others, however, including the UCL team, the sequence seems plausible and reasonable, and worth adopting as an investigative framework. Specifically, the UCL team wanted to profile the modes of knowledge exchange practices in each of the four 'SECI' modes of tacit–explicit knowledge interaction in project-based companies as opposed to investigating whether there was movement between the SECI quadrants.

Data collected from 64 respondents in eight organisations, three of which were construction firms, showed an overall bias towards the 'socialisation' mode, a strong bias against 'externalisation', a lesser bias against 'combination', and a smaller bias against 'internalisation' (quantitative data is given in Morris & Loch, 2004).

The importance of tacit–tacit knowledge interaction

Socialisation emerged as *the* most important of the four SECI modes. All three of the construction organisations showed a strong bias towards socialisation – the result of the deliberate person-to-person knowledge management strategies in two cases and of the adoption of a deliberately 'balanced' approach between the four SECI modes in the other. (Socialisation is being defined here in the very specific sense of the SECI tacit-to-tacit knowledge interaction. It is aligned with the broader usage of the word, however, as one where social relations are important and as such appears to resonate with the character of the construction industry, which, as we have already observed, is very much a people-orientated one.)

Of the three ICT-based organisations, two were weak in socialisation. The third was in fact the ICT arm of a global energy major, spending $2 billion a year on ICT projects. This organisation was strongly socialisation-orientated: heavy emphasis on person-to-person interaction, learning by narratives, and so on. Thus it would appear that ICT does not necessarily have to imply less tacit-to-tacit knowledge exchange: culture and intent would seem to be more important.

Barriers to intended learning across the project–enterprise interface

Most of the organisations studied experienced difficulties and frustration in implementing their chosen project–enterprise learning strategies. One organisation was unable to get the degree of socialisation it wanted, largely because of culture and its emphasis on technology; another was unable to change the focus of its knowledge creation from transaction support to real learning about projects; a third found difficulty getting people to focus on organisational learning. In general, culture, strategic intent and organisational constraints were found to inhibit the application of well-perceived knowledge management and organisational learning best practices.

Similarly, in all eight organisations we saw abstraction and sense-making (Weick, 1995) at the enterprise level of project learning but difficulty in getting this learning transferred to other projects – often because of the challenge of getting people to bother and a scepticism that such learning would actually lead to improved performance.

Most of the organisations used manuals, guidelines and other forms of 'systemic knowledge assets' (Nonaka & Takeuchi, 1995) as a core means of passing on knowledge to new projects. This is very different, however, from knowledge creation of the more cognitive type proposed by Argyris and Schön (1978) or Nonaka and Takeuchi (1995) – where the primary emphasis for learning was on expecting people to use these systemic assets there was little evidence of really effective learning.

The business driver for getting feedback and for the formalisation of a body of knowledge

All the organisations experienced difficulties in getting feedback from projects to the enterprise – in getting people to take the subject of project-based learning seriously. Of the two organisations that were best at this, one had knowledge managers and senior management support, and the other relied on the power of an IT portal to create a demand and to 'pull' information onto, and from, projects, thus emphasising the importance of socialisation. Both of these organisations were suppliers, however, and it is noticeable how strongly biased their project-based learning was towards transaction type knowledge – sharing, for example, technical or commercial information rather than project management learning.

Interestingly, it was largely the owner-client organisations rather than the suppliers who seemed to believe that project management 'good practice' could usefully be explicitly documented and formalised as a body of knowledge. While this is by no means true of all suppliers, one can see how organisations that take on a wide array of projects might find it harder and less worthwhile to structure their project management knowledge: it is more variable. (And this is particularly true of building, which as a construction sector is notable for the wide variability in its work – large numbers of different clients, varying types of projects, different supply chain configurations, and so on.) This may explain to some extent why construction, so long dominated by general contractors, has been slow to begin formalising knowledge on the management of projects, the initiative coming eventually from owners and external bodies such as the OGC, CIOB, BSI and APM. The owners, on the other hand, have less variability, greater contextual consistency, and perhaps crucially, a tighter linkage between their business processes and project processes. This echoes the importance of having a business reason for knowledge transfer – the finding that came out of the original LNG portal case study research.

Conclusions

Projects are important, and prima facie we ought to know how to manage them effectively. Construction, which is highly projects based, has only recently begun formally to structure its knowledge in this area with early initiatives in the process engineering sector now beginning to be matched in building and civil engineering. Latham and Egan led such efforts, though both without really attempting to formulate such knowledge systematically. Recently OGC has begun to do this though it has tended so far to do so – perhaps betraying its ICT provenance – initially in terms of tools, whereas much of construction's concerns have been more about people and organisations.

Too often project management is seen as synonymous with project execution, a view implicit in Egan and to a degree in Latham. To a regrettable extent the profession has only itself to blame: PMI's 'body of knowledge' entirely reflects this perspective. Other associations' BOKs, however, such as APM's, take a more holistic approach, explicitly including the front-end definitional stages of the project. This broader view better addresses the reasons behind projects failing or getting into trouble, as shown by academic research and by a whole series of industry studies, not least in construction.

Documenting such knowledge, as a social as well as a technical process, is only part of what needs doing to help people learn how to manage projects better. The knowledge needs contextualising and organisations need to draw on individuals' tacit knowledge as well as whatever explicit knowledge assets they possess. There are various means by which this can best be done: communities-of-practice, subject matter experts, and the use of IT tools that encourage an interaction between tacit and explicit knowledge, such as portals, for example. Our research suggests in fact that it is tacit-to-tacit knowledge interaction, 'socialisation', that is the preferred mode of knowledge interaction in projects and that this is more effective than merely getting people to refer to manuals and guides ('systemic knowledge assets').

Construction organisations seem particularly biased towards this form of knowledge sharing, which is not surprising really given the strong people-centred nature of the issues it faces in managing projects and the type of actions it believes are necessary to address them.

What would be sad, however, would be if construction as a whole – but particularly the building and civil engineering sectors – were to miss out on the formal knowledge that is available on the management of projects and that has been articulated elsewhere. We can understand how the variability of project type has made it more difficult for contractors and suppliers to formalise such 'bodies of knowledge' – a challenge that has been compensated with the socialised character of construction learning (partially itself a consequence of such variability). Although a lead in doing this is being taken by some owners and institutions, the challenge is to find ways of doing this across the whole supply chain.

References

Argyris, C. & Schön, D. (1978) *Organisational Learning: A Theory of Action Perspective.* Addison-Wesley, Reading, MA.

Association for Project Management (2000) *Project Management Body of Knowledge* (4th edn). APM, High Wycombe.

Association for Project Management (2005) *Body of Knowledge for Managing Projects and Programmes* (5th edn). APM, High Wycombe.

Ayas, K. & Zeniuk, N. (2001) Project-based learning: building communities of reflective practitioners. *Management Learning*, **32** (1), 61–76.

Boisot, M. (1998) *Knowledge Assets*. Oxford University Press, Oxford.

Brander-Löf, I., Hilger, J-W. & André, C. (2000) How to learn from projects: the work improvement review. *IPMA World Congress 2000*. IPMA, Paris.

Bresen, M. (2005) Conflicting and conflated discourses? Project management, organisational change and learning. In: D. Hodgson & S. Cicmil (eds) *Making Projects Critical*. Palgrave, London.

British Standards Institute (2002) *BS 6079–1 Guide to Project Management*. BSI, London.

Caupin, G., Knöpfel, H., Morris, P., Motzel, E. & Pannenbäcker, O. (1998) *ICB IPMA Competence Baseline*. International Project Management Association, Zurich.

Central Communications and Telecommunications Agency (1990) *PRINCE*. National Computing Centre, London.

Chartered Institute of Building (2002) *Code of Practice for Project Management*. CIOB, Ascot.

Cooke-Davies, T. (2004) Project Success. In: P. W. G. Morris & J. K. Pinto (eds) *The Wiley Guide to Managing Projects*. Wiley, Hoboken, NJ.

Crawford, L., Hobbs. J. B. & Turner, J. R. (2005) Project categorization systems and their use in organisations: an empirical study. In: D. P. Slevin, D. L. Cleland & J. K. Pinto (eds) *Innovations: Project Management Research 2004*. Project Management Institute, Newton Square, PA.

Davies, A. & Hobday, M. (2005) *The Business of Projects*. Cambridge University Press, Cambridge.
Dixon, N. M. (2000) *Common Knowledge: How Companies Thrive by Sharing What They Know*. Harvard Business School Press, Boston, MA.
Easterby-Smith, M. (1997) Organizational learning: contributions and critiques. *Human Relations*, **50** (9), 1085–1113.
Egan, J. (1998) *Rethinking Construction: The Report of the Construction Task Force*. Department of Trade and Industry, London.
Flyvberg, B., Bruzelius, N. & Rothengatter, W. (2003) *Megaprojects and Risk: An Anatomy of Ambition*. Cambridge University Press, Cambridge.
Griseri, P. (2002) *Management Knowledge: A Critical View*. Palgrave, London.
Hodgson, D. & Cicmil, S. (2006) Are projects real? The PMBoK and the legitimation of project management knowledge. In: D. Hodgson & S. Cicmil (eds) *Making Projects Critical*. Palgrave, London.
Koskela, L. & Howell, G. (2002) The underlying theory of project management is obsolete. *Proceedings of PMI Research Conference 2002*, 293–302. Project Management Institute, Newton Square, PA.
Latham, M. (1994) *Constructing the Team*. HMSO, London.
Lave, J. & Wenger, E. (1991) *Situated Learning: Legitimate Peripheral Participation*. Cambridge University Press, Cambridge.
Levinthal, D. A. & March, J. G. (1993) Exploration and exploitation in organizational learning. *Strategic Management Journal*, **14**, Winter, 95–112.
Miller, R. & Lessard, D. R. (2000) *The Strategic Management of Large Engineering Projects*. MIT Press, Cambridge, MA.
Miner, A. S. (1990) Structural evolution through idiosyncratic jobs: the potential for unplanned learning. *Organization Science*, 195–210.
Morris, P. W. G. (1994) *The Management of Projects*. Thomas Telford, London.
Morris, P. W. G. (2001) Updating the project management bodies of knowledge. *Project Management Journal*, **32** (3), September, 21–30.
Morris, P. W. G. (2003a) Science, objective knowledge, and the theory of project management. *Proceedings of the Institution of Civil Engineers*, **150** (12641), 82–90.
Morris, P. W. G. (2003b) The (ir)relevance of project management. *Proceedings of International Association for Project Management World Congress*, June, Moscow. IPMA, Zurich.
Morris, P. W. G. (2004a) Project management in the construction industry. In: P. W. G. Morris & J. K. Pinto (eds) *The Wiley Guide to Managing Projects*. Wiley, Hoboken, NJ.
Morris, P. W. G. (2004b) The validity of knowledge in project management and the challenge of learning and competency development. In: P. W. G. Morris & J. K. Pinto (eds) *The Wiley Guide to Managing Projects*. Wiley, Hoboken, NJ.
Morris, P. W. G. & Hough, G. H. (1987) *The Anatomy of Major Projects*. Wiley and Sons, Chichester.
Morris, P. W. G. & Jamieson, H. A. (2004) *Translating Corporate Strategy into Project Strategy*. Project Management Institute, Newton Square, PA.
Morris, P. W. G. & Loch, I. C. A. (2004) Knowledge creation and dissemination (organizational learning) in project-based organizations. In: D. P. Slevin, D. L. Cleland & J. K. Pinto (eds) *Innovations: Project Management Research 2004*. Project Management Institute, Newton Square, PA.
Morris, P. W. G. & Pinto, J. K. (eds) (2004). *The Wiley Guide to Managing Projects*. Wiley, Hoboken, NJ.

Morris, P. W. G., Crawford, L., Hodgson, D., Shepherd, M. & Thomas, J. (2006) The role of the bodies of knowledge in defining a profession – the case of project management. *International Journal of Project Management.*

Morris. P. W. G., Deason, P. M., Ehal, T. M. S., Milburn, R. & Bloomfield, D. (2003) IT support for knowledge management in designer and contractor briefing. *International Journal of IT in Architecture, Engineering and Construction*, **1** (1), 9–24.

Murray, M. & Langford, D. (2003) *Construction Reports.* Blackwell Science, Oxford.

National Audit Office (2001) *Modernising Construction.* Report by the Comptroller and Auditor General HC 87 Session 2000–2001, London.

National Audit Office (2004a) *Major IT Procurement: The Impact of the Office of Government Commerce's Initiatives on Departments and Suppliers in the Delivery of Major IT-enabled Projects.* Report to the Comptroller and Auditor General HC 877: Session 2003–2004, London.

National Audit Office (2004b) *Ministry of Defence: Major Projects Report 2004.* Report to the Comptroller and Auditor General HC 1159–l: Session 2003–2004, London.

National Audit Office (2005) *Improving Public Services through Better Construction.* Report by the Comptroller and Auditor General HC 364–1 Session 2004–2005, London.

National Economic Development Office (1991) *Guidelines for the Management of Major Construction Projects.* NEDO, London.

Nonaka, I. & Takeuchi, H. (1995) *The Knowledge Creating Company.* Oxford University Press, New York.

Office of Government Commerce (1999) *Managing Successful Programmes.* The Stationery Office, Norwich.

Office of Government Commerce (2002) *Managing Successful Projects with PRINCE 2.* The Stationery Office, Norwich.

Office of Government Commerce (2003) *Building on Success.* The Stationery Office, Norwich.

Packendorff, J. (1995) Inquiring into the temporary organisation: new directions for project management research. *Scandinavian Journal of Management*, **11** (4), 319–33.

Polanyi, M. (1958) *Personal Knowledge.* Routledge, London.

Project Management Institute (2004) *A Guide to the Project Management Body of Knowledge* (3rd edn). Project Management Institute, Newton Square, PA.

Project Management Professionals Certification Center (2002) *A Guidebook of Project and Program Management for Enterprise Innovation (P2M) – Summary Translation* (rev. edn). PMCC, Japan.

Reiss, G. (1996) *Programme Management Demystified.* E & F N Spon, London.

Senge, P. (1990) *The Fifth Discipline.* Doubleday, New York.

Shenhar, A. J. & Dvir, D. (2004) How projects differ, and what to do about it. In: P. W. G. Morris & J. K. Pinto (eds) *The Wiley Guide to Managing Projects.* Wiley, Hoboken, NJ.

Shenhar, A. J., Levy, O. & Dvir, D. (1997) Mapping the dimensions of project success. *Project Management Journal*, **28** (2), 5–13.

Söderlund, J. (2004a) Building theories of project management: past research, questions for the future. *International Journal of Project Management*, **22** (3), 183–91.

Söderlund, J. (2004b) On the broadening scope of the research on projects: a review and a model for analysis. *International Journal of Project Management*, **22** (8), 655–68.

The Standish Group (1994) *The CHAOS Report*, www.standishgroup.com

Strategic Forum for Construction (2002) *Accelerating Change*. Strategic Forum for Construction, London.

Thiry, M. (2004). Program management: a strategic decision management process. In: P. W. G. Morris & J. K. Pinto (eds) *The Wiley Guide to Managing Projects*. Wiley, Hoboken, NJ.

Thomas, J. L. (2005) Problematizing project management. In: D. Hodgson & S. Cicmil (eds) *Making Projects Critical*. Palgrave, London.

Thomas Telford (2005) *New Engineering Contract*. Thomas Telford, London.

Walker, A. (2002) *Project Management in Construction*. Blackwell Science, Oxford.

Wateridge, J. (1998) How can IS/IT projects be measured for success? *International Journal of Project Management*, **16** (1), 59–63.

Weick, K. E. (1995) *Sense-making in Organizations*. Sage Publications, Thousand Oaks, CA.

Wenger, E. (1998) *Communities of Practice: Learning, Meaning and Identity*. Cambridge University Press, Cambridge, MA.

Wenger, E., McDermott, R. & Snyder, W. M. (2002) *Cultivating Communities of Practice: A Guide to Managing Knowledge*. Harvard Business School Press, Cambridge, MA.

Williams, T. (2004) Assessing and building on the underlying theory of project management in the light of badly overrun projects. *Proceedings of the PMI Research Conference*, London. Project Management Institute, Newton Square, PA.

Winch, G. M. (2002) *Managing Construction Projects*. Blackwell Science, Oxford.

Yardley, D. (2002) *Successful IT Project Delivery: Learning the Lessons of Project Failure*. Addison Wesley, London.

Yeo, K. T. (2000) Critical failure factors in information system projects. *International Journal of Project Management*, **20**, 241–6.

3 Applying emotional intelligence in project working

Vanessa Druskat and Paul Druskat

Productive relationships in a business environment are developed and maintained through the use of interpersonal skills. In the last decade, research in a number of disciplines has revealed that these interpersonal skills are rooted in the ability to recognise, interpret and manage emotion during interpersonal interactions (Ashkanasy *et al.*, 2000; Lopes *et al.*, 2003). These emotion-focused skills have been labelled emotional intelligence (EI) and, because of its importance to building productive relationships, research links EI to positive life outcomes as diverse as personal well-being and job performance (for reviews, see Bar-On & Parker, 2000; Druskat *et al.*, 2006). In project management, we argue, the need for emotional intelligence is even more pronounced than in most business environments. This is because in project management relationships must develop more quickly and interpersonal interactions often occur across organisational and professional cultures. Also, research reveals that emotional intelligence is critical to the performance of those who, like project managers, must lead others towards a shared goal (Goleman, 2001a).

The notion of emotional intelligence is not new, but its specific definition is. Definitions of types of intelligence other than cognitive intelligence (measured via the IQ test) have emerged in the last two decades, in part, because IQ predicts, at best, 15–25% of individual differences in important life outcomes such as work performance or career success (Schmidt & Hunter, 1998). One reason that cognitive intelligence is not the sole predictor of work success is that human beings are not predominantly rational. Neuroscientists have discovered that our decisions and actions are as much influenced by emotion as by logic (Damasio, 1994, 1999). Therefore, research suggests, successfully adapting to life's emergent circumstances, coping with setbacks and dilemmas, and effectively interacting with others requires a form of intelligence that combines both cognitive and emotional capacities (Goleman, 1995; Salovey & Mayer, 1990). This form of intelligence is labelled *emotional intelligence*.

Emotional intelligence is defined as the ability to monitor one's own and others' emotions, to discriminate among emotions, and to use emotional information to guide thinking and actions (Mayer & Salovey, 1993). Specifically, it involves four distinct sets of abilities:

(1) The ability to perceive, appraise, and express emotion accurately
(2) The ability to access and generate feelings when they facilitate cognition

(3) The ability to understand emotion-laden information and make use of emotional knowledge
(4) The ability to regulate emotions to promote emotional and intellectual growth and well-being (Mayer & Salovey, 1997)

In this chapter we argue and discuss why the abilities that constitute emotional intelligence are critical in project management environments. Specifically, we argue that EI is important because, by definition, projects are:

- Temporary
- Unique
- Progressively elaborated (Project Management Institute, 2000)
- Generally conducted by coalitions of members drawn from different organisations, industries and disciplines, each with its own culture

Each of these characteristics helps create an environment in which quickly building relationships and trust with a diverse group of people is fundamental to job success.

We begin the chapter with a discussion of EI competencies. In that discussion we present research that reveals how and why EI influences job performance. In the second section we discuss how EI is particularly important for project management and how EI competencies can have a positive influence on projects. We end the chapter with some specific recommendations for applying and developing EI.

Emotional intelligence competencies

Scientists have discussed the possibility of a social or emotional form of intelligence for over a century (see Cronbach, 1960; Darwin, 1965; Thorndike & Stein, 1937). However, the first to formally define and measure EI were Salovey and Mayer (1990). It is important to emphasise that EI is not the absence of cognitive intelligence. Emotional intelligence involves noticing and understanding emotion and its implications and using this understanding to improve cognitive thinking including the quality of actions and decisions. Because emotion pervades every human interaction (Kemper, 1978), emotional intelligence is particularly useful when actions and decisions involve others (Salovey *et al.*, 2000).

Daniel Goleman was among the first social scientists to take an interest in how EI affects work performance. Goleman analysed data on hundreds of organisations and collaborated with Richard Boyatzis, an expert on the job competencies that underlie superior work performance, to produce a model of work-related emotional competencies (Goleman, 1998). They defined emotional competencies as learned capabilities based on emotional intelligence that result in outstanding performance at work (Goleman, 1998). Emotional

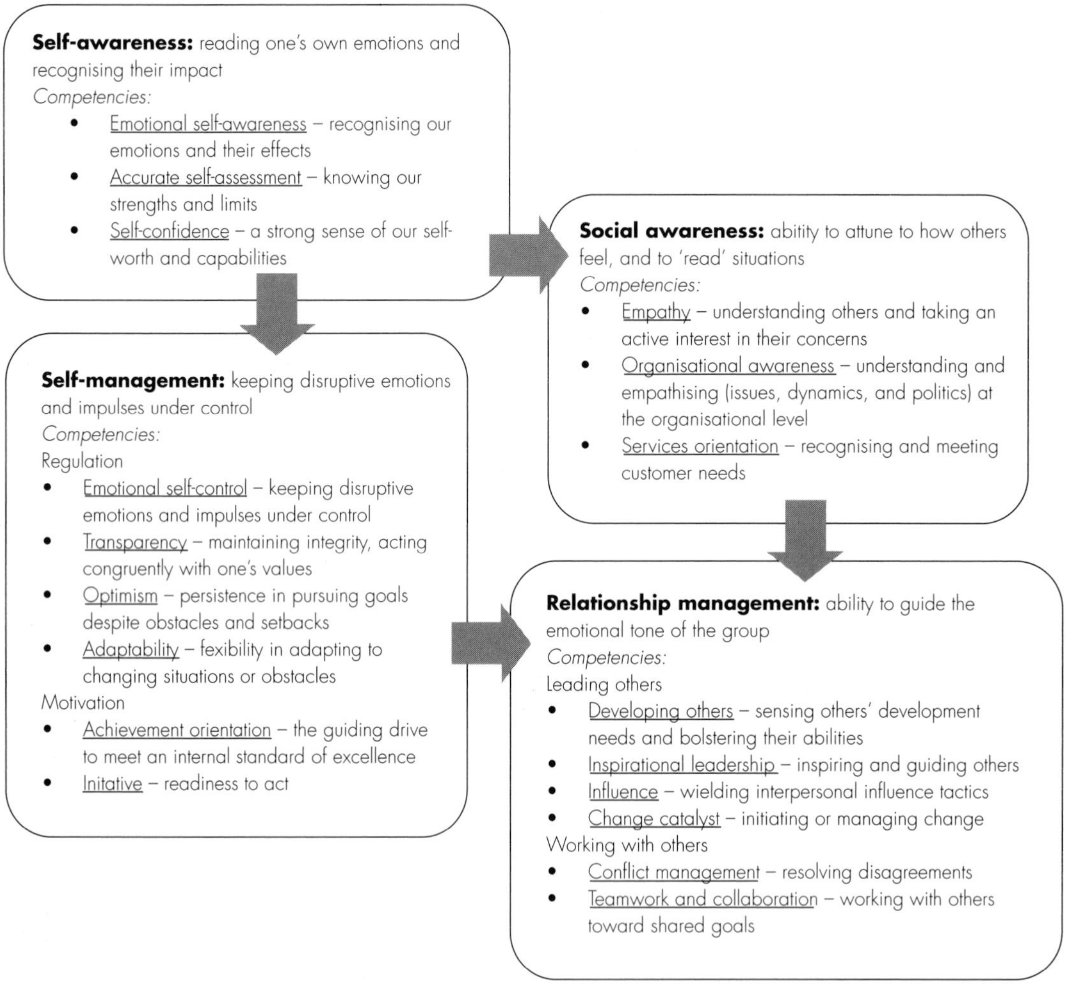

Self-awareness: reading one's own emotions and recognising their impact
Competencies:
- Emotional self-awareness – recognising our emotions and their effects
- Accurate self-assessment – knowing our strengths and limits
- Self-confidence – a strong sense of our self-worth and capabilities

Social awareness: abitity to attune to how others feel, and to 'read' situations
Competencies:
- Empathy – understanding others and taking an active interest in their concerns
- Organisational awareness – understanding and empathising (issues, dynamics, and politics) at the organisational level
- Services orientation – recognising and meeting customer needs

Self-management: keeping disruptive emotions and impulses under control
Competencies:
Regulation
- Emotional self-control – keeping disruptive emotions and impulses under control
- Transparency – maintaining integrity, acting congruently with one's values
- Optimism – persistence in pursuing goals despite obstacles and setbacks
- Adaptability – fexibility in adapting to changing situations or obstacles

Motivation
- Achievement orientation – the guiding drive to meet an internal standard of excellence
- Initative – readiness to act

Relationship management: ability to guide the emotional tone of the group
Competencies:
Leading others
- Developing others – sensing others' development needs and bolstering their abilities
- Inspirational leadership – inspiring and guiding others
- Influence – wielding interpersonal influence tactics
- Change catalyst – initiating or managing change

Working with others
- Conflict management – resolving disagreements
- Teamwork and collaboration – working with others toward shared goals

Fig. 3.1 The Goleman and Boyatzis emotional intelligence competency model. Adapted from Boyatzis *et al.* (2000) and Goleman *et al.* (2002).

intelligence determines potential for developing these competencies, which have all been linked to work success in a wide variety of contexts (see Druskat *et al.*, 2006; Goleman, 1998).

The Goleman and Boyatzis model presents 18 individual emotional competencies (see Figure 3.1), which fall into the following two categories and four subcategories of competence (Boyatzis *et al.*, 2000):

(A) Personal competence
  (i) Self-awareness
  (ii) Self-management

(B) Social competence
   (i) Social awareness
   (ii) Relationship management

As discussed in the following sections, the competencies in all four subcategories are meaningful for project management work.

Self-Awareness

Self-awareness involves the ability to be aware of one's own emotional states and conscious of one's impact on others. Self-awareness is considered the most fundamental subcategory of EI competencies because the self-awareness competencies are fundamental to the development of the competencies in the other three subcategories. For example, without awareness of one's own emotions, that is emotional self-awareness, it becomes less possible for an individual to engage in empathy, a social awareness competency that involves understanding others' emotions. Without a good understanding of emotion, it also becomes less possible for an individual to engage in inspirational leadership, a relationship management competency that involves engaging others' emotional commitments.

There are three specific self-awareness competencies. The first of these is *emotional self-awareness*, which involves knowing what one feels. It is formally defined as recognising one's own feelings and how they affect performance. The importance of this competency was highlighted in a study of successful financial planners who were having problems selling life insurance (Cherniss & Caplan, 2001). The study discovered that the financial planners were unaware of their own emotional discomfort with issues of mortality. This discomfort affected the way they discussed the topic of life insurance with their clients and resulted in poor sales of life insurance. To resolve the problem, the Fortune 500 Company, for which these financial planners worked, developed a training programme that focused on increasing the planners' emotional intelligence and, specifically, a module on increasing emotional self-awareness. The result was a sharp increase in sales of life insurance.

The second self-awareness competency is *accurate self-assessment*. Individuals who hold this competency understand their strengths and the areas in which they need to improve. They seek feedback to learn from their mistakes and know when to work with others whose strengths complement their own (Goleman, 2001a). In a study by Boyatzis (1982) of over 300 managers in 12 organisations, it was found that accurate self-assessment was consistently linked to managerial performance. Without this competency, managers tend to have blind spots that harm their performance. Common blind spots include unrealistic goals, relentless striving, driving others too hard, a hunger for power, the need for recognition and the need to seem perfect (Kaplan *et al.*, 1991).

The third self-awareness competency is *self-confidence*, defined as having a strong sense of self-worth and capabilities. It can be seen when presenting oneself in an assured manner. Presenting oneself with confidence is essential for garnering support for one's ideas and inspiring others to follow. For example, a major beverage firm was having a hard time hiring effective division presidents; 50% of their newly hired presidents were leaving within two years mostly because of their poor leadership performance (see Goleman, 1998). When the company started hiring based on emotional competencies including self-confidence, influence and inspirational leadership, turnover was reduced to 6% leaving within two years of hire. Those division leaders with emotional competencies performed at higher levels than their peer division presidents.

Self-management

Self-management involves managing one's emotions to control those that may be disruptive, to display emotions in productive ways, and to focus one's emotional drive to achieve standards of excellence. There are six specific self-management competencies. The first of these is *emotional self-control*, which involves the ability to keep emotions under control and to restrain negative actions when tempted or when working under stressful conditions. Self-control is referred to as an invisible competency (Goleman, 1998) because, when being utilised, there is no evidence of stress, anger, unhappiness, or any emotion that might disrupt the events occurring at the moment. Classic examples are seen when an angry or frustrated leader continues discussions in a fair and calm manner, when a manager who is experiencing a great deal of stress is able to manage that stress, think clearly and stay focused under pressure, and when a manager controls strong emotions and responds constructively to complaining employees. A number of studies have found that managers able to manage their emotions perform at higher levels than those who score lower on this competency (Boyatzis, 1982; Druskat *et al.*, 2006). It is important to note, however, that sometimes it may be most productive and honest to exhibit anger or frustration. When appropriate, individuals who demonstrate emotional self-control are able to exhibit anger or frustration in a constructive rather than a destructive manner. This leads us to the next competency in this subcategory: *transparency*.

Transparency involves the ability to maintain integrity and to behave authentically and to act congruently with one's values. Employees evaluate transparent managers as being trustworthy, and to honestly share their opinions and views, and act on their values even when it results in personal costs. Individuals who display this competency are honest about their own mistakes and problems, and are honest in their discussions with others about their mistakes and problems. When working with someone holding this competency there needs to be little guesswork in determining their values, opinions and evaluations. An increasing amount of research in organisational studies has linked transparency, trustworthiness and integrity to leader effectiveness (Dirks, 2000). Transparency results in trust that increases employees' willingness to cooperate, collaborate, and engage themselves in their work (Druskat & Wolff, 2001a).

The third self-management competency is *optimism*, which is defined as the ability to be positive and persistent when pursuing goals despite obstacles and setbacks. Recent research evidence confirms the long-suspected idea that emotion is contagious (Hatfield *et al.*, 1994). Thus, when a manager exhibits optimism it spreads to the others with whom he works and enables an entire team to remain optimistic in the face of setbacks (Barsade, 2002). Remaining optimistic despite obstacles requires managing one's emotions and focusing on the positive. Goleman (2001a) discusses it as a key ingredient in the ability to achieve goals and meet deadlines because of the inevitability of bumping into barriers along the way. Optimism keeps a team working hard in the face of those barriers. It is not surprising that research suggests optimism is often linked to success in sales (Seligman, 1991). Convincing others to purchase a product or to accept an idea requires persistence and a clear and transparent expectation of positive outcomes.

The fourth competency in this subcategory is *adaptability*, which is defined as the ability to adapt to change and to work effectively within a variety of situations and fluid circumstances. Adaptability requires managing oneself to remain flexible when priorities shift and to change rapidly when circumstances indicate that change is necessary. Adaptability requires that an individual view a situation objectively and without excessive emotional attachment to any one way forward. The performance relevance of adaptability was seen in a study of South African customer call centre agents in an insurance company (Sala, 2005). Those agents who scored high on the adaptability competency also scored higher in their performance ratings, which were calculated based on objective ratings of outcomes such as the quality of their conversations with clients and how quickly they were able to complete transactions with clients. Thus, adaptability enables one to more quickly and easily adapt to the specific needs of specific clients.

The fifth and sixth competencies in this subcategory are both related to motivation, which involves focusing one's emotional energy proactively and persistently to meet or surpass standards of excellence. These competencies are *achievement orientation* and *initiative*. Both represent a bias towards action and they often go hand in hand. The achievement orientation competency is defined as a drive to meet an internal standard of excellence or to improve performance. Initiative is defined as readiness to act in order to seize opportunities. These two competencies are practically synonymous with motivation and a great number of studies have linked them to performance. In Mount's study (2005) of competencies linked to the effectiveness of five critical business roles in a major international petroleum corporation (strategists, business developers, negotiators, business service managers, and project managers), he found that one of two competencies that predicted high performance in all five roles was achievement orientation. Also, in a study of 358 managers in Johnson & Johnson's Consumer and Personal Care Group, the highest performing managers were found to exhibit both achievement orientation and initiative significantly more often than the average performing managers (Gowing *et al.*, 2005).

Social awareness

Social awareness involves the ability to attune to how others feel and to 'read' the emotional tone in interpersonal situations. Social awareness is considered a prerequisite for the fourth subcategory of EI competencies, which is labelled *relationship management*. The social awareness competencies allow one to be attuned to how others feel, which improves one's competence at managing relationships and at guiding the feelings and emotional tone of a group. Goleman and Boyatzis (Boyatzis *et al.*, 2000) place three competencies in the social awareness subcategory. The first of these is *empathy*, which is defined as sensing others' feelings and perspectives and taking an active interest in their concerns. Empathy implies the desire to understand someone including his or her emotions, concerns and needs. Self-awareness is essential to empathy. It would be difficult to understand or take interest in the emotions experienced by another if we were perpetually unaware of our own emotions. Empathy is linked to performance in most jobs where sensitivity to others' needs and circumstances are important. In the Johnson & Johnson competency study (Gowing *et al.*, 1995), high performing managers were rated by their leaders and their direct reports as showing significantly more empathy than average performing managers. Also, in Mount's study (2005) of leaders in an international petroleum corporation, empathy was found to be related to the performance of international strategists, international deal makers and leaders in the international businesses services area. It makes great sense that empathy would be linked to performance in an international position that requires interacting with individuals from different societal cultures.

The second competency in the social awareness subcategory is *organisational awareness*, which is defined as the ability to read the currents of emotions and political realities in an organisation or group. It requires the desire to understand issues, dynamics and politics at the group or organisational level. In his study of over 300 managers in 12 organisations, Boyatzis (1982) found that superior performing managers displayed organisational awareness significantly more often than average performing managers. They were able to objectively view and understand the dynamics in groups and organisations, which enabled them to manage situations in ways that were constructive and appropriate to the situation.

The last competency in the social awareness subcategory is labelled *service orientation*. It is defined as the ability to anticipate, identify and meet clients' or customers' often unstated needs. It is also exhibited when an individual takes a long-term perspective and trades off short-term gains for a long-term customer relationship.

Relationship management

Relationship management includes six competencies that are often referred to as social skills. The use of 'relationship management' has a specific conceptual function within EI. While there are strong links with the 'relationship management paradigm' referred to elsewhere in the book, the application in EI is distinct. The competencies in this subcategory involve actions that are made effective because of one's self-awareness, self-management, and social awareness (Goleman, 2001a). The competencies in this subcategory focus on either:

- Leading others, or
- Working with others

Four specific competencies focus on leading others. The first of these is *developing others* and is defined as sensing others' development needs and taking actions to bolster their abilities. Developing others is a fundamental leadership role. Research indicates that it is associated with leader effectiveness. For example, in the Johnson & Johnson study (Gowing *et al.*, 2005) the developing others competency was found to be one of the most powerful differentiators between high performing and average performing leaders. Effective leaders use their competence at being in tune with their employees' needs (empathy) and in tune with the organisations' needs (organisational awareness) to take appropriate and effective actions to develop their employees in ways that help them succeed.

It is not surprising that the other three competencies in this subcategory focused on leading others were also found to be powerful differentiators between the high performing and average performing leaders at Johnson & Johnson. They are *inspirational leadership*, *influence* and *change catalyst*. *Inspirational leadership* involves recognising that leading and guiding individuals and groups requires inspiring them emotionally. Inspirational leadership matters because emotion is contagious (Hatfield *et al.*, 1994), and to be most successful a leader must display high levels of positive energy and enthusiasm that spreads to others. This is especially true because leaders are in highly visible positions and all eyes are often attuning to and 'reading' the leader's verbal and non-verbal emotions (Goleman, 2001a).

The *influence* competency is rooted in the intention to persuade or convince others and is defined as having an impact on others. Goleman (2001a) proposes that effective influence requires sensing others' reactions and fine-tuning one's response to move interaction in a constructive direction. He has found that those who exhibit this competency draw from a wide range of influence tactics including impression management, dramatic arguments or action, and appeals to reason. In Mount's study (2005) of competencies linked to the effectiveness of five critical business roles in a major international petroleum corporation (strategists, business developers, negotiators, business service managers, and project managers), the second competency that predicted high performance in all five roles was *influence*.

The *change catalyst* competency is defined as initiating or managing change. It involves recognising the need for change, removing barriers, and enlisting others in the pursuit of change. Like the inspirational leadership competency, it involves turning on others' excitement about new possibilities. Research on effective leaders has long shown that those who are most effective in the leadership role are adept at recognising the need for change and effective at managing the transformation process (Bass, 1990).

Two competencies under the relationship management subcategory focus on working with others. The first of these competencies is *conflict management* and is defined as negotiating and resolving disagreements. The conflict management competency requires one to be able to recognise when conflict is developing, and then take action to resolve issues or head off the conflict. Creating win–win situations through conflict management requires social awareness competencies such as empathy and organisational awareness, which enable one to understand the emotions and circumstances that fuel conflict.

The second emotional competency in this subcategory focused on working with others is *teamwork and collaboration*. It is defined as working with others towards a shared goal and creating group synergy in pursuing collective goals. It implies genuine intent to work cooperatively, as opposed to working separately or competitively, and the intent to influence others to do the same. When leaders demonstrate interest in working with others towards a shared goal, teamwork and collaboration are infectious and have an important influence on the behaviour of members (Druskat & Wolff, 2001b). In his study of the relevance of the emotional competencies displayed by managers in the international petroleum industry, Mount (2005) found that for his sample of worldwide project managers, the competency of teamwork and collaboration differentiated high performing from lower performing project managers. This makes sense: the willingness to cooperate and collaborate with others is fundamental to the project manager's role. Doing this well would certainly boost performance.

Summary

The 18 emotional competencies in the Goleman and Boyatzis model all contribute to managerial performance success in unique ways. Research conducted two decades ago revealed the importance of emotional competencies to manager success (Boyatzis, 1982). A growing number of studies confirm those early findings (see Druskat *et al.*, 2006). This research shows that emotional competencies matter in a wide variety of positions and in a wide variety of locations (see Druskat *et al.*, 2006).

Group emotional intelligence

A number of researchers have begun to discuss and test the performance benefits of developing emotional intelligence at the team level (Druskat & Wolff, 2001a; Elfenbein, 2006; Jordan & Ashkanasy, 2005). Druskat and Wolff define *group emotional intelligence* (group EI) as a group culture created by a set of norms that facilitate a productive social and emotional environment that leads to high performance. A productive social and emotional environment is one that includes trust, group identity, emotional capability, group efficacy and networks. They identified nine emotionally competent group norms that build such a culture and that have been found to lead to high team performance (Wolff *et al.*, 2005). Three of these norms are focused on awareness and regulation of emotions in individual team members: *interpersonal understanding, caring behaviour* and *confronting members who break norms*. These norms ensure that team members build an understanding of each other's strengths, weaknesses and idiosyncrasies, treat one another with respect, and are honest with one another. Together they build trust and a sense of group identity. Four group EI norms are focused on awareness and regulation of group-level emotions: *team self-evaluation, creating resources for working with emotion, creating an affirmative environment* and *proactive problem solving*. These norms ensure that a group steps back occasionally to discuss its progress, discusses, rather than avoids, tensions and difficult issues, and approaches challenges with a proactive and optimistic attitude. Together, they build emotional capability, defined as the ability to work through difficult issues, and group efficacy, defined as the sense that the group is stronger together than apart. The final two norms in the group EI model are focused on the group's external stakeholders: *organisational awareness* and *building external relationships*. These norms enable a group to recognise its interdependencies with others outside the group and help the group build networks of relationship that can help it perform at its best.

Emotional intelligence in project management

In a study specifically related to project management, Mount (2005) assessed the skills related to the success of 74 worldwide project managers in an international petroleum corporation. These project managers were a service group that partnered with business units from supply chain actors – upstream, downstream and petrochemical – to evaluate the feasibility of capital asset projects, and then managed the construction to bring them on-stream in a cost-efficient, safe and reliable manner. Successful project management contributed to the competitive advantage of the business unit partners. The study determined that, of the skills that predicted project manager success, 69% were emotional competencies – achievement orientation, influence, self-confidence, and teamwork and coordination – 0% were cognitive skills, such as analytical thinking or conceptual thinking, and the final 31% were related to business expertise.

These findings support those of Spencer (2001) who studied 28 engineering construction project managers. Spencer also found that emotional competencies were most strongly related to their job success. The construction project managers demonstrating emotional competencies saved $27 million through avoiding cost and time overruns and selling additional engineering change orders.

These studies highlight the importance of emotional intelligence competencies for project managers. As mentioned in our introduction, we argue that emotional intelligence is even more important in project management than in traditional business environments because by definition projects are temporary, unique and progressively elaborated (Project Management Institute, 2000); and projects are generally conducted by coalitions from different organisations, industries and disciplines, each with its own culture.

As projects are temporary, project-based relationships often exist only for the length of the project, a subproject, or even a checkpoint or progress meeting. Project managers and team members frequently move quickly from one project onto the next. Therefore, building new relationships and establishing trust swiftly is a constant part of the project manager's job. Similarly, the project manager needs to facilitate swift trust and relationships among relevant stakeholders in a project. Emotional intelligence competencies such as *self-confidence, emotional self-control, transparency* and *empathy* can help a project manager develop trust swiftly and create a working environment that becomes quickly productive.

By definition, each project and its product are also unique, which means that each project will involve factors that will not have been considered before and whose challenges cannot be known or anticipated. Without constructive relationships between the customer, team and other stakeholders, these emergent challenges will be more difficult to overcome, increasing risk to the project. Thus, emotionally intelligent competencies such as optimism, adaptability, initiative and organisational awareness can be helpful to a project manager's ability to confront and overcome unexpected challenges. Another way to ready the project team for unexpected challenges is to build group emotional intelligence norms in the project team. Group EI norms such as *creating an affirmative environment, proactive problem solving* and *organisational awareness* can all enable a group to approach unexpected challenges optimistically and effectively.

Projects are also progressively elaborated, which means that product design concepts and specifications are broad at project initiation and, by progressive steps through the planning and execution phases, the project and its product conceptual design are elaborated upon to produce greater and greater detail. In other words, by its very nature a project is meant to invoke and involve change. Project managers and other leaders must develop and exercise the emotional intelligence competencies needed to ensure that the individuals and groups and the relationships among them can thrive in this constantly evolving and emerging environment. The EI competencies that would help one to manage this environment involve the social awareness subcategory of competencies including *empathy, organisational awareness* and *service orientation*. Also, it would be helpful if the project manager had competence in the relationship management subcategory including *inspirational leadership, influence* and *change catalyst*.

A consequence of project and product uncertainty in the early stages of a project is that stakeholders will have differing, often changing, views of the 'product'. In these early stages, it may be difficult for stakeholders to verbalise concerns because the project and product outlines are still relatively vague. At this stage, members of the project team must manage their own emotions, and demonstrate empathy and awareness towards the verbal and non-verbal signals of other stakeholders. EI competencies in the self-management subcategory such as *self-control* and *adaptability* may be particularly necessary when a project manager discovers retrospectively that stakeholders thought something different had been approved.

Finally, projects differ from most other workplace environments in that work relationships in project management are frequently cross-cultural – client-to-designer-to-contractor, industry-to-industry, company-to-company, and country-to-country. The differences across the cultures of the various client, design and implementation and contractor organisations carry risk and provoke the natural human tendency to stereotype people who are different. Emotional intelligence competencies such as *empathy, organisational awareness* and *service orientation* must be strong in order to overcome this tendency (see Goleman, 2001a). These competencies are needed throughout the project coalition and therefore must be driven by project managers and other leaders with good relationship management competencies. The tendency to stereotype can also be dissipated when the project manager develops group EI norms in the project team. Norms such as *interpersonal understanding* and *caring behaviour* reinforce the need to get to know the backgrounds and perspectives of team members and to treat each member with caring and respect. These norms build the kind of group trust that enables members from different backgrounds to work together effectively.

Let's now consider each EI competency subcategory for its relevance to project management.

Self-awareness competencies – *emotional self-awareness, accurate self-assessment, self-confidence* – are the basis for the other competency subcategories. Self-confidence is particularly important for projects because of the high levels of uncertainty in the environment.

Self-management competencies of regulation – *emotional self-control, transparency, optimism, adaptability* – and motivation – *achievement orientation, initiative* – are all important in managing projects, transparency being especially important for the development of healthy relationships.

Optimism can be important for similar reasons. Mixing of organisational cultures can cause adversarial attitudes to develop. One new project manager told us that it took time to recognise and overcome the wastefulness in conflicts surrounding the project. Common negative workplace banter was rooted in and promoted further job stereotyping. He realised everyone involved in the project was qualified and competent. However, all were overloaded and all had their own unique stresses and worries, and the banter made matters worse. Thus, he chose to consciously adopt an optimistic view of the project participants and stakeholders. Controlling his emotional responses had a positive impact on the team and resulted in the team adopting a positive tone; meetings became more productive and efficient, and relations across groups and with the stakeholders changed from the mode of defending turf to that of resolving issues as common concerns.

The *adaptability* competency also continues to grow in importance as the pace of change in business environments increases (Goleman, 2001a). Project management continues to grow as a profession and organisations continue to transform their structures to 'management-by-project' models. Adaptability and project management go hand in hand.

Social awareness competencies – *empathy, organisational awareness, service orientation* – are important because projects span organisations, disciplines and professions. Empathy can help break down natural barriers among groups and develop project member organisational awareness. This in turn can help the project coalition develop a common services orientation or clarity of approach.

This type of competency progression occurred during a high-speed digital network implementation project. Halfway through the project, a major network hub went down causing connectivity outages to a large US region of about 12 states. The outage lasted almost two days during which millions of dollars of sales could not be processed and thousands of person-hours in productivity were lost. The project team's regional project manager was pressed to bring the network services provider representative to a meeting with the regional VP and to reprimand her and her company. However, the project team had been aware from the start that the network providers seemed to have evolved a culture in which, to protect themselves, network design details and problem diagnosis and resolution details were not shared with clients. Network services were seldom even noticed until problems occurred. Hence, communications between clients and network services providers were usually under a negative and defensive cloud. The network services providers seemed to have become skilled at placating bad feelings and hiding mistakes.

In this particular case the project team and network services provider had been working for years to develop a partnership to help improve cooperation during network projects. Thus, the regional project manager empathised with the network services provider support team's situation and decided against perpetuating the old traditional adversarial paradigm, and instead to behave as a partner on the same larger team. Thus, rather than accept the overly simplified, defensive explanation for the outage from the network service provider help desk, the project manager insisted that, *as a partner*, the support representative involve the network services provider design team in determining the problem's root cause. Then, as a team, the client regional project manager and network services provider's designer and manager took the overall problem diagnosis and resolution report to the regional executive management team. The project manager took the blame for the delayed response, but the network services provider was shown to be capable, competent and conscientious. This approach strengthened the partnership and the teamwork through the remainder of the project. Empathy and understanding in relationships with members of the services provider team led to better organisational awareness of the services provider's organisation and industry. This organisational awareness led to strengthening of the partner relationship in the interest of a common service orientation.

The relationship management competencies of leading others – *developing others, inspirational leadership, influence, change catalyst* – and working with others – *conflict management, teamwork and collaboration* – are key in project management. For example, one emotionally intelligent project manager studied by Mount (2005) scheduled a meeting with the intent of using it as a vehicle for developing relationships and trust. This project manager said:

> *'I decided to include the contractor's team (about 100 people) with our company team for the team meetings. To get everyone's involvement, I scheduled the meetings during lunchtime and bought them lunch. I used that time to inform everyone of what's going on and to do some team building. It worked in that it created a sense of belonging and togetherness, which was my intention.' (Mount, 2005 p.117)*

Another example comes from a project involving the construction of a new site filtration system at a major chemical plant. Regulators gave the company six months to build and install the new system to satisfy new regulatory requirements, otherwise the entire plant would shut down until the system was up and running. The site facilities management team believed that the project team had little chance of meeting the deadline, so they began planning to negotiate fines to avoid a shut down. However, because of the emotional intelligence of the project manager, as the project progressed the entire project coalition began to work as a team, to develop and share a common sense of urgency, and to strive to meet the challenge. The project coalition included the plant's engineering team, the architectural design firm, the contractors and subcontractors, and the OEM and custom equipment providers.

The project manager brought the engineers and contractors together to discuss their issues and concerns and to identify their suggestions about how he could tighten the schedule and avoid delays. Next, he and his procurement specialist visited the equipment providers to learn about their operations. Through his influence and inspirational leadership, the usual conflicts were replaced with teamwork and collaboration. The mechanical and electrical contractors, for example, made design suggestions for the engineers and architects and, most importantly, pointed out potential pitfalls. The project was completed within one week of the regulatory agency's deadline. No fines were charged and the plant was not shut down.

Summary

The purpose of this section has been to show the importance of EI and EI competencies and their application in the project management environment with an emphasis upon establishing constructive relationships. EI competencies and EI group norms are especially important because project managers and teams do not have the luxuries of time or background similarity that occur more often in traditional business environments. Also, projects inherently involve uncertainty and interdependencies.

In addition to the more general need for EI competencies, some competencies can be located in specific PM processes for example, relating EI competencies to human resource development policies. Organisational planning and staff acquisition should emphasise EI competencies in choosing and training project managers and team members. The team-building process should begin with development of all the EI competencies for the project manager and other leaders and should then include development and awareness of EI competencies for other members of the team.

Project managers, along with their managers, sponsors and teams, need to establish an emotionally intelligent mindset – a mindset that recognises the value of developing and strengthening strong emotional intelligence competencies and that sets the mind to recognise the influence of emotion in every interpersonal interaction. As we have discussed, the importance of emotion in social interactions has been recognised by social scientists and neuroscientists alike (Ashkanasy *et al.*, 2000; Damasio, 1994; Salovey & Mayer, 1990). It is not a question of whether we *should* attend to emotion, but rather *when* we will begin treating emotion seriously in our training and development programmes and project management processes so that we help build EI and its development into the project management mindset. We could, for example, put EI on the agenda for project meetings and practice using it in managing stakeholder issues when working with head office. Below, we discuss some basic information about the development of EI competencies.

Developing emotional competencies

Two questions are frequently asked about emotional competencies:

- Does one need to be competent in all 18 competencies in order to enjoy the benefits of emotional intelligence?
- Can one develop EI competencies, and if so, how?

It is not necessary to be effective in all emotional competencies. Research indicates that demands for emotional competencies are context specific. For example, in environments where employees are encouraged to hold their tongues, the empathy competency, including the ability to read non-verbal communication, is of great necessity. In environments where employees are encouraged to speak up, the need for self-management and the regulation of emotion would be greater. Thus, organisations can and should choose to build their own unique competency profiles. A good place to start is by determining the emotional competencies demonstrated by top performing employees and using it to identify the competencies that support success in a particular job context. The process can also start by using a panel of experts to discuss specific organisational roles and agree on which competencies matter most. In these ways a target list of competencies can then be identified for hiring and training purposes (see Boyatzis, 1982).

Research suggests three important considerations to determine the competency profile and reap the benefits of emotional intelligence:

(1) *Self-awareness* is a foundation for developing the rest of the emotional competencies in the model and is a prerequisite for recognising emotions in others.
(2) Competencies are not absolutes and can be developed on a continuum. Research suggests that small changes in the amount of a competency can make a large difference (Goleman, 1998).
(3) Nine competencies developed to a reasonable level can 'tip the balance'. In a study of partners in a consulting firm, Boyatzis (1999) determined that nine emotional competencies tipped the balance and enabled the team led by a consulting partner to show significantly superior financial performance.

Emotional competencies can be developed (see Cherniss & Adler, 2000; Cherniss & Goleman, 2001; Druskat *et al.*, 2006; see also the Consortium for Research on Emotional Intelligence in Organizations (CREIO), www.eiconsortium.org). Boyatzis has had great success developing EI competencies through a process of 'self-directed change and learning' (Boyatzis *et al.*, 2002; Goleman *et al.*, 2002), which involves measuring current exhibition of EI competencies, identifying which competencies need development (based on job demands and current level of demonstration), and developing a personalised competency development plan. The personal plan includes goals and means, for example through training, coaching, or learning through taking risks and seeking feedback. There are several sources for measuring EI competencies, and some of the best include:

- Bar-On Emotional Quotient Inventory (EQ-I) (Bar-On, 1997)
- Emotional Competence Inventory (ECI) (Sala, 2002)
- Emotional Intelligence Test (MSCEIT) (Mayer *et al.*, 2003)

Each test is slightly different: the EQ-I is a self-report only test, the ECI is a 360 degree competency feedback test, and the MSCEIT is the most like an IQ test as it does not involve self-report.

Conclusion

In the project management environment relationships must often be developed quickly, and important interpersonal interactions occur frequently in situations where relationships are still being established. Furthermore, relationships must withstand constant change as the project and its product progress. Perhaps most important is that projects cross organisational, job and professional cultures. Missing verbal or non-verbal signals during interpersonal interactions or failing to manage one's own emotion can result in missed information, adversarial relationships and increased risk.

Establishing EI competence in the mindset of the project managers, administrators and team is basic to establishing and maintaining constructive relationships across organisations and groups.

We have argued that project leadership must develop strong EI competencies and set the tone for the rest of the project coalition. However, everyone involved should develop some EI competencies for relationships to be constructive. Some competencies such as self-awareness and adaptability are essential and a critical mass of nine competencies is desirable. There are also advantages to developing Group EI norms within the project team (Elfenbein, 2006; Wolff *et al.*, 2005). In fact, most EI research has focused on the emotional intelligence of managers and leaders. A fruitful area for future research is further examination of EI within groups and project teams.

We have also recommended that EI competency development be part of project manager and team training. It is important to understand that EI and EI workplace competencies can be learned and developed.

As project management is a widely adopted model across many industries, there are further opportunities for research and the identification of the best approaches for mobilising EI competencies in specific project environments.

References

Ashkanasy, N. M., Hartel, C. E. J. & Zerbe, W. J. (2000) Emotions in the workplace: research, theory, and practice. In: N. M. Ashkanasy, C. E. J. Hartel & W. J. Zerbe (eds) *Emotions in the Workplace*. Quorum, Westport, CT.

Bar-On, R. (1997) *The Emotional Quotient Inventory (EQ-I): Technical Manual*. Multi-Health Systems, Toronto.

Bar-On, R. & Parker, J. D. A. (2000) *The Handbook of Emotional Intelligence: Theory, Development, Assessment, and Application at Home, School, and in the Workplace*. Jossey Bass, San Francisco.

Barsade, S. G. (2002) The ripple effect: emotional contagion and its influence on group behavior. *Administrative Science Quarterly*, **47**, 644–75.

Bass, B. M. (1990) *Bass & Stogdill's Handbook of Leadership: Theory, Research, and Managerial Implications* (3rd edn). Free Press, New York.

Boyatzis, R. E. (1982) *The Competent Manager: A Model for Effective Performance*. Wiley, New York.

Boyatzis, R. E. (1999) *The financial impact of competencies in leadership and management of consulting firms*. Unpublished paper, Department of Organizational Behavior, Case Western Reserve University, Cleveland, OH.

Boyatzis, R. E., Goleman, D. & Rhee, K. (2000) Clustering competence in emotional intelligence: insights from the emotional competence inventory (ECI). In: R. Bar-On & J. D. A. Parker (eds) *The Handbook of Emotional Intelligence: Theory, Development, Assessment, and Application at Home, School, and in the Workplace*. Jossey Bass, San Francisco.

Boyatzis, R. E., Stubbs, E. C. & Taylor, S. N. (2002) Learning cognitive and emotional intelligence competencies through graduate management education. *Academy of Management Journal on Learning and Education,* **1** (2): 150–62.

Cherniss, C. & Adler, M. (2000) *Promoting Emotional Intelligence in Organizations.* American Society for Training and Development, Alexandria, VA.

Cherniss, C. & Caplan, R. D. (2001) Implementing emotional intelligence programs in organizations. In: C. Cherniss and D. Goleman (eds) *The Emotionally Intelligent Workplace.* Jossey-Bass, San Francisco.

Cherniss, C. & Goleman, D. (2001) *The Emotionally Intelligent Workplace.* Jossey-Bass, San Francisco.

Cronbach, L. J. (1960) *Essentials of Psychological Testing* (2nd edn). Harper & Row, New York.

Damasio, A. (1994) *Descartes' Error: Emotion, Reason and the Human Brain.* Avon, New York.

Damasio, A. (1999) *The Feeling of What Happens.* Harcourt Brace, New York.

Darwin, C. (1965 [1872]) *The Expression of Emotion in Man and Animals.* University of Chicago Press, Chicago.

Dirks, K. T. (2000) Trust in leadership and team performance: evidence from NCAA basketball. *Journal of Applied Psychology,* **85**, 1004–12.

Druskat, V. U., Sala, F. & Mount, J. (2006) *Linking Emotional Intelligence and Performance at Work.* Lawrence Erlbaum Associates, Mahwah, NJ.

Druskat, V. U. & Wolff, S. B. (2001a) Building the emotional intelligence of groups. *Harvard Business Review,* **79** (3), 81–90.

Druskat, V. U., & Wolff, S. B. (2001b) Group emotional competence and its influence on group effectiveness. In: C. Cherniss & D. Goleman (eds) *The Emotionally Intelligent Workplace.* Jossey-Bass, San Francisco.

Elfenbein, H. A. (2006) Team emotional intelligence: what it can mean and how it can impact performance. In: V. U. Druskat, F. Sala & J. Mount (eds) *Linking Emotional Intelligence and Performance at Work.* Lawrence Erlbaum Associates, Mahwah, NJ.

Goleman, D. (1995) *Emotional Intelligence.* Bantam Books, New York.

Goleman, D. (1998) *Working with Emotional Intelligence.* Bantam Books, New York.

Goleman, D. (2001a) An EI-based theory of performance. In: C. Cherniss and D. Goleman (eds) *The Emotionally Intelligent Workplace.* Jossey-Bass, San Francisco.

Goleman, D. (2001b) Emotional intelligence: issues in paradigm building. In: C. Cherniss and D. Goleman (eds) *The Emotionally Intelligent Workplace.* Jossey-Bass, San Francisco.

Goleman, D., Boyatzis, R. & McKee, A. (2002) *Primal Leadership.* Harvard Business School Press, Boston, MA.

Gowing, M. K., O'Leary, B. S., Brienza, D., Cavallo, K. & Crain, R. (2006) A practitioner's research agenda: exploring real-world applications and issues. In: V. U. Druskat, F. Sala & J. Mount (eds) *Linking Emotional Intelligence and Performance at Work.* Lawrence Erlbaum Associates, Mahwah, NJ.

Hatfield, E., Caccioppo, J. T. & Rapson, R. L. (1994) *Emotional Contagion.* Cambridge University Press, New York.

Jordan, P. & Ashkanasy, N. M. (2006) Emotional intelligence, emotional self-awareness, and team performance. In: V. U. Druskat, F. Sala & J. Mount (eds) *Linking Emotional Intelligence and Performance at Work.* Lawrence Erlbaum Associates, Mahwah, NJ.

Kaplan, R. E., Drath, W. H. & Kofodimos, J. (1991) *Beyond Ambition: How Driven Managers Can Lead Better and Live Better*. Jossey-Bass, San Francisco.

Kemper, T. D. (1978) *A Social Interactional Theory of Emotions*. John Wiley & Sons, New York.

Lopes, P. N., Salovey, P. & Straus, R. (2003) Emotional intelligence, personality, and the perceived quality of social relationships. *Personality & Individual Differences*, **35**, 641–58.

Mayer, J. D. & Salovey, P. (1993) The intelligence of emotional intelligence. *Intelligence*, **17**, 433–42.

Mayer, J. D. & Salovey, P. (1997) What is emotional intelligence? In: P. Salovey & D. J. Sluyter (eds) *Emotional Development and Emotional Intelligence: Educational Implications*. Basic Books, New York.

Mayer, J. D., Salovey, P., Caruso D. & Sitarenios, G. (2003) Measuring emotional intelligence with the MSCEIT V2.0. *Emotion*, **3**, 97–105.

Mount, J. (2006) The role of emotional intelligence in developing international business capability: EI provides traction. In: V. U. Druskat, F. Sala & J. Mount (eds) *Linking Emotional Intelligence and Performance at Work*. Lawrence Erlbaum Associates, Mahwah, NJ.

Project Management Institute (2000) *A Guide to the Project Management Body of Knowledge: PMBOK Guide* (2000 edn). PMI, Newton Square, PA.

Sala, F. (2002) *Emotional Competence Inventory (ECI): Technical Manual*. Hay/McBer Group, Boston, MA.

Sala. F. (2006) The international business case: emotional intelligence competencies and important business outcomes. In: V. U. Druskat, F. Sala & J. Mount (eds) *Linking Emotional Intelligence and Performance at Work*. Lawrence Erlbaum Associates, Mahwah, NJ.

Salovey, P., Bedell, B. T., Detweiler, J. B. & Mayer, J. D. (2000) Current directions in emotional intelligence research. In: M. Lewis and J. M. Haviland-Jones (eds) *Handbook of Emotions* (2nd edn). Guilford Press, New York.

Salovey, P. & Mayer, J. D. (1990) Emotional intelligence. *Imagination, Cognition, and Personality*, **9** (3), 185–211.

Schmidt, F. L. & Hunter, J. E. (1998) The validity and utility of selection methods in personnel psychology: practical and theoretical implications of 85 years of research findings. *Psychological Bulletin*, **124**, 262–74.

Seligman, M. E. P. (1991) *Learned Optimism*. Knopf, New York.

Spencer, L. M. (2001) The economic value of emotional intelligence competencies and EIC-based HR programs. In: C. Cherniss & D. Goleman (eds) *The Emotionally Intelligent Workplace*. Jossey-Bass, San Francisco.

Thorndike, R. L. & Stein, S. (1937) An evaluation of the attempts to measure social intelligence. *Psychological Bulletin*, **34**, 275–84.

Wolff, S. B., Druskat, V. U., Koman, E. S. & Messer, T. E. (2006) The link between group emotional competence and group effectiveness. In: V. U. Druskat, F. Sala and J. Mount (eds), *Linking Emotional Intelligence and Performance at Work*. Lawrence Erlbaum Associates, Mahwah, NJ.

4 Measuring, developing and managing trust in relationships

Hedley Smyth

Over the last decade 'trust' has become a word in common use in project industries. Trust is particularly used in relation to alliances, including supply chain management, partnering and partnerships, such as Public Private Partnerships (PPP) and their forerunner, the Private Finance Initiative (PFI). Trust is a central element in UK government reports, such as the Egan Report (1998) and subsequently in *Constructing Excellence*. In fact, trust is important in any market relationship including traditionally adversarial industries where work is delivered to contract and uncertainties are high, providing conditions in which mistrust and opportunism can prevail.

In order to provide fertile ground for the development and management measurement of trust, it is necessary that we know what *trust* is. Attending a series of M4i meetings of the London Cluster, which was a best practice forum for demonstration projects in the wake of Egan's Report on *Rethinking Construction*, many listened at successive meetings to conversations that went like this: 'Trust is very important, and that means open communication.' Well intentioned such comments may be, but they fundamentally miss the point. If communication is open you do not need trust (Smyth, 2003; Swan *et al.*, 2001). Everything is transparent. Trust is needed over things you cannot see, things not known. Good communication arises out of trust and, as evidence of trust, may then be used to develop further trust. It is not the cause of trust development *per se* or, as O'Neill (2002) stated in her BBC Reith Lectures on trust:

> *'Where we have guarantees or proofs, we don't need to trust. Trust is redundant.'*

We need to know more about trust. The financial 'bottom line' is well understood. We can analyse it from many angles: gross profit, net profit, profit margin, return on capital employed, 'windfall' profits and the absence of profits in the forms of breakeven or loss. We need a similar understanding of trust, the social 'bottom line'. A detailed analysis is needed to begin to develop trust, in order to manage trust as a *competency*. We need metrics so that managers can understand areas of weakness and take systematic action to improve performance. Using a financial analogy, trust is a 'stock' that is earned, built up over time as an asset on one side of the balance sheet, relating to concepts such as *social* and *human capital*, and trust is a 'currency' that is spent in the outworking and development of relationships, and thus accrues as a liability on the other side of the balance sheet, remaining such until a return is earned and the stock of trust increases.

This chapter draws together new unpublished data, alongside data used in previous publications (Edkins & Smyth, 2006; Smyth & Edkins, forthcoming; Smyth & Thompson, forthcoming) that has been reconfigured for the purposes of this chapter, on the measurement of trust. The purpose is to analyse trust across a series of relationships within different client types concerned with the management of projects. The relationships are:

- The client and design team
- The client and project coalition of professional firms
- The client and contractor
- The client and project coalition of contracting firms

Therefore, the analysis focuses upon 'supply chains', or more precisely supply and work clusters. The format of the chapter is to first define trust. Three 'building blocks' of trust will be analysed: *conditions of trust*, derived from the work of Butler (1991), *characteristics of trust*, derived from the work of Lyons and Mehta (1997), and the *components of trust*, derived from the work of Smyth (2003). Second, the chapter will present the empirical findings, and then provide an analysis of trust from the findings. The analysis will consider how these elements of trust are linked together and finally will place these elements into a broader picture of trust. The measurement will provide guidance for evaluating how trust helps the management of projects in a relationship approach for the development and management of trusting relationships. The chapter will conclude with a summary and an agenda for future action in practice and for research.

Definitions and dissection of trust

Trust

The financial analogy posited trust as an asset and a liability. The liability is incurred in order to build up the asset, providing that there is a return on the trust expenditure. In other words, to trust individuals and to trust an organisation as represented by individuals involves being prepared to be vulnerable. As O'Neill stated:

> *'Since trust has to be placed without guarantees, it is inevitably sometimes misplaced: others let us down and we let others down. When this happens trust and relationships based on trust are both damaged. Trust, it is constantly observed, is hard earned and easily dissipated. It is valuable social capital and not to be squandered.' (2002)*

Defining trust precisely has proved elusive. In the same way a number of concepts have spawned many definitions, such as marketing. The lack of agreed definitions for marketing has not prevented academics and practitioners from formulating marketing strategies. Indeed, it could be argued that the definitional differences are potential sources for market differentiation. The same can be said of trust: there is no agreed definition. Definitions of trust accord with the bundle of relationship experiences encountered, recognising that every relationship is different in character.

Diversity of definitions does not provide a reason for failing to define trust. Definitions are not only important for seeking agreement, but also help in understanding differences. Therefore definitions provided in this chapter may not accord with other definitions, yet will help others to understand the position adopted here as a basis for interpreting and applying the analysis in other contexts. This reflexive process helps synthesis over a period, with the end result perhaps being an agreed definition in the future.

There is general agreement in the literature that trust involves a willingness to be vulnerable (Mayer *et al.*, 1995; Mishra, 1996; in construction see Smyth, 2003; Wood & McDermott, 1999). This willingness to trust also implies an expectation of mainly positive outcomes (Rousseau *et al.*, 1998). In a business context expectation is ultimately for positive relationships against financial performance, although for individuals there may be mixed expectations and motives. Trust is intangible (cf. Fukuyama 1995; Ganesan, 1994; McAllister, 1995; Misztal, 1996). It is an attitude (Flores & Solomon, 1998; Luhmann, 1988) as a *noun*, and a disposition (Fukuyama, 1995) as a *verb* that informs action.

We may believe that those on whom we depend will meet our expectations of them (Shaw, 1997). Trust is observed indirectly as evidence in behaviour (Currall & Judge, 1995; Mayer *et al.*, 1995; Moorman *et al.*, 1993; Smith & Barclay, 1995; Smyth & Thompson, 1999, forthcoming), and using this indirect evidence we evaluate whether we shall continue to trust. Thus there are two steps. First, the willingness to be vulnerable, where trust is not in evidence on either side, but at least one party has indicated the preparedness to trust the other. Second, and following from this, is the opportunity to explore whether the other party takes the opportunity to be trusting or to take advantage. How the other party responds provides indirect evidence through their behaviour.

On this basis a definition has been proposed as:

> *'Trust is a disposition and attitude, giving rise to a belief, concerning the willingness to be vulnerable in relation to another party or circumstance.' (Smyth, 2003)*

In this definition 'belief' embodies the expectations arising in particular contexts, the behaviour in context providing the evidence. This definition was developed to take account of business-to-business relationships characterised by project contracts:

> *'Trust is a disposition and attitude concerning the willingness to rely upon the actions of or be vulnerable towards another party, under circumstances of contractual and social obligations, with the potential for collaboration.' (Edkins & Smyth, 2006)*

These definitions are socio-psychological with cultural overtones in character. Economic or socio-economic views vary. From a transaction perspective Williamson (1985) originally argued that 'trust' does not exist in the market, occurring as 'personal relations' against which calculations and measures cannot be assigned. As such, trust is indirectly expressed through the concept *atmosphere*, which is a summation of what individuals find satisfying in the way transactions are conducted. Korcynski (2000) criticises Williamson as essentially neo-Hobbesian, following neo-classical orthodoxy in respect of denying trust in the market. Korcynski builds upon Durkheim, citing that any economic contract formed in the market gives rise to social obligations that are not stated in the contract terms. The motivation to fulfil such obligations is relationship based, mediated by trust. Barney and Hansen define trust in economic relations as:

> *'an attribute of a relationship between exchange partners.' (1995, p.2)*

Since Williamson consigned trust to personal relations and by implication atmosphere, a whole body of management literature has emerged, called *relationship marketing* and *relationship management*, that has posited how business-to-business relations between firms are articulated through relationships (Christopher *et al.*, 2001; Grönroos, 2000; Gummesson, 2001; Ford *et al.*, 2003). This constitutes *social capital*, potentially yielding returns on investment. Trust is therefore an important foundation in economic relationships. In addition, Dasgupta (1988) has stated that trust in a market helps to increase stability, whereas its absence can lead to market failure. Trust adds value through incentives or governance, increases the value of personal or corporate relations, adds value through all forms of scarce knowledge, and through the perceived reliability of corporate systems in the exchange process (Korcynski, 2000). In contracting, it is demonstrated that trust is linked to financial issues through better business performance from working in teams across organisational boundaries (Swan *et al.*, 2001; cf. Cherns & Bryant, 1983). Therefore, the original claim of Williamson that personal relations are not germane to the market is unsustainable. Paradoxically, Williamson acknowledged the importance of team relationships – *relational teams*:

> *'The firm here will engage in considerable social conditioning to help assure that employees understand and are dedicated to the purposes of the firm, and the employees will be provided with considerable job security, which gives them assurance against exploitation. Effective adaptation in a cooperative team context is especially difficult and important to achieve.' (1985 p.247)*

The focus of Williamson in this quote is within organisations; however, many project teams are temporary and multi-organisational, purportedly acting in unison (Cherns & Bryant, 1983). Thus there is a link between relational teams and market governance. Williamson (1985) identified three forms of business-to-business governance:

- Classical contract law
- Neoclassical contract law
- Relational contracting

In *relational contracting* Williamson correctly identified that many transactions are not discrete and some or even the entirety of the relations have to be taken into account, including norms that have been established. While relational contracting concerns governance in the market, the conduct of that governance embraces personal relations and relational teams, and hence trust.

However, there is a subtle distinction between *relational contracting* and *relationship management*. The minimum requirement for *relational contracting* is reactive market management, although it may embrace proactive management. The minimum requirement for *relationship management* is proactive market management, involving relationship development as a competence for firms to systematically invest in and develop.

Williamson (1985) also argued that trust cannot be calculated in economic terms. This may have been accurate at the time, yet is becoming increasingly untenable. At one time it would not have been possible to calculate the value of commodities. It was only with the development of money as a representation of value that measurement became possible. The current position regarding trust is that ways of measurement are being developed, as represented through behaviour. Tracing the impact of trust in relationships so that it can be attributed to relationship revenue in economic terms (Storbacka *et al.*, 1994) still has some way to go, but no more so than tracing *transaction costs* in practice.

The primary obstacle to trust development is a lack of willingness to be vulnerable towards another party, arising from fear that others will not look after your best interests and will selfishly pursue opportunistic behaviour. Economists have more recently embraced trust and such fears, especially through game theory (see Kreps, 1990). Ive (1995) explored the outworking of trusting and adversarial relations through game theory in construction using the Nash equilibrium, identifying four potential outcomes for the two parties:

(1) Win–win
(2) Win–lose
(3) Lose–win
(4) Lose–lose

Classical economics assumed a party would look after its own interests. Game theory states that if one party looks after both their own interests *and* the interests of the other party, trust is present and a win–win outcome results. The win–win outcome is based upon mutual self-interest. On the other hand, if one party opportunistically takes advantage of the trust the other party has towards them, then a win–lose outcome arises and vice versa. Where one party intends to play by the rules, being willing to trust in principle, yet in practice mistrusts the other party, they will tend towards opportunism, which invokes a similar response from the other party, the result being lose–lose.

The win–win outcome derived from mutual self-interest is possible under a single exchange, but is more likely where there is a series of exchanges. Such 'repeat games' increase the value of pursuing mutual self-interest, gain from learning from relationship experience and the opportunity to solicit evidence about the relationships, and hence increase trust. In construction, strategic partnering provides a repeat game context. 'One-off' games can have the same effect where the game or contract is long enough for feedback, in other words where the contract induces cycles or patterns of relationship behaviour. In construction, some project partnering provides such a context, as do PPP/PFI concession contracts.

Win–win outcomes may vary. The size of the benefits or pay-offs may be different for the parties. There may also be some variance in the perceived level of trustworthiness in the parties' regard for each other. However, there is a level of trust in the relationship, therefore a maximum pay-off that may be distributed between the parties in a number of ways. Although trusting relations yield the greatest probability of inducing win–win outcomes, there is an underlying assumption that win–win is the optimum outcome. It is possible to go beyond a simple win–win for one-off and repeat games.

The proactive management of relationships, for example using *relationship management* or *emotional intelligence*, can be used to develop trust. Proactive management can increase the level of trust in the relationship; therefore the maximum pay-off is also increased. This induces an 'enhanced win–win', which in economic terms is somewhat analogous to 'added value' in a transaction. Alliances, such as partnering or supply chain management, try to promote the benefits of collaboration, which induce an enhanced win–win for both parties – added value for the client and increased profitability, repeat business or referrals for contractors. In practice, experience has been patchy. Trust can only contribute to an enhanced win–win where parties understand how trust is developed and managed. This requires definition and articulation of the *characteristics of trust*, the *components of trust* and the *conditions of trust*.

Characteristics of trust

Lyons and Mehta (1997) identified two stages of trust, the character changing as trust develops:

- *Self-interested trust* – a willingness to trust with minimal or no evidence for trust, but where it is estimated that there is mutual short-term advantage in trusting another party. The risk is small and so is the initial reward, yet there may be potential to build rewards beyond the initial willingness to trust.

- *Socially orientated trust* – generated through obligations in a social network, and coming through relationships, plus reputation and advocacy. It is sustained through experience, leading to a preparedness to 'go the extra mile' for another trustworthy party. In this sense it is sacrificial.

Self-interested trust is therefore predicated upon behaviour where it is in mutual self-interest to trust each other. This type of trust requires minimal evidence, probably at most an intuitive sense or knowledge of company reputation. It could be summarised as being prepared to accept the other party as trustworthy until proved otherwise, which, if successful, is the classic 'win–win'. In summary the motive to trust is: 'What can the other party do for me?' This is the character of self-interested trust (see Edkins & Smyth, 2006; Smyth, 2003). Self-interested trust provides an initial stage for developing trust, although if evidence of trust is not forthcoming a relationship may be sustained at this level, even for 'repeat games' in the game theory sense.

Socially orientated trust is a deeper level of trust that is giving in character, hence more philanthropic. The definition of 'service' is taken literally, and requires that one party 'goes the extra mile' for the other. Socially orientated trust is built with care through relationships, as each party must be aware of the willingness of the other party to be equally trustworthy. In a business context, there must also be investment on both sides in the relationship. While there must be a return on the investment in the long run, each act of investment does not necessarily require a short-term return. This is a second stage for developing trust, assuming parties proceed with caution in being vulnerable towards another party.

Socially orientated trust potentially increases the maximum pay-off, inducing an enhanced win–win. The performance pay-off may be unevenly distributed, yet must be greater for both parties than under the win–win of self-interested trust. From the deeper level of trust in the relationship, each party is not only benefiting in terms of the performance pay-off, but is also benefiting on the social bottom line, that is, building trust as an asset – the social capital of trust (O'Neill, 2002, cf. Swan *et al.*, 2001 in construction). In terms of motivation, socially orientated trust asks: 'What can I do for the other party?' (cf. Edkins & Smyth, 2006; Smyth, 2003). On occasions, socially orientated trust may emerge, yet development depends upon investment and proactive management through the governance and systems of *relationship management*. Socially orientated trust will be most easily developed and maintained in repeat games or where contracts are long enough for cycles or patterns of relationship behaviour to provide feedback.

Components of trust

The *components of trust* are a family of related concepts: expectations and confidence. *Expectations* arise in two forms: *faith* and *hope* in the future performance of the other party based upon current assessment. Faith is determined as the *unseen* capabilities of other parties to perform. Faith is a disposition

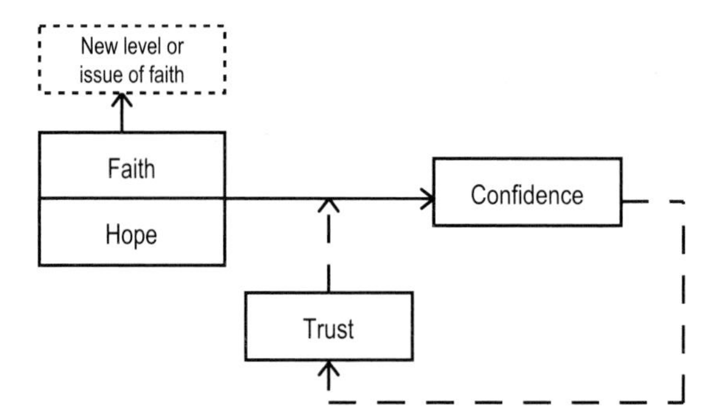

Fig. 4.1 Components of trust (Edkins & Smyth, 2006). Used with permission of ASCE.

and relates as a *noun* to the disposition of trust. Hope is formed through the *seen* capabilities of the other parties to perform. Hope is a disposition and relates as a *noun*, yet is also an intangible activity and relates to trust as a *verb*. The other component is *confidence* in the other party, derived from evidence of recent past performance within relationships, thus being an evaluation akin to a mathematical probability statement.

Williamson (1993) stated that 'trust' is used as an indicator of probability (cf. Gambetta, 1988), which is inaccurate. Where probability exists there is 'calculated risk', which concerns *confidence* and not trust, as confidence embodies evidence, which is measurable. The role of trust is to work with hope and faith through actively being vulnerable, so that evidence for trustworthiness in the other party may come to light. Trust mediates between these expectations and confidence, with evidence aiding mediation (see Figure 4.1).

If trusting another party brings forth evidence to induce confidence, then faith and hope can rise to new levels. Expectations rise until trust reaches the level of the management 'comfort zone' in terms of investment and transaction costs of, and strategic commitment to, trust, or until trust is broken and either expectations are lowered or all goodwill is spent.

Conditions of trust

Trust is engendered and developed in fertile conditions. In this context trust is related to behavioural intent (Thompson, 2003): the intention to be willing to be vulnerable and the judgments concerning the behavioural intent of the other party. A decision to trust carries a socio-psychological risk, the liability on the social balance sheet. The intention has a primary focus on inputs, hence relationship investment at a practical level, and how that is harnessed to competencies or abilities to yield behavioural outputs. While establishing *conditions of trust* can be circumstantial, the conditions can also be managed within paradigms, for example *relationship management* or *emotional intelligence* (Goleman, 1998; see **Chapter 11** by Cox & Ireland and **Chapter 3** by Druskat & Druskat).

Butler (1991) developed *conditions of trust* for consumer markets, including a 'conditions of trust inventory' of attributes and attitudes, converting related

Table 4.1 Ranked client perceptions of required conditions of trust by input–output contractor behaviour (Thompson, 2003; Smyth & Thompson, forthcoming).

Rank	Input/intent dimension	Rank	Output/ability dimension
1	Integrity	1	Consistency
2	Receptivity	2	Promise fulfilment
3	Loyalty	3	Fairness
4	Discretion	4	Competence
5	Openness	5	Availability
	Alpha = 0.92		Alpha = 0.90

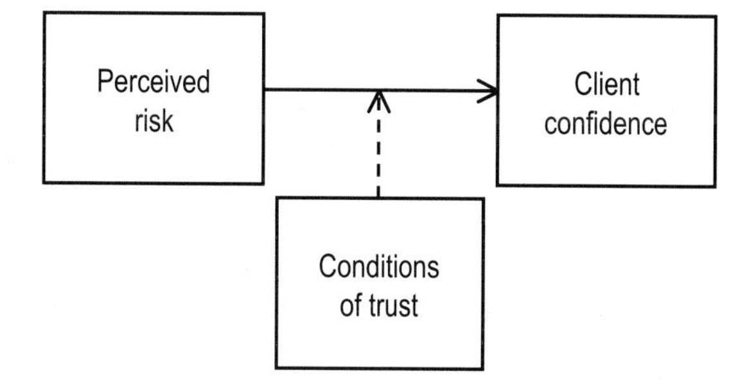

Fig. 4.2 Effect of conditions of trust upon perceived project risk.

behaviour patterns into an atmosphere and culture of trust. This is particularly relevant in business-to-business relationships extending over long project cycles (one-off games) and project programmes (repeat games), where framework agreements, PFI/PPP concessions or repeat business are prevalent.

An exhaustive study among leading construction clients established an inventory for construction (Smyth & Thompson, forthcoming; Thompson, 2003; Thompson *et al.*, 2003), and a ranking (see Table 4.1). These conditions are based upon client perceptions, thus have overlaps and grey areas. They reflect real conditions and expectations in building procurement (Smyth & Thompson, forthcoming). Clients perceive the following benefits:

- Reduced monitoring of providers
- Increased information disclosure
- Improved chance of overall project success (Smyth & Thompson, forthcoming; Thompson, 2003)

Establishing *conditions of trust* has the effect of increasing confidence in the other party, and consequentially the perceived project risks are reduced (see Figure 4.2).

While the assessed or 'objective' risk remains the same, the presence of trust means that clients are confident that timely and appropriate solutions will be identified; hence, the perception of risk, or subjective risk, is reduced. It is perceived risk that is most important, influencing behaviour, and hence trust. This links to the *components of trust*. Trust helps convert faith and hope into confidence as successful outcomes are achieved. Setting *conditions of trust* in place helps convert objective risk into confidence, the *components* acting intangibly, and the *conditions* acting at the tangible level of risk.

Winch (2002), using an information processing approach, stresses the need to make information available. Where project uncertainty is assessed as probability from the data, yielding an initial perceived risk, the *conditions of trust* can potentially reduce this risk by building confidence (Smyth & Thompson, forthcoming). Indeed, the presence of confidence through *conditions of trust* itself constitutes additional information. While this is not information that reduces 'objective' uncertainty, it is information that reduces perceived risk and in this sense is an objective contribution.

Trust in practice

Measurement of trust has been conducted across senior management in 13 separate organisations for a series of projects with a value in excess of £0.5 billion or almost exactly US$1.0 billion. The response rate was high:

- Client and design team 82%
- Client and project coalition of professional firms 87%
- Client and contractor 88%
- Client and project coalition of contracting firms 86%

The response represents a significant volume of public and private sector projects. The method was survey based, key managers being asked to rate their perceptions between late 2004 and early 2005. The components of trust utilised three-point Likert scales for faith, hope and confidence (cf. Edkins & Smyth, forthcoming), while the characteristics of trust utilised a five-point scale in order to include degrees of adversity or mistrust as well as self-interested and socially orientated trust (cf. Edkins & Smyth, 2006). The data should be considered qualitative, as further data for a broader project range would be needed to establish general patterns, and would need to be longitudinal to map the dynamics of trust over project life cycles.

The data provides indicative trust profiles of different project relationships and some pointers concerning projects for public and private clients. The client organisations are of three types:

- Private sector client – a leading property developer and investor
- Public sector client – PFI procurement, essentially representing design-build-finance-operate concession (cf. **Chapter 12** by Ive & Rintala)
- Public sector client – PPP procurement, essentially representing design-build-operate concession

Table 4.2 Perception rating of characteristics of trust in key relationships.

Client relationship	Trust	
	Self-interested (%)	Socially orientated (%)
Design team	53	40
Project coalition of professional firms	42	38
Contractor	42	13
Project coalition of contracting firms	46	27

Table 4.3 Perception rating of supply cluster of the characteristics of trust in the client relationship.

Client type	Trust	
	Self-interested (%)	Socially orientated (%)
Private sector developer-investor	42	38
Public sector PFI (with finance)	42	26
Public sector PPP (without finance)	37	11

Findings

The *characteristics of trust* in the key relationships are considered in Table 4.2. The data represents scores for self-interested and socially orientated trust (+1 and +2 on the Likert scale), expressed as a percentage of the maximum achievable. There is generally a reasonable level of socially orientated trust, particularly among the professional relationships at the design stages of a project and where the professions are directly 'representing' the client. The professional coalition includes specialist consultants as well as the design team consisting of architect, cost consultants, and structural and services engineers. Therefore the lower levels of socially orientated trust are due to these specialists, namely planning consultants and chartered surveyors responsible for letting. The client–contractor relationships are weak. They are strengthened when other immediate players in the supply cluster are brought into the coalition.

The private client–design team achieved 60% of the net score across the entire scale for the *characteristics of trust*, thus including adversarial behaviour, while the client–contractor achieved 15% – a very low level of trust.

Table 4.3 considers perceptions that supply team members have of the client. This establishes whether the client type affects the presence of the *characteristics of trust*, for clients occupy positions of market dominance (see **Chapter 11** by Cox & Ireland) and are regular procurers of buildings. It is found that the client type makes a distinct difference, especially in terms of socially orientated trust. The private sector client appears to induce circumstances in which self-interested trust is converted to socially orientated trust

Table 4.4 Perception rating of the components of trust in key relationships (%).

Client relationship	Faith	Hope	Confidence
Design team	100	100	93
Project coalition of professional firms	85	92	92
Contractor	29	46	43
Project coalition of contracting firms	60	81	65

in some relationships and in which self-interested trust can continue to develop. The PFI public sector client, where private finance is supplied for capital costs by the concession consortium, also induces reasonable levels of socially orientated trust. The PPP-type client, where the client supplies the finance directly for the capital costs of the projects, paradoxically seems to have the effect of inducing less trust, especially socially orientated trust.

There is also evidence of adversarial behaviour. It is found that the relationships amidst the entire private sector developer-investor supply cluster, including the client, scored 40% of the net score across the entire scale for the *characteristics of trust*. Some adversarial behaviour, especially among the letting agents, offset some benefits derived from socially orientated trust. The public sector PFI (with finance) scored 39%, a similar percentage to the private sector client set of relationships. The PPP (without finance) scored –2.6% net across the entire scale, where negative behaviour patterns prevailed. Therefore, the PPP public sector client body is creating a net liability on the social balance sheet.

The *components of trust* in the key relationships are considered in Table 4.4. The average perception rating per respondent is very high at the client–professions interface, whereas it is low between the main contractor and client. Even though some adversarial behaviour among the letting agents has been noted, the confidence for successful outcomes is very high indeed.

The client–contractor rating is moderate for hope and confidence and low for faith. The coalition of contracting firms does improve the rating considerably. This may be because clients have greater hope concerning other players within the supply cluster. This hope seems to exist despite the lack of trust characteristics. Therefore it must be assumed that the market and contract mechanisms help induce successful outcomes. Certainly the level of confidence is reasonable, suggesting that hope is being converted into confidence through evidence of performance. However, this confidence in performance is probably induced to a limited degree through trust (see Table 4.2), other factors, perhaps technical competence or contractual obligations, beyond the scope of this research creating these levels of hope.

The perceptions that the supply team members have of the client concerning *components of trust* are considered in Table 4.5. The private sector market environment induces higher levels of faith, hope and confidence. The public sector client faired reasonably well where financial risk had been transferred to the private sector, yet poorly where the public sector was holding the purse strings.

Table 4.5 Perception rating of supply cluster of the components of trust in the client relationship (%).

Client type	Faith	Hope	Confidence
Private sector developer-investor	85	92	92
Public sector PFI (with finance)	58	74	68
Public sector PPP (without finance)	33	47	32

Table 4.6 Perceived importance of conditions of trust in key relationships.

Client Relationship	Ranking of conditions of trust		
	Overall ranking (1–5)	Input/intent	Output/ability
Client-contractor (Thompson's ranking)	1 Integrity	1 Integrity	1 Consistency
	2 Consistency	2 Receptivity	2 Promise fulfilment
	3 Promise fulfilment	3 Loyalty	3 Fairness
	4 Receptivity	4 Discretion	4 Competence
	5 Loyalty	5 Openness	5 Availability
Client-design team	1 Integrity	1 Integrity	1 Competence
	2 Competence	2 Loyalty	2 Promise fulfilment
	3 Loyalty	2 Openness	3 Consistency
	4 Promise fulfilment	4 = Receptivity	4 Fairness
	5 Openness	4 = Discretion	5 Availability
Client-project coalition of professional firms	1 Integrity	1 Integrity	1 Competence
	2 = Loyalty	2 Loyalty	2 Promise fulfilment
	2 = Competence	3 Openness	3 Consistency
	4 = Promise fulfilment	4 Receptivity	4 Fairness
	4 = Openness	5 Discretion	5 Availability

Measuring the presence of the *conditions of trust* is shown in Table 4.6. Data was collected for the design team and coalition. Data was not collected for the client–contractor dyad, which had been the focus of Thompson's findings (2003; see Table 4.1). A direct comparison with previous data is limited as this survey did not replicate the study (Thompson, 2003) for reasons of complexity and duration; however, respondents were asked to say whether they perceived the range of conditions to be present and the ranking is derived from the summation for each condition. The objective is to establish whether the findings of Thompson are echoed here.

This research finds that there are differences in the importance ascribed to some conditions. Thompson (2003) found that clients considered *integrity* and *consistency* among contractors to be most important. *Integrity* remained most important, yet *consistency* was not considered so important among the professions. *Competence* was highly valued, and *openness* was a valued condition in relationships between clients and the professions. This can be explained by contractors not being directly responsible for design and specification, or only indirectly where they subcontract design under D&B-type arrangements.

Table 4.7 Perception of conditions of trust in key relationships.

Client relationship	Conditions of trust					
	Input/intent			Output/ability		
	Rank importance of condition	Instances present	Difference in rank	Rank importance of condition	Instances present	Difference in rank
Client-design team	1 Integrity	15	–	1 Competence	14	down 1
	2 Loyalty	13	down 3	2 Promise fulfilment	13	down 1
	3 Openness	14	–	3 Consistency	11	down 1
	4 = Receptivity	14	up 1	4 Fairness	15	up 3
	4 = Discretion	14	up 1	5 Availability	11	up 1
Client-project coalition of professional firms	1 Integrity	26	–	1 Competence	25	down 1
	2 Loyalty	22	down 3	2 Promise fulfilment	22	down 1
	3 Openness	23	down 1	3 Consistency	19	down 1
	4 Receptivity	25	up 2	4 Fairness	26	up 3
	5 Discretion	25	up 3	5 Availability	18	–

Their role is to align themselves with the design and specification of others through receptivity and consistency, and deliver accordingly via subcontractors and suppliers.

There is reasonable congruence between the importance ascribed to the *conditions of trust* and the extent of their presence. However, there are areas for improvement. In terms of behavioural intent, the design team fell short on *loyalty* – down three positions – although the number of instances scored reasonably high (Table 4.7). From the disaggregated figures it would appear that there is some concern over the *loyalty* of the client towards the design team and also between architects and cost consultants.

Analysis of findings

Comparing the results from the *characteristics* and *components of trust* in key relationships, confidence was 78% of the characteristic of trust Likert scale score for the design team. It was 114% for the professional coalition, especially high as trust was lower and confidence must also be derived from non-trust performance factors. This additional finding shows there is scope for building and maintaining confidence in the future. Confidence was 89% of the characteristic of trust Likert scale score for the contracting coalition, which is partly an expression of low trust levels with confidence either arising from other factors or being the result of trust that has been subsequently dissipated. However, trust-confidence was merely 42% for the client–contractor relationship indicating that a great deal more investment into building trusting relations is needed in order to induce confidence in the future.

Comparing the results on the *characteristics* and *components of trust* from the viewpoint of the supply cluster, it is found that hope is a factor that is particularly strong in the PFI client–supplier cluster. While this may be an anomaly, it could also be the case that this is a positive aspect for developing trust. Trust is clearly low for what could be expected to be favourable, given the partnership nature of the policies and the long-term project life cycles.

The analysis clearly demonstrates that there are differences in the strengths of relationships generally and the extent of trust in the relationships across two dimensions:

(1) The type of relationship between the client and the suppliers
(2) The client type

Concerning the type of relationship, it has been demonstrated that the client–design team and professional coalition fare positively in comparison with the contractors and the contracting coalition. It has been argued that contractors primarily manage transaction costs (Gruneberg & Ive, 2000; Winch 2002), which is to say they neither specify the product nor produce it, subcontracting the majority of the work. However, the letting agents, although more adversarial than the design team, maintain trust in the professional coalition, although they too manage transaction costs. While the levels of uncertainty may be lower (cf. Winch, 2002), the main difference is that agents are paid on a percentage fee basis, which helps induce a positive outlook, whereas contractors endeavour to reduce costs, which helps induce a negative outlook – at its worst manifested as a blame culture in behavioural terms (Smyth, 2000). As trust seems to be higher in public sector work where concession consortiums have more control over the finance, it is possible that incentive contracts could induce more positive behaviour among contractors. Positive incentives enable contractors to gain from effective and efficient management and may prove conducive to trust development.

It might be argued that gain share contracts are one option, incentivising cost reductions. Pain-gain share contracts may prove less conducive as the penalty element could provoke defensive behaviour, which is contrary to the willingness to be vulnerable that trust is predicated upon. Gain share contracts can be structured around the tender price, yet can also use the cost consultant's estimated price from the pre-tender stage as a benchmark for establishing the share of cost reductions gained.

Cost plus contracts, although closest to the mode of payment for letting agents, tend to prove expensive for clients as contractors have no incentive to control client cost – indeed, the reverse. Cost plus, with a guaranteed maximum price, is a moderated option. An alternative to cost plus contracts is target cost contracts, which essentially use the pre-tender estimates or an independently established schedule of rates to form targets. Contractors are appointed by negotiation centred on non-financial criteria, such as service quality, or the ability to manage relationships. The target cost can also act as a guaranteed maximum price, and then contractors are incentivised to build below the target cost on a gain share basis. Target cost was used regularly in the Second World War in order to achieve speed of construction by letting contracts as soon as possible and proved financially more robust than its predecessor, cost plus (Smyth, 1985).

Another measure, which may be used in parallel to target cost, is an *escrow account*. It has been found that supply contracts with penalties tend to induce negative and at worst adversarial behaviour in many industries. An escrow account is used in the context where either party in the dyad is perceived by the other to be unresponsive. A charge is levied and paid into the escrow account, which is spent to reinvest in the relationship, for example in new information systems, joint team education or travel to get people face-to-face more often. The results from using escrow in this way have been an increase in the level of trust (Birt, 2003). The effectiveness of these contract measures in relation to trust requires further research and almost certainly would require a programme of contracts, such as strategic partnering or a contract with a long project life cycle that is phased.

This research has found that trust characteristics and components of trust are higher in the private sector coalition than the public sector ones. The public sector projects involved concession contracts and thus involved cycles and patterns of project behaviours that echo features of repeat games. Therefore, the nature of the client organisations, and hence their ability or desire to learn from experiences, appears to be more important than whether the game is one-off or repeat in nature. Also the preparedness of the client to use their leverage in the market is also important, which this study demonstrates according to whether the client retains the purse strings (cf. **Chapter 11** by Cox & Ireland). Thus, PFI contracts, which increase financial risk, carry lower project risks as client leverage is reduced and relationships are more collaborative.

The public sector client is generally poor at managing relationships, resulting in poor trust levels. This might seem surprising, government being concerned to drive continuous improvement since 1994. PPP contracts imply a partnership, and hence collaborative working over long periods of the project life cycle, typically 25 years or more. While procurement has become much more sophisticated through the 'Gateway' process introduced by OGC through the Treasury department, managing relationships seems to lack any systematic approach (Edkins & Smyth, 2006, forthcoming).

Although this problem is less in the private sector, there is scope for improvement there too. Currently, responsibility for relationship development is left at the level of the individual and how teams operate on this basis, rather than there being a proactive organisational strategic system to support and manage relationship development (Smyth & Edkins, under review; cf. Edkins & Smyth, 2006, forthcoming).

Developing non-adversarial relations in construction has been advocated (Latham, 1994), with trust being important (Egan, 1998). However, moving from adversity can simply result in 'good relations' rather than proactively managed relations that improve performance.

Relational contracting and relationship management

The term *relational contracting* has been widely used in recent years in the context of alliances, partnering and collaboration (see, for example, **Chapter**

7 by Kumaraswamy & Rahman; Edkins & Smyth, 2006; Macneil, 1974; Rahman *et al.*, 2002; Rahman & Kumaraswamy, 2004; Rowlinson & Cheung, 2004). Within the field of transaction cost economics, *relational contracting* moves project practice away from the worst excesses of adversarial behaviour characterised by opportunism (cf. Ive 1995). However, *relational contracting* does not have the strategic strength of concept to improve performance in other areas of management as *relational contracting* is rather reactive.

Concepts within the *relationship management* paradigm (see Ford *et al.*, 2003; Grönroos, 2000; Gummesson, 2001; in construction see, for example, Rowlinson & Cheung, 2004; Smyth, 2000, 2004) have the strategic content and proactive requirements to match the strategic procurement approaches, such as supply chain management. It has been found that PFI/PPP contractors have yet to take on board the need to proactively manage relationships in systematic ways (Smyth & Edkins, under review). It has been argued that when the *relationship marketing* and *management* paradigm is harnessed to the procurement approaches in construction, there is potentially further enhancement (Smyth, 2005).

Public and private sector organisations tend to rely on responsibility and behaviour at the level of the individual. This chapter has demonstrated that the private sector has a more conducive culture. It is likely that public sector 'accountability' dominates other legitimate considerations, yet both the public and private sectors could benefit from considering management programmes and systems for improving relationship management competencies, either within the *relationship management* paradigm or through other paradigms of competencies, such as *emotional intelligence* (see **Chapter 3** by Druskat & Druskat) or *knowledge management* (see **Chapter 2** by Morris) or some combination (Smyth, 2004).

Figures 4.3, 4.4 and 4.5 build up the dynamics of how trust works in practice from the three building blocks – *characteristics*, *components* and *conditions of trust* – to show how trust operates for the management of projects. These figures do not consider the implications of levels of hierarchies within an organisation, nor project management tasks at head office and on site, nor the impact of the marketplace beyond transaction economic issues raised, all of which are schematically considered elsewhere (Smyth, 2003). They do provide the building blocks, from which managers can develop their approaches to develop, manage and measure trust.

Figure 4.3 shows how the *characteristics of trust* and *components of trust* interact, thus helping to deepen trust from a self-interested level towards a social orientation, in order to develop trust. Figure 4.4 adds the *conditions of trust* and its contribution to confidence in managing projects. Where the presence of trust is left to individuals or chance an overemphasis upon accountability and opportunistic behaviour can arise, eroding trust with opportunism, inducing mistrust and increasing uncertainty. Figure 4.5 adds links to the information processing approach of Winch (2002). The negative impact of both increased uncertainty derived from adversarial behaviour and the positive impact of increased confidence provide additional information, which is

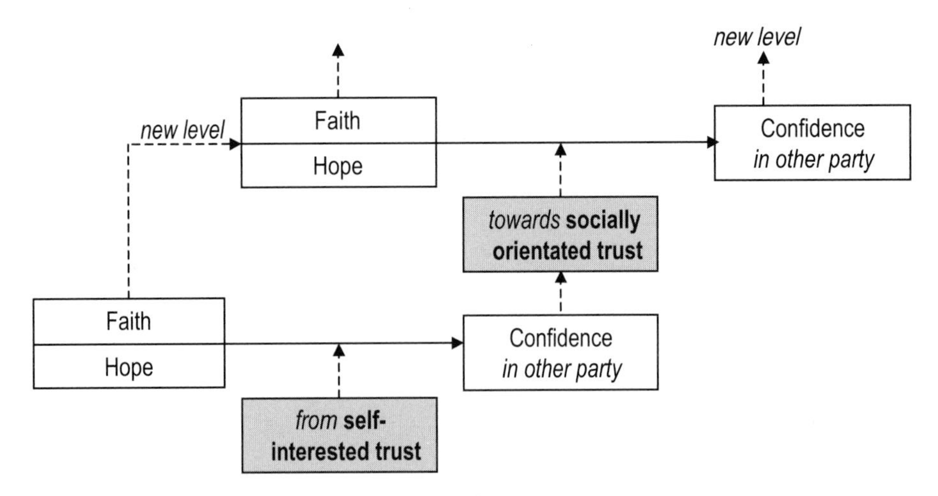

Fig. 4.3 The characteristics and components of trust dynamics.

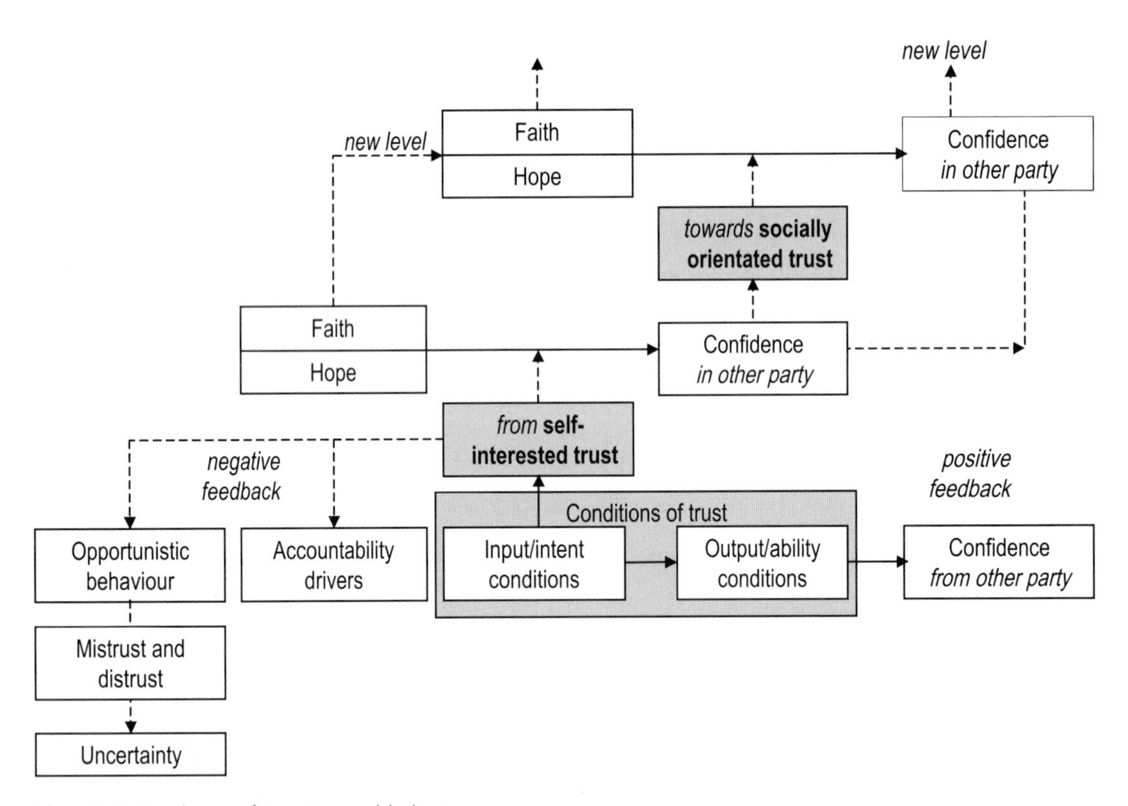

Fig. 4.4 Conditions of intention and behaviour.

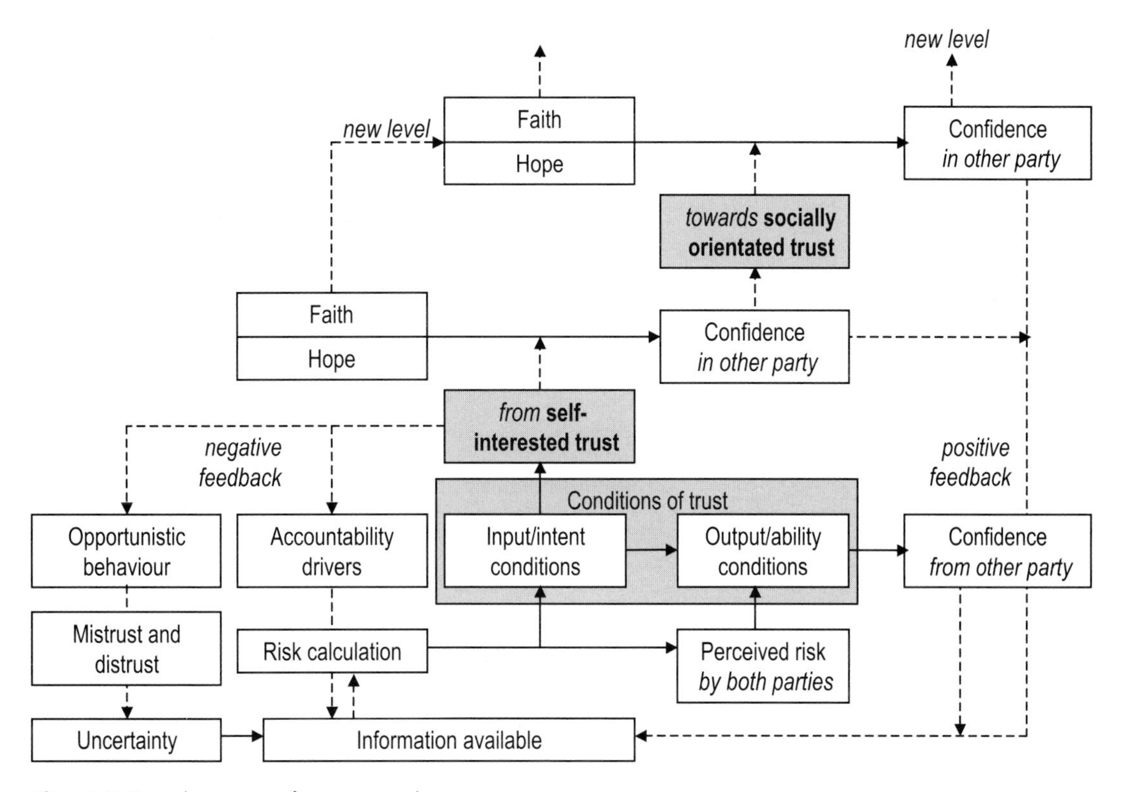

Fig. 4.5 Trust dynamics, information and uncertainty.

reprocessed into risk calculations and trades off opportunism against confidence as a probability statement.

It has been shown that trust can in practice go beyond the simple 'win–win' envisaged in game theory towards 'enhanced win–win'. It has also been shown that the nature and leverage of the client seems more decisive for developing trust than whether a project is one-off or a repeat game or behaviour pattern. The analysis has shown that there are differences between key relationships and client types. While the extent of trust is largely reactive and minimal in terms of proactive management, there is sufficient evidence of individuals developing trust to show that there is scope for managers to develop trust proactively beyond *relational contracting* towards *relationship management*.

Conclusion

Three 'building blocks' have been analysed: the *characteristics of trust*, the *components of trust* and the *conditions of trust*. This chapter has demonstrated that it is possible to break down the concept of trust and measure the extent to which it is present. The method used a relatively simple Likert scale questionnaire that can be replicated by organisations and project coalitions without incurring high costs. The key relationships investigated were the client–design team, the client–project coalition of professional firms, the client–contractor, and the client–project coalition of contracting firms.

Trust is generally agreed to be very significant in developing good relations with other parties. The evidence shows there are specific areas that managers can identify in which to develop trust, the dynamics of which have been presented in Figures 4.3–4.5. A proactive stance to management raises trust as a strategic competence. Trust is located within transaction cost economics and *relational contracting*; it is also located as part of proactive management within the *relationship management* paradigm (Smyth & Edkins, under review), or within another paradigm, such as *emotional intelligence*. Developing trust from a self-interested basis to a socially orientated one is an important foundation and requires proactive management (cf. Edkins & Smyth, 2006).

This analysis has therefore confirmed the ability to measure trust and has argued the need for managing trust within the management of projects. Apart from linking the analysis to information availability, uncertainty and risk, *how* trust is developed and managed operationally between the head office and site and across projects is beyond the scope of this chapter and would benefit from further investigation. There are several related points to make.

First, it would be useful to link the measurement approach developed here to benchmarks provided through key performance indicators (KPIs). This is especially the case for the KPI of client satisfaction as current practice renders it rather innocuous – satisfaction is not formally evaluated against expectations, which is a shortcoming given the lack of personnel continuity on many projects (see **Chapter 6** by Wilkinson), and the duration of project life cycles (Winch, 2002). Project specific expectations around time-cost-quality, the briefing documentation and selection criteria can all be used, and can be enriched as expectations pertaining to trust, that is faith and hopes concerning project relationships.

Second, the means for developing trust may be context specific both in terms of organisational culture, systems and procedures and in terms of any resultant benefits accruing as competitive advantage. Related to this are market restraints on firms to keep investment costs and transaction costs low. However, these exist in any situation. Management will be aware of these risks compared to the benefits that accrue as organisational competitive advantage, and accrue as barriers to entry when pursued by the majority of players in any tier of the market.

Third, there is the need to establish a clear link between trusting relations and project performance. The link between trust and KPI measures, measures of profit margins and return on capital employed need to be established. Further research is therefore needed, on the one hand to extend the breadth of projects covered by this work, and on the other to conduct a longitudinal study during project life cycles.

A longitudinal study would also provide firmer evidence concerning the points of intervention management can target and what actions are likely to yield improvements to trust levels both within and across organisational boundaries. Costa (2003) analysed trust in non-project teams within organisations, finding that trust enhanced attitudes of commitment to team tasks, but did not necessarily improve issues of continuity (cf. **Chapter 6** by Wilkinson).

Fourth, it has been suggested that contracts that positively incentivise contractors may help induce trust, or at least offer more scope for developing trust. This conclusion from the analysis is indicative. Establishing this in a more robust way requires further research over a programme of projects.

In summary, some dimensions have been developed to enhance understanding of the notion of *trust*. Trust has been measured across these dimensions. Analysis has been provided of both the data and concepts in order to demonstrate that trust can not only be measured in a detailed way, but also developed and managed for the management of projects. A research agenda has been set out so that the contribution that the presence and development of trust directly makes to project performance can be evaluated. An opportunity is present to develop relationship management in ways that could significantly improve project performance and added service value in the management of projects. While such measures may not be appropriate for all clients and coalitions, there is potential in certain market segments, particularly the upper tiers of the market and for niches within the project market.

References

Barney, J. B. & Hansen, M. (1995) *Trustworthiness as a Source of Competitive Advantage*. Paper given at the Australian Graduate School of Management, University of South Wales, Sydney.

Birt, D. N. (2003) Supply chain challenges: building relationships. *Harvard Business Review*, July, 64–73.

Butler, J. K. (1991) Toward understanding and measuring conditions of trust: evolution of conditions of trust inventory. *Journal of Management*, **17** (3), 643–63.

Cherns, A. B. & Bryant, D. T. (1983) Studying the client's role in construction management. *Construction Management and Economics*, **1**, 177–84.

Christopher, M., Payne, A. & Ballantyne, D. (2001) *Relationship Marketing: Creating Stakeholder Value*. Butterworth-Heinemann, Oxford.

Costa, A. C. (2003) Work team trust and effectiveness. *Personnel Review*, **32** (5), 605–22.

Currall, S. C. & Judge, T. A. (1995) Measuring trust between organisational boundary role persons. *Organisational Behaviour and Human Decision Processes*, **64** (2), 151–70.

Dasgupta, P. (1988) Trust as commodity. In: D. Gambetta (ed) *Trust: Making and Breaking Cooperative Relations*. Blackwell, Oxford.

Edkins, A. J. & Smyth, H. J. (2006) Contractual management in PPP projects: evaluation of legal versus relational contracting for service delivery. Special issue on Legal

Aspects of Relational Contracting. *ASCE Journal of Professional Issues in Engineering Education and Practice.*
Edkins, A. J. & Smyth, H. J. (forthcoming) The imperative of trust in PPPs: evaluations from the provision of 'full service' contracts. *Journal of Construction Procurement.*
Egan, J. (1998) *Rethinking Construction.* HMSO, London.
Flores, F. & Solomon, R. C. (1998) Creating trust. *Business Ethics Quarterly*, **8** (2), 205–32.
Ford, D., Gadde, L-E., Håkansson, H. & Snehota, I. (2003) *Managing Business Relationships.* Wiley, London.
Fukuyama, F. (1995) *Trust: The Social Virtues and the Creation of Prosperity.* Penguin Books, Harmondsworth.
Gambetta, D. (1988) *Trust: Making and Breaking Cooperative Relations.* Blackwell, Oxford.
Ganesan, S. (1994) Determinants of long-term orientation in buyer-seller relationships. *Journal of Marketing*, **58**, 1–19.
Goleman, D. (1998) *Working with Emotional Intelligence.* Bloomsbury, London.
Grönroos, C. (2000) *Service Management and Marketing.* John Wiley and Sons, London.
Gruneberg, S. L. & Ive, G. J. (2000) *The Economics of the Modern Construction Firm.* Macmillan, Basingstoke.
Gummesson, E. (2001) *Total Relationship Marketing.* Butterworth-Heinemann, Oxford.
Ive, G. (1995) The client and the construction process: the Latham Report in context. In: S. L. Gruneberg (ed.), *Responding to Latham: the Views of the Construction Team.* CIOB, Ascot.
Korcynski, M. (2000) The political economy of trust. *Journal of Management Studies*, **37** (1), 1–21.
Kreps, D. M. (1990) *Game Theory and Economic Modelling.* Oxford University Press, Oxford.
Luhmann, N. (1988) Familiarity, confidence, trust: problems and alternatives. In: D. Gambetta (ed.) *Trust: Making and Breaking Cooperative Relations.* Basil Blackwell, Oxford.
Lyons, B. & Mehta, J. (1997) Contracts, opportunism and trust: self-interest and social orientation. *Cambridge Journal of Economics* **21**, 239–57.
Macneil, I. R. (1974) The many futures of contracts. *Southern California Law Review*, **47** (3), 691–816.
Mayer, R .C., Davis, J. H. & Schoorman, F. D. (1995) An integrative model of organizational trust. *Academy of Management Review,* **20** (3), 709–34.
McAllister, D. J. (1995) Affect and cognition based trust as a foundation for interpersonal cooperation in organisations. *Academy of Management Review*, **38** (1), 24–59.
Mishra, A. K. (1996) Organizational responses to crises: the centrality of trust. In: R. M. Kramer & T. R. Tyler (eds) *Trust in Organizations: Frontiers of Theory and Research.* Sage Publications, London.
Misztal, B. A. (1996) *Trust in Modern Societies.* The Polity Press, Cambridge.
Moorman, C., Deshpande, R. & Zaltman, G. (1993) Factors affecting trust in market research relationships. *Journal of Marketing*, **54**, 81–101.
O'Neill, O. (2002) Lecture 1: Spreading suspicion. *A Question of Trust, Reith Lectures* (radio programme). British Broadcasting Corporation, London.
Rahman, M. M. & Kumaraswamy, M. M. (2004) Potential for implementing relational contracting and joint risk management. *ASCE Journal of Management in Engineering,* **20** (4), 178–9.

Rahman, M. M., Kumaraswamy, M. M., Rowlinson, S. & Palaneeswaran, E. (2002) Transformed culture and enhanced procurement: through relational contracting and enlightened selection. In: T. M. Lewis (ed) *Proceedings of the CIB Joint Symposium on W092 and W063, TG36 and TG23*, January 14–17, Trinidad & Tobago.

Rousseau, D. M., Sitkin, S. B., Burt, R. S. & Camerer, C. (1998) Not so different after all: a cross-discipline view of trust. *Academy of Management Review*, **23** (3), 393–404.

Rowlinson, S. & Cheung, Y. K. (2004) Relational contracting, culture and globalisation: value in project delivery systems facilitating a change in culture. *Proceedings of the CIB Symposium on Globalization and Construction*, CIB-W107, November 17–19, Bangkok.

Shaw, R. B. (1997) *Trust in the Balance: Building Successful Organizations on Results, Integrity, and Concern*. Jossey-Bass Management Series, San Francisco.

Smith, J. B. & Barclay, D. W. (1995) *Promoting effective selling alliances: the roles of trust and organizational differences*. Technical Working Paper, Report No. 95–100. Marketing Science Institute, Cambridge, MA.

Smyth, H. J. (1985) *Property Companies and the Construction Industry in Britain*. Cambridge University Press, Cambridge.

Smyth, H. J. (1999) Partnering: practical problems and conceptual limits to relationship marketing. *International Journal of Construction Marketing*, **1** (2), www.brookes.ac.uk/other/conmark/IJCM

Smyth, H. J. (2000) *Marketing and Selling Construction Services*. Blackwell Science, Oxford.

Smyth, H. J. (2003) *Developing customer-supplier trust: a conceptual framework for management in project working environments*. CRMP Working Paper, http://www.crmp.net/papers/index.htm

Smyth, H. J. (2004) Competencies for improving construction performance: theories and practice for developing capacity. *Journal of International Construction Management*, April, 41–56.

Smyth, H. J. (2005) Procurement push and marketing pull in supply chain management: the conceptual contribution of relationship marketing as a driver in project financial performance. Special issue on commercial management of complex projects (eds D. Lowe & R. Leiringer). *Journal of Financial Management of Property and Construction*, **10** (1), 33–44.

Smyth, H. J. & Edkins, A. J. (under review) Relationship Management in the management of PFI/PPP projects in the UK. Submitted to a journal, 2006.

Smyth, H. J. & Thompson, N. J. (1999) Partnering and conditions of trust. In: P. Bowen & R. Hindle (eds) *Customer Satisfaction: A Focus for Research & Practice*. CIB W55 & W65 Joint Triennial Symposium, 5–10 September, Cape Town.

Smyth, H. J. & Thompson, N. J. (forthcoming) Managing conditions of trust within a framework of trust. *Journal of Construction Procurement*.

Storbacka, K., Strandvik, T. & Grönroos, C. (1994) Managing customer relationships for profit: the dynamics of relationship quality. *International Journal of Service Industry Management* **5** (5), 21–38.

Swan, W., Wood, G. & McDermott, P. (2001) *Trust in Construction: Achieving Cultural Change*, http://www.scpm.salford.ac.uk/trust/publications.htm

Thompson, N. J. (2003) *Relationship marketing and client trust toward contractors within the large private building sector of the UK construction industry*. PhD thesis, Oxford Brookes University.

Thompson, N. J., Raftery, J. & Thompson K. E. (2003) Conditions affecting the development of trust within virtual organisations: the client perspective in the UK construction industry. *Proceedings of the Academy of Marketing Conference*, Aston Business School, Aston University.

Williamson, O. E. (1985) *The Economic Institutions of Capitalism*. Free Press, New York.

Williamson, O. E. (1993) Calculativeness, trust, and economic organization. *Journal of Law and Economics*, **36**, April, 453–86.

Winch, G. M. (2002) *Managing the Construction Project*. Blackwell, Oxford.

Wood, G. & McDermott, P. (1999) *Searching for Trust in the UK Construction Industry: An Interim View*. CIB W92 Procurement Systems Conference, Thailand.

Section III

Relationships at the Client–Design Team–Contractor Interface

Context

Albert Einstein said: 'Imagination is more important than knowledge.' Managing relationships is no mean feat. Nor is it a 'magic wand'. As **Chapter 1** demonstrated, not all relationships are healthy ones. The adversarial culture of blame and personality has dominated project firms that endeavour to manage transaction costs at minimum levels. Moving away from adversarial behaviour has been advocated in many countries, but success has been mixed. Some clients, design teams and contracting firms have tried with varying success; others have not tried. Firms that try face both internal critics that wish to return to the 'bad old days' and external critics that have indirect yet vested interests in identifying failure.

Successful change depends upon developing successful relationships. **Section III** examines a selection of important relationships, contexts and processes. It broadens out the relationship approach, taking the exploration and discourse further beyond psychology and organisational behaviour, through what we have called 'soft side of organisational ergonomics', and links into organisational interfaces for project teams and the links with other mainstream issues of managing projects. In addressing these issues, an overview of the contribution made by each chapter is first set out and then the chapters are located in the framework.

Contributions of chapters in this section

In **Chapter 5** Cova and Salle look at *Communications and stakeholders*. They put forward the idea that the project comprises a type of network involving relationships between project stakeholders, and stakeholders and project suppliers (suppliers in the broadest sense here) embedded within a dynamic environment. This unit of analysis is referred to by Cova and Salle as the *milieu*. They contend that communications between the actors within the milieu aim to 'construct, develop and maintain extra-business relationships with the stakeholders' and that this 'concerns the logic of friendship or ritualising what will be developed (t)hereafter'.

The chapter looks at definitions of the stakeholder and inconsistencies in such definitions. The authors contrast Turner's rather restrictive definition (1995) with the rather broader and inclusive definition adopted by Pinto and Rouhiainen (2001). The chapter moves on to deal with some social network analysis terms and applies these terms to the milieu. A case study involving a subsidiary of a major construction group working in the Seine-Maritime *département* of France is dealt with as an example of the milieu in operation. We see the way in which a network of project stakeholders is identified and managed in order to achieve a successful construction project outcome. The importance of friendship and mutual debt in the operation of the milieu is examined.

In **Chapter 6** Wilkinson tackles *Client handling models for continuity of service*. She takes a fresh look at *continuity of service*, building on continuity achieved through *account handlers* and *relay teams* (Smyth, 2000). An account handler is one person responsible from cradle to grave for the project and client relations. Others have examined this within relationship marketing and management paradigms using the term *key account manager* (KAM).

Wilkinson examines an 'old chestnut', the most appropriate organisation to manage the interests of the client, especially from the viewpoint of service continuity. In other words, taking the client handling models *within* a firm, she explores how these models apply *across* organisations. She examines the traditional client's representative, architect or engineer, considers the independent project manager, and considers the contractor too. She applies the account handler and relay team models to these options, examining under what circumstances the models are most suited to client interests.

Applying a relationship approach provides new management insights. Drawing upon the results of recently completed research and additional analysis of new research data, Wilkinson develops an analysis of *roles* to be performed at each stage of the project life cycle, which highlights the needs for managers in respective organisations handling client interests and service continuity to have or develop particular competencies to improve project performance. Similarly, organisations representing the client need to have or develop capabilities at the level of strategic management to ensure continuity of service for the client. This is necessary in order to ensure the representative organisation delivers a consistent service to its clients, that they manage the project teams or coalitions consistently and that the coalition provides continuity as far as it is in the power of the representative at each relevant stage to do so.

It cannot be assumed by representatives that this is achieved by pursuing 'business as usual', and Wilkinson identifies some of the problem issues and recommends areas for strategic and policy focus. In the context of how contractors provide continuity of service using either the account handler and KAM model or the relay team model, one of the first issues representatives need to address is how they can oversee the contractor.

In **Chapter 7** Kumaraswamy and Rahman address *Applying teamworking models to projects*. This chapter brings together some of the work developed elsewhere by the authors, both integrating and extending it. They look at teambuilding by objectives as a key input and value as a key output within the remit of *relational contracting*. The main areas addressed are:

- Relationally integrated supply chains (RISCs)
- Relationally integrated project teams (RIPTs)
- Relationally integrated value networks (RIVANs)

Relational contracting is conceptually located in the work of Williamson (1985), classical contract law and neoclassical contract law being the two other main forms of business-to-business governance. As mentioned in **Section I**, this chapter poses interesting challenges because given the implications of the way in which RISCs, RIPTs and RIVANs are construed, it is arguable that their development goes beyond the accepted remit of *relational contracting* and moves towards *relationship management*. This raises two important issues that are worth exploring further. First, relational contracting primarily addresses what is possible within the market as constituted by the laws of economics, particularly transaction costs as a branch of economics. Relationship management starts at the other end, recognising how people actually behave and operate in the market and developing that in order to improve business transactions (Grönroos, 2000; Gummesson, 2001). The result is a possibility that the market itself changes and that current economic analysis is unable to effectively embrace the market changes, assuming that it has done so up to now.

This raises the second issue, which is how this relates to the relationship approach in this book. It was said at the outset that there are different approaches and there is importance in recognising this. This chapter comes from a relational contracting perspective and begins to cross boundaries of definitions. In contrast, in the next section both Cox and Ireland in **Chapter 11** and Ive and Rintala in **Chapter 12** stay within the boundaries of transaction cost economics and relational contracting, yet draw different and somewhat contrasting primary conclusions.

This compare and contrast exercise is posed as Kumaraswamy and Rahman having been pushing the boundaries in their work and here begin to encroach on other territories, which is exactly what one would expect in a discourse and debate that is trying to develop a field, in this case a relational approach to managing projects.

Chapter 8 by Loosemore concerns *Managing project risks*. He takes issue with the rationalist approach to risk management, arguing that it has largely failed to correctly identify risks in practice because rational assessment does not accord with the individual and social perceptions of problems and resultant behaviour. While he acknowledges a role for rational assessment, greater importance is ascribed to the management of risks through involvement and hence communication with stakeholders.

Locating the chapters within a relationship framework

As we stated in **Section II**, each chapter for **Section III** is located within the framework outlined in **Section I**. The framework is provided to anchor the work as it is developed in order to improve our understanding. It is hoped that in future the framework will develop or be reconstituted as understanding progresses.

The *objectives* provided in **Section I** are addressed in this section as follows. Cova and Salle focus upon interpersonal and inter-organisational relationships at the very front end of the project from the contractor perspective: selling the services directly to the client and to influencers across other stakeholders. Wilkinson looks at bringing success for projects by determining the best options for service continuity and Kumaraswamy and Rahman consider the constitution of project teamworking for the contracting organisation. Finally, Loosemore considers risk and the role of relationships in affecting risk perceptions. These are all interface issues that cover the objective concerning interpersonal and inter-organisational relationships. All argue that managing these relationships has significance for project success. While all have a strong tactical thrust to them, the approach to selecting and implementing the tactics is strategic. Thus, while some decisions may be context specific for certain projects, the marketing, human resource and financial plans of the firm set out the project strategy. This must be strategic for the firm, at the front end of the management of projects and thus relatively stable and enduring if the syndrome of 're-inventing the wheel' is to be avoided and the personality and blame cultures of adversity have truly been buried.

The relationship context is predominantly business-to-business or organisation-to-organisation in this section. Cova and Salle focus upon the front end of projects, indeed the front end of marketing and sales. In addition to predominantly business-to-organisation relations, there are also individual-to-individual relationships of significance where key influencers are courted, not only motivated by task and social obligation but also through friendships. Wilkinson also has a business-to-business focus, having taken the largely organisation-to-individual issue within the construction firm as to how to promote continuity of service and transposing that into the context of independent project management by a client representative or specialist project management company.

Kumaraswamy and Rahman also have the same context although the detailed analysis about creating teamworking is more focused upon the individual, unless the team is also conceived as an organisational entity in its own right. However, the focus is a task orientation, with authority relations of management and leadership guiding how team formation and teamworking are engendered.

Finally, Loosemore is concerned with external relationships of stakeholders, drawing them in through organisation-to-organisation/individual relations, not simply in order to serve their needs, but also to be better able to manage the associated risks posed by unwarranted or unexpected stakeholder intervention.

All these contributions are located in the operational context according to the system and culture both within firms and in the wider environments and networks within which the individuals and organisations work. **Chapter 6** by Wilkinson and **Chapter 7** by Kumaraswamy and Rahman are primarily concerned with classic market relationships, whereas Cova and Salle in **Chapter 5** and Loosemore in **Chapter 8** are concerned with both *classic market* and *special market relationships*. There is a secondary concern with *nano relationships* for both Kumaraswamy and Rahman and for Loosemore in their respective chapters. Table III.I examines these categories further, the allocation and weighting being somewhat subjective; however, it provides guidance.

Table III.I Contributions within market relationships. Developed from Gummesson (2001).

Relationships	Chapter 5 Cova and Salle	Chapter 6 Wilkinson	Chapter 7 Kumaraswamy and Rahman	Chapter 8 Loosemore
Classic market relationships				
R1. The dyad (supplier-customer)	Direct and indirect	Direct	Direct	Indirect
R3. The network	Direct		Indirect	Direct
Special market relationships				
R4. Relationships via full-time and part-time marketers	Direct	Direct	Indirect	Very indirect
R5. The service encounter	Very indirect	Direct	Direct	Direct
R6. The many-headed customer and many-headed supplier	Direct		Indirect	Direct
R7. The relationship to the customer's customer	Direct			Direct
R8. Close versus distant relations	Direct	Direct	Direct	Direct
R9. The dissatisfied customer	Indirect	Direct	Direct	Indirect
R13. Parasocial relationships	Direct			Indirect
R14. The non-commercial relationship	Direct			Direct
R15. The green relationship	Very indirect			Indirect
Mega relationships				
R18. Personal and social networks	Direct			
R19. Mega marketing	Direct			Indirect
R20. Alliances that change market mechanisms	Direct			
R22. Mega alliances that change the market structure	Indirect			
Nano relationships				
R26. Quality and customer orientation		Direct	Direct	
R27. Internal marketing			Indirect	Indirect

In **Chapter 1** a wider range of *nano* or internal relationships was recognised. Table III.II addresses these internal relations for this section. This analysis can then be located within the context of portfolios, programmes and the social environment for projects (Figure III.I).

Figure III.I shows the players immediately involved with project delivery – client, design team and contractor. In reality there is a distinction between the contracting organisation as a firm and its involvement in any project as part of its own programme of projects, and the management of the firm rather than projects. Figure III.II tries to address this, applying the relationship management paradigm derived from Figure 1.10.

This introduction to **Section III** has focused on the context, contributions and framework of the three chapters to a relationship approach with particular emphasis on relationships located at the client–design team–contractor interface.

Table III.II Contributions for internal relationships.

Relationships	Chapter 5 Cova and Salle	Chapter 6 Wilkinson	Chapter 7 Kumaraswamy and Rahman	Chapter 8 Loosemore
Internal structural relationships				
R31. Line management relations		Direct	Direct	
R33. Peer relations	Indirect		Direct	
R34. System relationships and communication	Indirect	Direct	Direct	Direct
Internal functional relationships				
R35. HR relations		Indirect	Direct	
R36. Immediate job function relations			Indirect	Direct
R37. Other departmental relations	Direct	Indirect	Direct	Direct
R38. Interdepartmental relations	Indirect		Indirect	Indirect
R39. Circumstantial relations				Indirect
Strategic relationships				
R40. Formal relationships	Indirect	Indirect	Direct	Direct
R41. Career and organisational politics		Indirect	Direct	Direct
R42. Embedded relationships	Direct	Direct	Direct	
R43. Cultural relations	Indirect			Direct
R45. Strategy identification			Direct	Direct
R44. Strategy implementation-tactical response	Direct	Indirect	Direct	Direct
Tactical relationships				
R45. Heroes, champions and role models	Indirect		Direct	Direct
R46. Nurturers			Indirect	
R47. Supporters, members and fan clubs			Indirect	
R47. Surrogate relationships			Indirect	Indirect
R46. Relationships of dissent and blame			Direct	Indirect
Personal relationships				
R49. The affirming-orientated relationships	Indirect	Indirect	Direct	Indirect
R50. The serving-orientated relationships	Direct		Direct	Direct
R51. The performance-orientated relationship		Direct	Direct	Direct
R50. The appearance-orientated relationships	Indirect			
R51. The blame-orientated relationships				Indirect

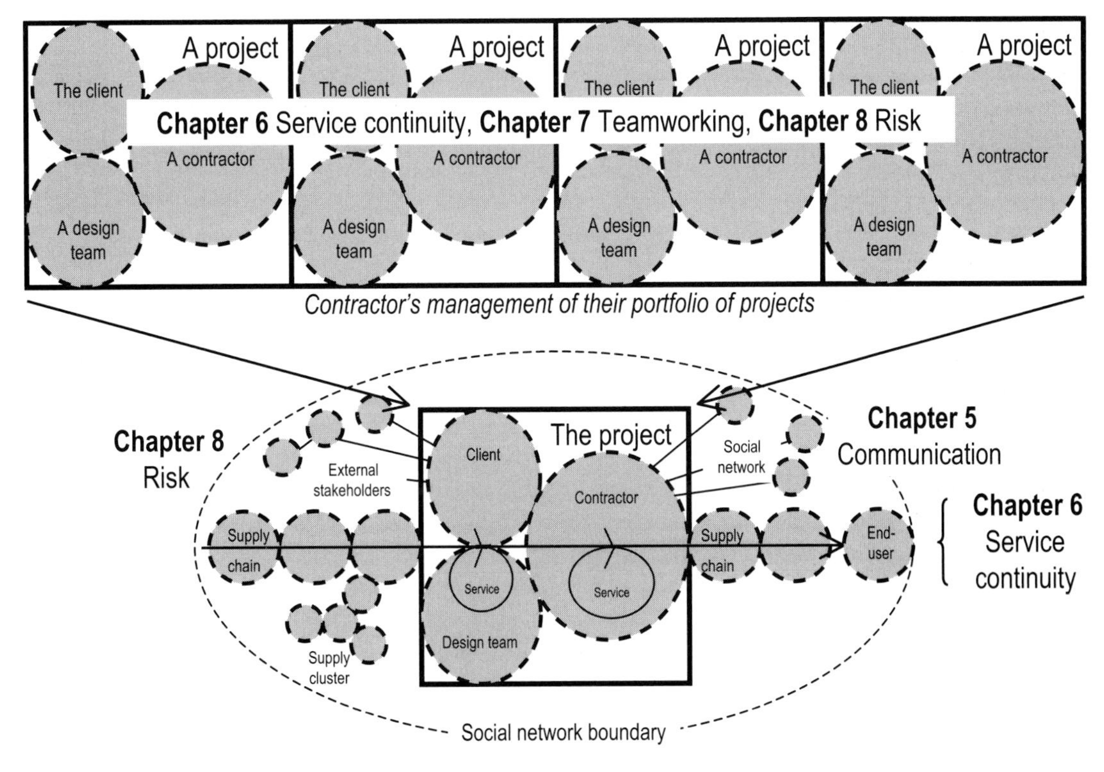

Fig. III.I Contributions in relation to portfolios, programmes and the social environment for projects.

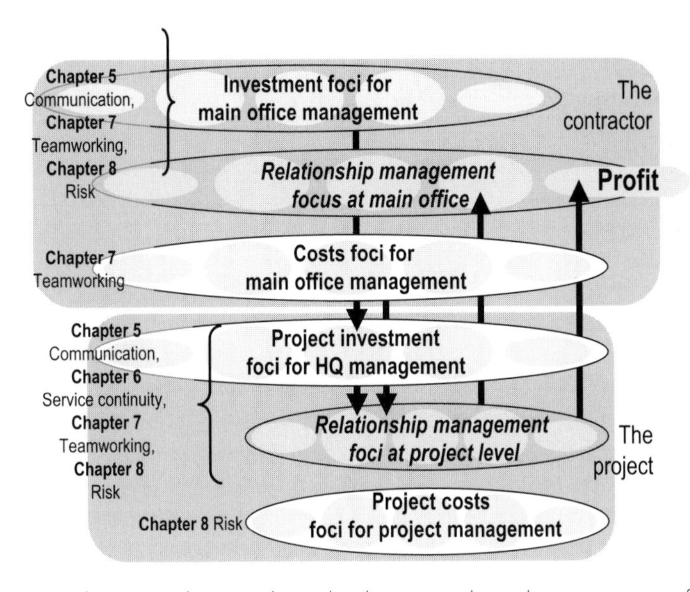

Fig. III.II Contributions to the interrelationship between relationship management for projects and the main office.

References

Grönroos, C. (2000) *Service Management and Marketing*. John Wiley and Sons, London.

Gummesson, E. (2001) *Total Relationship Marketing*. Butterworth-Heinemann, Oxford.

Pinto, J. K. & Rouhiainen, P. (2001) *Building Customer-based Project Organizations.* John Wiley & Sons, New York.

Smyth, H. J. (2000) *Marketing and Selling Construction Services*. Blackwell Science, Oxford.

Turner, J. R. (1995) *The Commercial Project Manager.* McGraw-Hill, London.

Williamson, O. E. (1985) *The Economic Institutions of Capitalism*. Free Press, New York.

5 Communications and stakeholders

Bernard Cova and Robert Salle

Over the last decade, the notion of stakeholders has penetrated the field of project management thinking (Turner, 1995; Pinto & Rouhiainen, 2001). This has followed on from the movement of project re-embedment in a wider context: the success of a project is more and more conditioned by the capacity to positively orientate a group of players who are involved in diverse phases of the project cycle (Söderlund, 2004). When we consider the point of view of a project supplier, spotting stakeholders and communication with them is an obligatory part of the marketing process. To do this, it is necessary to perceive stakeholders as more than simply players organised in homogeneous groups, or segments, in marketing language. For the supplier, it is a question of developing a network approach that is capable of both taking into consideration the global dynamic of the context – that is, the *milieu* (Cova *et al.*, 1996) – and detailing the characteristics of each particular relationship, in other words the relationships among stakeholders and the relationships between stakeholders and the supplier.

It is also necessary to reintegrate the notion of stakeholders in a temporal perspective: stakeholders are very varied and act differently throughout the project cycle phases. During the project inception phase, a myriad of potential stakeholders can be spotted. Not all will be active in the development and management of the project. For the supplier, it is therefore a matter of conceiving a communication approach targeting certain potential stakeholders while there is not yet any project in view. Communication with stakeholders is therefore principally localised in the inception phase of projects and based on a unit of analysis known as the *milieu*. This communication action aims to construct, develop or maintain extra-business relationships with the stakeholders; it concerns the logic of friendship or ritualising what will be developed hereafter.

From stakeholders to milieus

Business and non-business actors

The management of project stakeholders represents a challenge that most project managers are only just beginning to acknowledge. According to Turner:

'It is only since the late 1960s, that investors in projects have begun to consider the stakeholders in a project, and it is only over the last decade, since the 1980s, that truly adequate notice has been given to their requirements . . . If more effort and expense is put in at the start of the project managing the needs of the stakeholders, then the overall cost can be lower and the duration of the project shortened.' (1995)

In fact, according to Pinto and Rouhiainen:

'Little is known about the nature of the various project stakeholders: Who they are, what their drivers and separate agendas are, and how to understand the nature of project stakeholder trade-offs.' (2001 p.158)

In view of the shortcomings that have been encountered until now, there is distinctly more desire in project management today to take stakeholders into consideration.

But the conceptualisation of the stakeholder is inconsistent, ranging from a limited conception to a very wide approach. A restrictive conception adheres to the understanding that stakeholders are:

'A group of people who are often involved without their prior agreement, sometimes against their will, and who often view the project as being a disbenefit because it somehow distracts from their local environment.' (Turner, 1995)

Stakeholders are therefore essentially perceived as *non-business actors*, who react positively or negatively to the project and who can therefore condition the social acceptability (Miller & Lessard, 2000). Calvert states:

'Although appearing as indirect players in the project environment, stakeholders can have a major impact on project success. Failure to recognize their existence and their potential power at a project's strategic level may lead to serious problems at the advanced stages of project planning and implementation.' (1995 p.214)

In the same limitative manner, for mega-projects stakeholders frequently comprise 'key institutional actors, such as NGOs, various levels of government, industrial interests, scientific and technical expertise and the media' (Flyvbjerg *et al.*, 2003 p.7).

Conversely, certain authors adopt a very inclusive approach to the notion of stakeholders. Pinto and Rouhiainen (2001) identify several types of stakeholders that project managers must consider: internal stakeholders such as top management, accountants, project team members and other functional managers and external stakeholders such as customers, competitors, suppliers and intervener groups. The latter are defined as 'any environmental, political, social, community-activist, or consumer-groups that can have a positive or (more likely) a detrimental effect on the project's development and successful launch' (Pinto & Rouhiainen, 2001 p.147).

By taking the perspective of a supplier wishing to optimise its chances of pulling off a project, the inclusive approach of stakeholders seems the best adapted (Cova *et al.*, 2002). This is the view of what is called *project marketing*: a specific approach developed by contractors and projects-to-order companies in order to create a competitive advantage and win the contract (Cova & Holstius, 1993). For project marketing, stakeholders include all players grouped along two lines:

- *Business actors*: consultants, financial backers, agents, engineering companies, sub-contractors
- *Non-business actors*: governments, syndicates, lobbies, unions, pressure groups, activists

Players and relationships

For traditional project management approaches, the focus is on the stakeholders themselves without taking into account relationships between them. Indeed, the project management view might regard stakeholders as a group of people who do not interact or form relationships but share the same characteristics or the same roles: for example investors, owners, sponsors, consumer-groups. Conversely, the theory of industrial networks (Håkansson, 1987; Mattsson, 1985) deals with stakeholders from the standpoint of 'markets as networks', that is, stressing the relationships between all rather than the immediate players in the exchange. This approach has enabled project management to be developed as a strategic approach based on relationships with key stakeholders (Cova *et al.*, 1996). According to Wasserman and Faust:

> *'Rather than focusing on attributes of autonomous individual units, the associations among these attributes . . . , the social network perspective views characteristics of the social units as arising out of structural or relational processes or focuses on properties of the relational systems themselves.' (1994 p.7)*

Thus, social network theory describes the collective body of relevant stakeholders in a company's project management activities. Project marketers (Cova *et al.*, 1996) call this collective body a milieu because it acts as a microcosm in which network forces are at play, reinforced and empowered by spatial and cultural proximity.

For suppliers, stakeholders are therefore identified and analysed in a logic of markets as networks, which brings them to focus not on the project but on the milieu, where the milieu is defined as a *socio-spatial entity* that is *geographically bound* and in which, through the frequency of socio-economic exchanges, business and non-business players are intertwined. They share a common vision of business and a set of tacit rules – the 'law of the milieu'. This milieu represents a collective group that cannot be categorised or defined as within the manner of the segmentation approach cited previously (Turner, 1995). The milieu (Cova *et al.*, 2002) is characterised by four elements:

- Territory
- *Network of heterogeneous players*, such as business people, governmental bodies, civil society organisations related to each other within this territory
- Representation constructed and shared by these groups
- *Set of rules and norms* regulating interaction between these groups

An example of milieu analysis in project business

To illustrate the pertinence of the milieu concept in the identification and analysis of stakeholders for the purpose of project firm communication, an example has been developed: a firm in the construction sector that works in a logic of markets as networks. In 1996 the construction group 'Constructa', a pseudonym for reasons of confidentiality, covered the entire French territory through local subsidiaries. These local subsidiaries were not created by Constructa but were the result of their buying local firms. In order to preserve the local character of its subsidiaries in local markets in which this represents an important competitive advantage, Constructa kept the original name of each subsidiary. In the Seine-Maritime *département*, the subsidiary is called Garaud. Garaud is a middle-sized company specialising in local public projects, but is also interested in the construction of premises for foreign industrial firms wishing to develop in the Seine-Maritime. Garaud does not wait for a foreign company to decide the details of its settlement in the Seine-Maritime; it would have zero chance of winning the project. The construction of private industrial premises does not follow the advertisement rules of public markets. In general, a call for tender is issued and sent by the foreign company to some engineering firms selected for their reputation or their relational proximity to the buyer. Then, the selected engineering firm chooses a construction company to implement the project.

With more than 1.2 million inhabitants, the Seine-Maritime is ranked just after the *départements* of the largest French metropolises – Paris, Lyon, Marseille and Lille. The Seine-Maritime borders the most frequented shipping lane in the world, the Channel, its culture being anchored in related trade. The *préfecture* of the *département* is Rouen, a town of 150 000 inhabitants. The principal cities are the port of Le Havre, Dieppe, Fécamp and Yvetot.

In 1973, the first oil shock plunged the region into a recession of long lethargy. A new way of running Rouen was instigated under Jacques Canu, who became Mayor of Rouen in 1993. As an entrepreneurial mayor he opened in November 1993 a new division, the ADEAR (the Association for the Economic Development of Rouen) run by a former salesman from Schneider, Alain Mathieu. He introduced a policy of prospecting for and communicating with foreign companies setting up plants in Europe. From 1993 onwards, Rouen began to attract several companies through a network of international consultants of the Normandy Development Institute, particularly Monsieur Lorme, a consultant in Chicago assigned to the Institute.

A network of key players was built up in the following way. The *Conseil Général* – the 'county council' – of the Seine-Maritime *département* was active (a *département* being a subdivision of France administered by a prefect who is located in the *préfecture*). At the beginning of the 1990s Blaise Arribart, the centrist president of the county council, used his staff to attract companies from abroad, taking the risk of echoing the message of the ADEAR or even those of the Chamber of Commerce and Industry of Rouen (CCI). He was assisted by, among others, Philippe Rouard, the socialist mayor of Dieppe. In 1994, one year after the launching of ADEAR, Blaise Arribart set up, with the support of the county council, Seine-Maritime Expansion (SME – the agency for the economic development of the Seine-Maritime). He appointed Xavier Alduy, a former tax inspector in the *préfecture* of the region, to the position of general delegate of the SME. Blaise Arribart ran the county council like a company board. Xavier Alduy planned expenditure as investments, 'reasoning in terms of revenue for the department as much as expenses for development'. This included the divisions of the *préfecture* that functioned simultaneously, thus making it possible to speed up administrative work. According to Blaise Arribart, 'consensus is the secret of our success'.

The Seine-Maritime is a *département* that wastes no time transforming projects into actions. Here, converting land at its own expense; there, buying a factory that was about to close down; elsewhere proposing funds and credits or accepting to pay a high proportion of the infrastructure costs. Xavier Alduy states that these were 'actions which are only normal for a rich region and which offer no fiscal advantage'. Indeed, companies come to the Seine-Maritime for two major reasons: the workforce is qualified and stable, which is not the case in the Paris region, and there are two major harbours, one on the sea, one on the Seine river, that ease the access. Unlike other *départements* that are economically underprivileged, the Seine-Maritime hardly benefits from aid from the DATAR (the national agency for territorial development). This agency tends to direct large-scale industrial development towards other zones using financial and fiscal incentives.

The municipality and county council, represented by Alain Mathieu and Xavier Alduy, worked in close collaboration on important international dossiers, creating the Rouen dynamo and linked to the economic hinterland of the Seine-Maritime. This collaboration also included the CCI and MEDEF Rouen-Dieppe (the local association of CEOs). The ADEAR and the SME of the county council were careful to keep local companies at a distance during the initial phase of prospecting. The public infrastructure services, like the ANPE (National Employment Agency) and the EDF (*Electricité de France*) were associated early on if projects looked viable. In order to further develop the setting up of foreign companies, Rouen needed more sites for construction. Rouen is looking for partners in the communes that are able to provide land. One example is the commune of Fécamp, in the Pays des Hautes Falaises, a zone in which a large area was created for industrial development.

The stakeholders presented in Figure 5.1 are mostly circumscribed in a clearly defined geographical space. They are heterogeneous and interlinked. This milieu has its central actors and its peripheral actors characterised by their business and non-business nature, and by their value-creating or non-value-creating links. In 1996, the analysis of the milieu enabled Garaud to

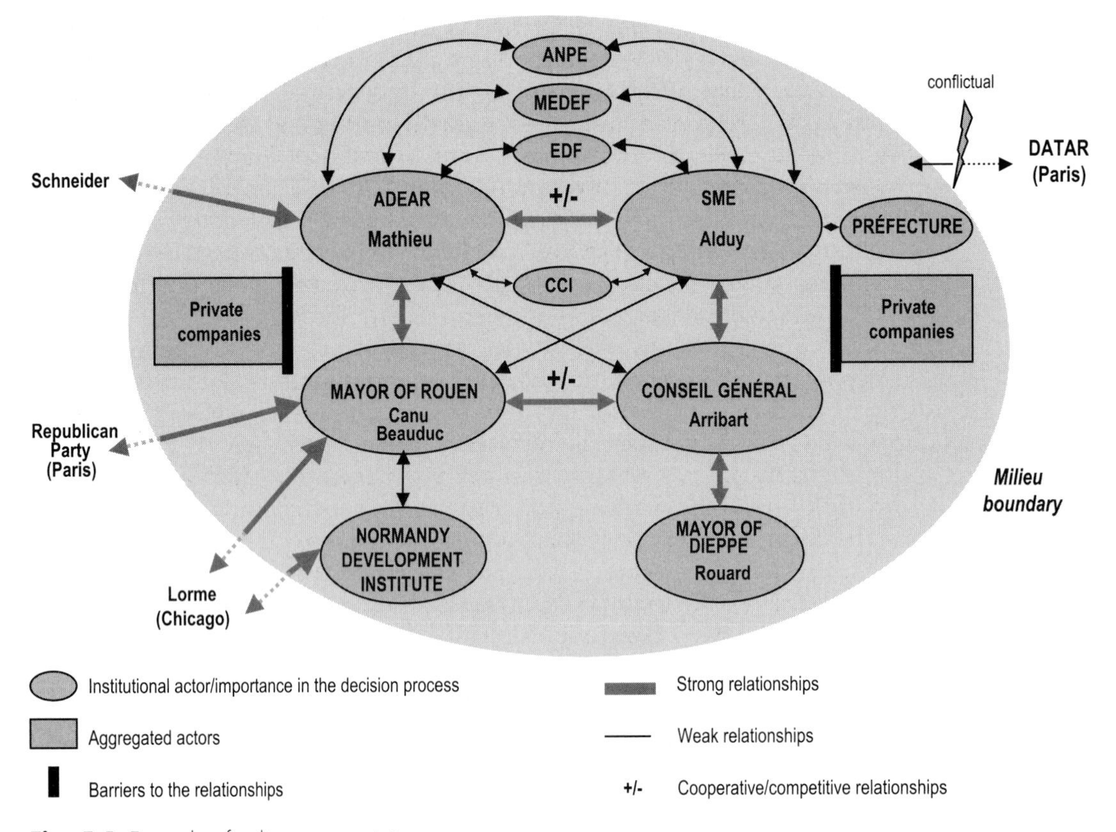

Fig. 5.1 Example of milieu representation.

represent and analyse relationships and thus a network of stakeholders, in a well-defined geographical framework. In a given territory and for a given object, the analysis aimed at identifying and qualifying both the links and the subjects of the links (the stakeholders). Garaud was now able to target directly or indirectly – through other players – several stakeholders that could have an impact on forthcoming projects and to design the best way to communicate with them.

Towards an outward project focus

Being aware of the stakeholders in a project leads us to question when the right time to communicate with them is. Determining when a project starts is problematic. The dominant idea today is to communicate even earlier, when the project is at the 'very beginning' of its 'pre-inception' or *generation* phase. Flyvbjerg *et al.* state:

> *'Stakeholder groups and civil society should be invited to participate from an early stage and throughout feasibility studies and decision-making. Who, exactly, should participate should vary from project to project. But, in principle no party affected by a given decision should be excluded and all participants should have equal possibility to present and criticise validity claims. Participation should be as representative as possible. This is best ensured by the government, or parties acting on behalf of government, taking an active role of identifying, inviting and balancing stakeholder and civil groups, so that all relevant groups get an opportunity to participate, and no one group gets to capture and dominate this aspect of the process.' (2003 p.111)*

It is possible to have even more foresight and seek to communicate with certain stakeholders even when there is not yet any 'project'. This is the approach that is more and more followed by suppliers who wish to increase their chances of success. Companies try to indirectly anticipate customers' projects through links with stakeholders within the milieu concerned. Project marketing then focuses on a firm's relational investment in a milieu, independent of any project opportunity (Hadjikhani, 1996). This approach totally reverses the existing perspective; the focus of communication to stakeholders is not at the phase 'inside the project' but at the phase 'independent of any project' when there is no business opportunity.

It is therefore pertinent to consider three phases for efficient stakeholder communication:

- *Project generation* – the project does not yet exist.
- *Project inception* – the project is in the process of definition.
- *Project purchase* – the project officially exists in the form of a market consultation by the customer (invitation to tender) calling for an offer by the supplier.

Communication towards stakeholders is to be understood as being a continuous action throughout three phases within the large group of actors of the milieu. In the following pages, we set out two original modalities of communication with stakeholders: the friendship approach at a dyadic level and the ritual approach at a group level.

Friendship and mutual debt

Project stakeholder communication can be based around relationships, which have been created, developed and maintained by company members with a number of socio-economic players acting either directly or indirectly, in the manner in which the milieu companies operate. These contacts are the result of the work not just of internal salespeople or project managers, but of all the members of the company (Cunningham & Turnbull, 1982). The managers' personal networks (Dubini & Aldrich, 1991), in fact all the members of the staff, have a part to play in the company's communication process. The term 'extended staff' is even used to designate staff relations (friends, groups, family) who also participate in the project stakeholder communication.

Communication with project stakeholders starts with interpersonal relations. The study of personal networks in project business raises the question of whether relations in the milieu can come in different forms, from the interested contact to the uninterested purely social relationship. Indeed, inter-individual relationships in a professional context are controlled less by opportunism for short-term egoistic interest than by confidence for a mutuality of long-term interest. The basis of a long-term business relationship between two individuals is less reciprocal than an equality of spin-offs resulting from common effort (redistribution of profit), which can be called mutual interest or mutuality of interest. In this case, the social ties between two individuals serve the economic link.

However, there are a number of cases where the economic link serves the social link (Godbout & Caillé, 1992). The most obvious example is the case of the shared professional project that we design simply so as to be with our friends more often. Indeed, some information circulates each day primarily to safeguard social ties. The interpersonal relationship is therefore not controlled by some functional and utilitarian demand, but rather by itself. In this case, the episodes of the relationship are more favour exchanges, just like the countless and daily favours we do for those we love and to whom we wish to remain close.

These regular favours nourish the idea of debt that exists between two individuals. This is what is known as 'mutual debt' and is the opposite of 'reciprocity'. If one party wishes to reduce or even break the social tie, that party should settle the debt and reinstate reciprocity. Therefore a continuum of the systems of interpersonal business relations exists. This is based upon the balance between utility on the one hand and resonated agreement, called *syntonism*, on the other. The giving of such favours is undertaken as a series of discrete transactions, creating mutual debt via reciprocal processes and mutuality of interests (Cova & Salle, 1992). Equally, there are not only pure systems; sometimes the utility and syntonism are closely related, one conditioning the other. A business relation that was initiated under the system of reciprocity can become one of mutual debt, just as a relationship based on mutual debt can turn into mutual reciprocity. A friendly business relationship can swing between moments of utility and moments of syntonism.

Business relations use everyday words, allowing the sharing of favours to create mutual debt without ever settling it, so social ties continue: information, availability, sharing company, food and social events together. During these social exchanges something important for a project activity might be said, but this is not the goal of the relationship. However, relationships controlled more by mutual debt than by interest can lead to important contracts and major project success for companies interested in a leading position in a milieu.

There are two schools of thought concerning the reasons behind personal relations and networks in a milieu (Belk & Coon, 1993). The first school, utilitarianism, considers reciprocity and mutual interest as the basis to all relationships. It believes that a simple diagnosis of the relations within the milieu followed by an evaluation of the entrance process to the milieu leads to a form of stakeholder management and communication. The manager can then target relations. Each relationship is examined to see which allow information to be gained concerning a specific project or to influence decisions concerning a specific project. In this situation, both parties are aware they are beginning a relationship that could be mutually interesting in the short term with potential for the mid/long term based on mutuality of interests. This type of approach is fairly common in sectors such as the construction industry or public works where the mutuality of interests can facilitate mutually beneficial agreements and conversely can lead to criticisms relating to probity, particularly where public funds are concerned.

The second, non-utilitarian, school of thought sees the problem of the creation and management of personal networks as being more complex. It considers the very notion of network management to be contradictory, the very strength of an individual's network being based on elements that are neither planned nor managed. In other words, they can be effective because they are not organised for precise reasons, such as collecting information or influence. The social ties are vital and sacred in this approach. For some top engineers, it is strictly out of the question to use their personal relations to obtain information concerning a project. This is particularly the case in the aerospace and energy sectors.

If we continue with this train of thought, communication with stakeholders is based principally on networks of interpersonal relations that could not be organised without a risk of destroying them, which takes us to the idea of the impossibility of managing relational investment (Godbout & Caillé, 1992). As a solution, some propose that the person who is (or who will be) in a relationship with stakeholders in the milieu does not intentionally manage his action to remain in the non-utilitarian dimension of the relation. However, it is up to the manager to 'use' the person by 'exposing' him to help develop a network. There could therefore be a double structure in a company with people dominated by a non-utilitarian logic – *givers* – who work on relationships and social ties, and those dominated by a utilitarian logic – *takers* – who work on easing and directing the relationships of the former.

In reality, the same people often swing between the two approaches, utilitarian and non-utilitarian. An engineer can go from a state of latent consciousness of what they are doing (in the relationship to pass on information) to a state of unconsciousness, generally during a one-to-one occasion, especially when they are spontaneous, enjoying the relationship, acting with a clear conscience and in a non-utilitarian way that could be useful. A form of debt has been created with the other person, a debt that will strengthen social ties and thereby guarantee a source of information and influence. These engineers and project managers are key players in the animation of networks, giving a lot of themselves, creating an atmosphere of confidence and a contagious need to remain part of the process.

Box 5.1 Mutual debt in project business

With the following case based on the aerospace industry, we shall try to demonstrate how the concept of mutual debt can throw a different light on the exchanges between individuals involved in business relationships.

A German aerospace manufacturer was competing against an English counterpart for the supply of an important part for a new Italian aircraft. A short while before the Farnborough Air Show, the German marketing department learnt from the Italian purchasing director that the English supplier was to be chosen because of their better offer. Following this announcement, the German marketing department admitted defeat. However, during the inauguration of the Farnborough Air Show, the director of the German company had the opportunity to dine with the Italian boss, someone he had known for a number of years. The Italian company belonged to the Europaero group, in which the German director used to have an important position before changing jobs (ten years previously). The German boss had fought long and hard to have the Italians in the group, alongside the Germans, English and French. That evening, the Italian boss passed on some information to his German counterpart and friend who had not asked for anything: 'This is the price to offer if you want to beat the English.' The following day, the marketing department of the German firm quickly made another offer, and before the show was over the contract was signed between the Germans and the Italians. This is bad bidding practice and can be located within the criminal relationship (see Gummesson, 2001). We propose three ways to interpret it:

- A first utilitarian interpretation could find elements of an agreement or even corruption.
- A second utilitarian interpretation, using the notion of give and take reciprocity, shows how the Italian boss, ten years on, returns the favour. This is a good example of reciprocity between two individuals who trust each other.
- A third, non-utilitarian interpretation, using the concept of mutual debt, shows that the Italian boss, who sees his German counterpart less frequently than in the past (new position, from Europaero to industry) uses the opportunity of a long-term economic exchange to recreate the social ties. He is not looking to settle his debt, but rather to start the mutual debt system up again, which at the same time may lead to more frequent social exchanges.

Source: Interviews by the authors

Rituals and extra-business relationships

In the milieu, communications that are not purely dyadic (two parties in the exchange) involve relationships that are mainly based on rituals. Indeed, today's project business is made up of a host of actions intended to maintain the atmosphere of the relationship with stakeholders, which are often misunderstood and sometimes considered anecdotal. This social dimension often escapes the attention of project management theorists although it is of major importance in the management of stakeholders. Consequently, the question of how to communicate with stakeholders outside any project opportunity can then be formulated in the following way: *Is there a structure for organising actions taking place in the context of extra-business relationships that makes it possible to control and support a part of these actions?* Bearing in mind that a long and costly investment is necessary for building up a relationship, it is easy to understand that maintaining it in the period 'pre-inception' or *generation* is a crucial aspect of the project stakeholder communication. Ritual being the basic social act, the following ritualisation framework aims at facilitating this task by providing a structure for organising communications with stakeholders that take place in the context of extra-business relationships.

In sociology, it has been known for a very long time that ritual is a means by which social phenomena mark and ensure their permanence. Indeed, for the human sciences all social relations need rituals in order to develop and to persist, and all social groups need rituals to assert and reassert their existence and the adherence of their members. To sum up, the arguments that are used in the functional analysis of rituals can be reduced to three:

- Ritual renews and invigorates adhesion to beliefs
- Ritual has a psychologically soothing function
- Ritual helps the integration of the individual in the group

The functions of belief, of uncertainty reduction and of social integration are therefore the three facets of ritual that are traditionally qualified as being religious. But today there are many sociologists (Bell, 1992; Rivière, 1995; Segalen, 1998) who consider that these facets are to be found in what they call profane rituals or contemporary rituals, that is to say the sacred dimension of our post-modern societies.

Indeed, the sacred is not less present than it used to be but a transfer has taken place. The sacred is in good health but it is different. That which, in the course of history, had found its abodes and its rituals in quite specific religious images and gestures has moved on since the beginning of modernity. It has multiplied and scattered. Thus our era has not been characterised so much by its materialism as by the dispersal of the experiences of the sacred, closer perhaps to the polytheism or even the animism of our ancestors than to the monotheism on which our modern civilisation has been built. It can be said that today we are witnessing a proliferation of apparently desacralised rituals, which are the result of a profane re-ritualisation – a sort of 'revenge of rituals' (Pitt-Rivers, 1987).

Recognised as a general form of expression of society and of culture, ritual has emancipated itself from the religious context in which it was automatically perceived in the past. The functioning of profane rituals is linked to social utility; execution is imperative in order to recreate periodically the moral being of society. Ritual brings together discontinuous elements and closes gaps. By the adoption of rules and roles, in the framework of an order that it expresses, ritual strengthens the integrating social link. It is both a way of exalting collective meaning or integration and a means of regulating social relations, and hence order.

For sociologists, a ritual is an ordered sequence of behaviour that is more rigid and predictable than ordinary action (Segalen, 1998). It is a meaningful but scarcely conscious procedure of dramatised roles, of values and finalities, of real and symbolic means, of communication through coded systems (Bell, 1992). Every ritual has its recurrent temporalities, its models marked by history and its divisions of space to serve as a setting. According to Rivière:

> *'The force of the ritual can be measured in part by the emotion it gives rise to: an emotion favoured by the attention it wins from the actors of the ceremony, from the onlookers, from the participants involved in this type of communication. It is an emotion affected by the metaphors transmitted by the ritual and which move the psyche all the more intensely as they refer to vital situations.' (1995 p.66)*

But, 'between the classical habit of restricting the ritual to the domain of the sacred – religious – and the temptation to call ritual all routine behaviour, the spectre of rituals to be identified remains immense' (Rivière, 1995 p. 10), and it is therefore necessary to clarify the notion of profane ritual.

Apart from mainstream sociology, the study of rituals is highly related with the symbolic interactionist heritage and the study of human group life and conduct (Blumer, 1969). The Chicago-trained interactionists and especially Goffman (1959, 1967) forged a framework for the reading of everyday life with a concern for manners, ritual, secrecy, deference, proper demeanour, the presentation of the right self . . . the micro-sociological or ethnomethodological viewpoint (Garfinkel, 1967) adopted by these researchers emphasising the negotiation of personal identity or the self-meaning of the person in each situation of interaction. These situations of interaction may be routinised, ritualised or highly problematic.

In fact, Goffmann (1967) is somewhat conflicting with Durkheim (1912) and classical sociology when he develops his notion of ritual. He changes scale, as noted by Rivière (1995), doubling the ritual concept. Goffman adds the small everyday venerations to the institutionalised ceremonies. For him, the analysis of interaction rituals, as he names them, has to explore intermediary forms of socialisation that are placed between the two extremes of routinised social behaviour and special collective events (Joseph, 1998). Thus, side by side with big ceremonies there are for Goffman a host of situations that have to be considered as rituals, such as when we ask someone what time it is: we are not only formulating a demand, we solicit a favour, we apologise for the disturbance, we thank for the assistance, which is then reciprocated. This is a small, daily and ordinary veneration that is typical of the micro-ceremonies Goffman analyses as rituals in his micro-sociology.

Consequently, rituals can be roughly divided into two types:

- *Rituals of integration* – macro-rituals, which are elaborate procedures (ceremonies) for consolidation of community belonging (Durkheim, 1912; Van Gennep, 1909). The origin of the integration ritual is to be found in the notions of gift, of spending, of sacrifice . . . which come from the sphere of the sacred, and which link together (*relie* in French as in *re-ligion; re-ligare* in Latin) a community, and are opposed to the utilitarian.
- *Rituals of interaction* – micro-rituals or ritualised behaviour, which are simple procedures for facilitating everyday contacts and negotiating personal identity (Goffman, 1967; Picard, 1995). The basis of the interaction ritual is to be found in the notion of respect for the other: respect for

his/her face (Goffman, 1959) and respect for his/her territory (Picard, 1995 p.37); 'ritual used out of respect for face and territory thus plays the role of defence of identity and relational defence'. This is seen in the tact and the reserve that order (put order into) human interactions. Interaction rituals are not useless old relics of politeness. They answer specific needs in everyday situations: the psychological need for self-protection and the communicational need of facilitating social contacts.

Macro-rituals or rituals of integration, because of their main function of fostering belief, of uncertainty reduction and of social integration, seem capable of providing a framework adapted to the needs of boosting confidence, of maintaining interdependence and socio-economic integration between the company and different stakeholders in the milieu over time. These macro-rituals, that could be of major interest for project management, are said to operate for a period (Picard, 1995): annual festivals put rhythm into social life to make of it a long repetitive movement; other ceremonies emphasise major steps in life such as birth, celebrating adulthood and funerals. Macro-rituals involve a community as a whole: every ritual event is an opportunity to celebrate commonality. Macro-rituals can be classified in four main categories all of which refer to time:

- Initiation rituals or rites of passage (such as the ragging of first-year students)
- Calendar rituals or commemorative rituals (such as Bastille day)
- Cyclical rituals (such as the every Monday morning get-together in the office)
- Occasional rituals (rituals for special occasions such as a marriage)

A case study illustrating these four types of macro-ritual is shown in Box 5.2. It is to be noted that the functioning of macro-rituals is based on a number of principles (Rivière, 1995):

- Meeting – every ritual implies a situation in which the actors are in each other's presence
- Delegation – all the members are not necessarily present; it is enough for some to perform the ritual behaviour as representatives who are delegates of the community
- Shared emotion – solidifies the link

But the way rituals work is never clearly expressed and expressible: 'ritual procedures are certainly significant but often actors are not aware of the ritual nature of these procedures' (Rivière, 1995 p.70).

Based on the previous investigation of the ritual concept and especially on the classification of rituals into four categories (initiation, calendar, cyclical, occasional), it is possible to build a comprehensive framework for all the actions taken to manage the extra-business relationship with stakeholders. We can see this in Box 5.2.

Box 5.2 Rituals in project business

The company Xava manages complex projects of revalorising chemical industry sub-products. These sub-products being delicate to handle and to treat require the use of precise technologies. These projects have the distinctive feature of not often involving one particular customer but involving many customers of the same country and even other players – local prescribers. But what is striking is that the 'extra-business' relationship is made up of a multitude of small actions carried out by different people at Xava without their necessarily consulting each other. One person will go to the conference in September in the USA and will meet numerous American and other players, another will go to another conference in June in Europe and will meet some of the same players, yet another will go to a student reunion. Although it is more the technical aspects of these conferences that trigger people to attend, they ritualistically use the opportunity to meet such and such a player. We can also see a unity by country and region, with the organisation of an annual meeting with all the North American chemists, or similarly of an annual meeting with all the German chemists. More precisely, we can say that relationships are kept up through big events such as the party organised for the transition of factory C to ISO 9002 or the one organised for the start up of a new revalorisation unit: large numbers of major national and foreign stakeholders are invited to share these important occasions. Factory visits are very highly prized by Xava whenever they wish a potential customer or prescripters in a country to adopt the idea of sub-product revalorisation: this is part of the plan of action designed to arouse and organise the interest of customers themselves and also interest within their country as a whole, going so far as to bring about a public debate. This is not only communication, but is also the construction of the milieu and the players' mentalities. Finally, Xava has its own in-house festivities to which certain stakeholders are invited: the 100th tonne valorised, the last dispatch to a particular destination.

The interaction actions of Xava and of stakeholders are organised on several levels and present ritual characteristics of a social dominance. We can therefore find four categories of integration rites:

- Rites of passage or initiation rituals: for example, the party for the transition of factory C to ISO 9002
- Calendar rituals or commemorative rituals: for example, the September annual conference in the USA
- Cyclic rituals: for example, the annual encounters with German chemists
- Occasional rituals: for example, on the occasion of the 100th tonne valorised.

Source: Survey by the authors.

Today's rites, which are in considerable expansion, include several characteristics that make them actual events of our times. They correspond to more relaxed and even funnier behaviours than those required for religious rituals and they are to be found in sports and politics as well as in business. Accepted and shared, they are vectors of new forms of identity that the socio-economic actors are trying to build (Segalen, 1998). Project firms are trying to set up extra-business rituals that are rituals of integration, designed to support the identity building of people in stakeholder organisations. Through certain rituals at different levels of socialisation, they build and develop the identity of their organisations (and their personal identity). At the same time, the company is limiting stakeholder relationship discontinuity. The final aim is to provide a ritualistic platform that enables the company to support the construction, the development and the maintenance of interpersonal contacts, not to organise them! All people in the project company's organisation, such as sales people, technicians and managers, may use this platform according to their objectives. The platform can encapsulate different types of ritual actions such as clubs, events and more, to provide a global framework for the communication activities outside any project opportunities.

Conclusion

The notion of stakeholders has recently penetrated the field of project management thinking. Who are they? How and when to communicate with them? These are today's key questions. In this chapter, we have defined stakeholders as all actors, whether business or non-business actors, that can play a role in the *milieu* where a certain type of project can be generated – the *milieu* being the collective body of stakeholders that acts as a microcosm in which network forces are at play. Network analysis enables representation and analysis of stakeholder relationships and thus a milieu. Communication towards stakeholders is to be understood as being a continuous action throughout three phases (project generation, project inception and project purchase). It is not an action that can be easily decided and directed. It goes through the channels of interpersonal relationships and rituals in order to develop and improve the network position of the contractor.

However, the major concern is about how to communicate with stakeholders outside any project opportunity in order to anticipate new projects. Bearing in mind that a long and costly investment is necessary for building up a relationship, one must recognise that developing and maintaining it in the project generation phase accounts for a crucial aspect of the project stakeholder communication. In this perspective, the concept of *milieu* seems less helpful in industries where the locus of projects is not geographically bound. The strength of the concept lies in its ability to integrate economic, social and cultural elements in specific locations and where these elements are heavily intertwined. As many project management situations take place on a global scale, one can surmise whether they still follow the logic of the milieu. Future development in the area of stakeholder communication would try to incorporate the worldwide dimension of some project industries.

References

Belk, R. W. & Coon, G. S. (1993) Gift giving as agapic love: an alternative to the exchange paradigm based on dating experience. *Journal of Consumer Research*, **20**, December, 393–417.

Bell, C. (1992) *Ritual Theory, Ritual Practice*. Oxford University Press, Oxford.
Blumer, H. (1969) *Symbolic Interactionism*. Prentice Hall, Englewood Cliffs, NJ.
Calvert, S. (1995) Managing stakeholders. In: J. R. Turner (ed.) *The Commercial Project Manager*. McGraw Hill, London.
Cova, B. & Holstius, K. (1993) How to create competitive advantage in project business. *Journal of Marketing Management*, **9** (2), 105–21.
Cova, B. & Salle, R. (1992) Project marketing and network theory: a gift approach. *Proceedings of the 8th IMP Conference*. September, Lyon.
Cova, B., Ghauri, P. N. & Salle, R. (2002) *Project Marketing: Beyond Competitive Bidding*. John Wiley, Chichester.
Cova, B., Mazet, F. & Salle, R. (1996) Milieu as a pertinent unit of analysis in project marketing. *International Business Review*, **5** (6), 647–64.
Cunningham, M. T. & Turnbull, P. (1982) Interorganizational personal contact patterns. In: H. Håkansson (ed.) *International Marketing and Purchasing of Industrial Goods: An Interaction Approach*. John Wiley & Sons, Chichester.
Dubini, P. & Aldrich, H. E. (1991) Personal and extend networks are central to the entrepreneurial process. *Journal of Business Venturing*, **6**, 305–13.
Durkheim, E. (1912) *Les Formes Élémentaires de la Vie Religieuse*. Alcan, Paris.
Flyvbjerg, B., Bruzelius, N. & Rothengatter, W. (2003) *Megaprojects and Risk: An Anatomy of Ambition*. Cambridge University Press, Cambridge.
Garfinkel, H. (1967) *Studies in Ethnomethodology*. Prentice Hall, Englewood Cliffs, NJ.
Godbout, J. T. & Caillé, A. (1992) *L'Esprit du Don*. La Découverte, Paris.
Goffman, E. (1959) *The Presentation of Self in Everyday Life*. Doubleday, Garden City, NY.
Goffman, E. (1967) *Interaction Ritual*. Doubleday, Garden City, NY.
Gummesson, E. (2001) *Total Relationship Marketing*. Butterworth-Heinemann, Oxford.
Hadjikhani, A. (1996) Project marketing and the management of discontinuity. *International Business Review*, **5** (3), 319–36.
Håkansson, H. (ed.) (1987) *Industrial Technological Development: A Network Approach*. Croom Helm, London.
Joseph, I. (1998) *Erving Goffman et la Microsociologie*. PUF, Paris.
Mattsson, L. G. (1985) An application of network approach to marketing: defending and changing market position. In: N. Dholakia & J. Arndt (eds) *Changing the Course of Marketing: Alternative Paradigms for Widening Marketing Theory*. Research in Marketing 2, JAI Press, Greenwich, CT.
Miller, R. & Lessard, D. R. (2000) *The Strategic Management of Large Engineering Projects: Shaping Institutions, Risks, and Governance*. MIT Press, Cambridge, MA.
Picard, D. (1995) *Les Rituels du Savoir-vivre*. Seuil, Paris.
Pinto, J. K. & Rouhiainen, P. K. (2001) *Building Customer-Based Project Organizations*. John Wiley, New York.
Pitt-Rivers, J. (1987) La revanche du rituel dans l'Europe contemporaine. *Les Temps Modernes*, **488**, Mars, 50–74.
Rivière, C. (1995) *Les Rites Profanes*. PUF, Paris.
Segalen, M. (1998) *Rites et Rituels Contemporains*. Nathan, Paris.
Söderlund, J. (2004) On the evolution of project competence: empirical regularities in four Swedish firms. *Communication IRNOP VI Project Research Conference*, August 25–27, Turku, Finland.
Turner, J. R., ed. (1995) *The Commercial Project Manager*. McGraw Hill, London.
Van Gennep, A. (1909) *Les Rites de Passage*. Nourry, Paris.
Wasserman, S. & Faust, K. (1994) *Social Network Analysis*. Cambridge University Press, Cambridge.

6 Client handling models for continuity of service

Suzanne Wilkinson

When a construction project is first initiated, clients must think through the ways in which they want the project to be managed. There are various ways to consider, from self-management to using an external company to manage the project. Whatever the configuration used, the client needs to be sure that some continuity of service is present. This requires the service by the companies involved in the project to follow a smooth path, and relationships, responsibilities, authorities and contractual arrangements to be conducted with continuity from one stage to another and ultimately to the successful conclusion of the project.

This chapter analyses the process of project management with particular reference to the relationships of project players at different stages of the project life cycle. It assesses the deliverables of the players at each stage of the project and then analyses how these deliverables might alter the relationships of the project team members and improve or disrupt the continuity of the service being offered. Various models of client handling are considered in this chapter, such as the account handler (or single point of contact) and the relay team proposed by Smyth (2000) within organisations, and the client representative (this term is used in the broadest sense, rather than as a specific contractual definition) system analysed by Scofield and Wilkinson (2001) primarily across organisations. The chapter examines the impact of relationships at life cycle interfaces so that a disruption of service is avoided. It develops research by Scofield and Wilkinson (2001) concerning the role of the client representative and the client at each stage in the life cycle, assessing how this relationship ensures continuity of service throughout. In particular, this chapter provides a new and extended interpretation of the client–client representative relationship and the changing roles of the client representative. Arguments presented in the chapter will lead to recommendations for improving relationships with the client, with a focus on the role of the client handler as key to project continuity and success.

Construction life cycle models

The Project Management Body of Knowledge (PMBOK, 2000) suggests that any project can be broken into five process groups. These are initiating processes, planning processes, executing processes, controlling processes and closing processes. Breaking down a project into generic stages is useful when managing construction projects. The number of stages and definition of stages can vary. For instance, the PMBOK uses five stages, or process groups. Shtub *et al.* (1994) also suggest five stages, these being conceptual design, advanced development, detailed design, production and termination. Meredith and Mantel (2000) suggest four stages: conception, selection, 'planning, scheduling, monitoring and control' and 'evaluation and termination'. These authors relate their life cycles to generic projects of any type and not just construction projects. For construction, Kwayke (1997) breaks the project into six stages: conception and briefing, design, documentation, tendering and estimating, construction, and commissioning. The RIBA Architects Plan of Work (RIBA, 2000) further breaks a construction project down into 11 stages under three main sections. These are: Section 1 – feasibility, which includes the stages appraisal and strategic briefing; Section 2 – pre-construction period, which includes the stages outline proposals, detailed proposals, final proposals, production information, tender documentation and tender action; and Section 3 – construction period, which includes mobilisation and construction to practical completion, and after practical completion.

Based on these life cycles, an equally valid way of representing the construction stages is:

- Inception
- Feasibility
- Design
- Tender
- Construction
- Commissioning

Relating this to the PMBOK and Kwayke's stages, inception and feasibility are linked to the PMBOK initiating processes or Kwayke's 'conception and briefing stage'; design to the PMBOK planning stage, or to design and documentation in Kwayke's stage; tender and construction to the executing and controlling processes in the PMBOK, or broken down into tendering and estimating and construction in Kwayke's scheme; and finally, commissioning can relate to the closing processes in PMBOK and is the same as commissioning in Kwayke's stages.

As a construction project progresses, the level of effort and resources required usually increases to a maximum and falls as the project reaches conclusion, or commissioning. The life cycle is an indication of the stages that a construction project travels through. It will obviously vary from project to project, with some projects necessitating the increase or decrease of one or more stages, and stages sometimes overlapping or their order being reversed. However, it can generally be said that in a construction life cycle the project starts with a few people and resources involved, gradually building up a momentum until peak activity is achieved at the construction stage. Then the project reduces to a conclusion after the commissioning stage.

As a project passes through these stages, relationships are formed between the parties undertaking the functions. The parties have to quickly form relationships if they are to work effectively. At interfaces of the stages, for instance at the design/tender interface, the change in personnel necessitates changes in roles and responsibilities of the parties, changes in lines of authority and changes in relationships. This means that parties will inevitably be in a state of flux at these interfaces. Careful change management is required for relationships at interfaces to operate smoothly.

Roles and relationships at each life cycle stage

Wilkinson and Scofield (2003) reported on the roles of each player at each life cycle stage. At the inception stage of the construction process the project initiator, usually the client, identifies a need and wishes to fulfil that need. At inception the client may appoint an adviser to assist with the decision-making. The adviser may design an initial brief with the client that can be used as a basis for a feasibility study. The adviser needs to be familiar with the client's organisation and the client in order to function well in the role. Often clients will use the same adviser for multiple projects (Scofield & Wilkinson, 2001). At the inception stage, relationship management for an adviser is focused on strengthening existing relationships with the client or establishing a rapport with a new client.

The feasibility stage requires the analysis of the proposed project to see whether it is possible within the required specifications of time, cost and quality (Ireland, 1995). If it is not, the feasibility study identifies the modifications required and how these might impact on the timing, cost and quality of the project. Depending on the size of the project, a feasibility study may be conducted by the client or by an adviser to the client, either one from within the client's organisation or an external consultant. A professional, architect or main designer may be selected to undertake some preliminary work. With extra personnel involved, and potentially different companies, the relationships begin to become more complex. The client, or the client's adviser, needs to establish clear lines of authority and responsibilities to make sure those relationships are operating well for the benefit of the project. At this stage, decisions to proceed are made and more detailed design begins.

At design stage, other parties will become more heavily involved, for example mechanical and electrical engineering personnel. This stage of the work will involve coordination between various design specialists, many of whom will be unfamiliar with how the other parties operate and the specific documentation to be produced. In addition, external parties begin to have an impact on the project, for instance local authorities for consents. It is at this stage that the web of relationships becomes more complex as different parties deal with each other for differing aspects of the work.

Establishing a project network at this stage is useful to ensure that every party in the project is aware of the roles, responsibilities and relationships (both contractual and personal) of each party to the project. Pryke's research (2004) shows the complexities of these networks for various construction projects and the importance of understanding these transitory relationships between actors whose roles are changing due to project-based and other external influences. (See also **Chapter 9**.)

To verify the cost of constructing the building, the client or client's adviser organises the tender process. There may be particular builders that the owner or the representative wishes to engage. These contracts may be offered because the client has built a relationship with the party and trusts that they can undertake the work. From a contractor's point of view, there may be many factors that influence their decision to work on the project including type of client and professionals involved. Likewise, selection of the contractor may be based partially on perceived management skills, which include the ability of the contractor to work well with other parties.

Walker (1998) suggested that the manner in which the contractor and the client's advisers and managers communicate at the construction stage could have an impact on the success of the project. Close working relationships need to be formed between the parties to ensure that this happens. Following construction, and as the project enters the commissioning stage, project relationships alter as players leave the project. Maintaining relationships becomes possible through more personal business relationship links (Blismas *et al.*, 2004).

Client handling models for continuity of service

Models for continuity of service help in understanding how clients are handled across life cycle intersections throughout a project. Smyth (2000) proposed two models for client continuity of service for a contractor managing their projects. The relay model proposed by Smyth (2000) allows for a different person handling the client's needs *within* the contractor's organisation at different stages, whereas the account handling model allows for continuity of service provided by one point of contact within the contractor's organisation. The *relay team* is typically favoured within contracting organisations.

These models can be applied equally throughout the whole project life cycle, where the relay model proposed by Smyth (2000) can allow for a different person to handle the client's needs *across* organisations at different life cycle stages and the account handling model can allow for continuity of service *across* organisations provided by one point of contact throughout the whole project life cycle. Typical examples of *organisational account handling* can be seen in traditional design and build contracts whereas an *organisational relay team* example would be novated design and build. In this case we see continuity of service by the architect in the context of contractors 'with design' procurement.

Research by Scofield and Wilkinson (2001) on the client–client representative relationship, and the roles that the client representative plays, suggests the typical form of client handling is the *account handling* method. This form can be across organisations throughout the whole project life cycle or within organisations at certain life cycle stages. This was confirmed recently in discussion with three senior managers in different New Zealand consulting companies. These discussions suggested that within organisations at senior level relationships are formed and developed with clients, and that these clients are the responsibility of one point of contact. This responsibility can be for only one aspect of the life cycle or for the whole life cycle. The senior manager will manage the relationship with the client, but the detailed running of the project may be delegated to staff at different stages of the project.

In effect what happens is that the client has a client's representative in the professional consultant company who acts as the account handler, but the detail of the project follows more the relay model, where responsibility for the representation is moved through various sections relating to the project life cycle. In this sense the relay baton (as proposed by Smyth, 2000) is passed across each life cycle interface as each team takes on their respective roles. For instance, the change from design to tendering necessitates a different team taking the lead role. This is in contrast to a single point of contact, for instance where there is a project management company managing the whole project with one team. However, if teams change at the interface of each stage, careful management is required for handing over the baton. Smyth (2000) reports that it is advisable to have some overlap between the interfaces so that there is some continuity in relationships. Signing off from one interface to another ensures smooth transition and shows the new team the limit of responsibility of the previous life cycle stage team. Criticism of the relay method stems from the lack of continuity of service (Smyth, 2000) and is supported by the author's discussions with senior managers.

As clients require certainty of service, client representatives or contractors can ensure continuity by one point of contact within the company. Smyth's account handler (2000) will be responsible to the client for all the client's needs on that project. In practice, as stated, this account handler may be external to all the companies involved: the client may choose a project manager or external client representative to act as project coordinator for the project and will use this person as a source for all the information about the project and as manager of the project. This client handler, or client representative, takes on different configurations, such as internal to the client's organisation, external to that organisation, part of the professional design organisation or part of a project management company. Ideally, all companies involved echo the management process concerning account handling.

Whatever the system in use, the client's need for certainty and continuity of service should become a priority. Assigning one person to look after the project account ensures this. The common theme of both models proposed by Smyth (2000) within organisations and other types of client handling across organisations is the need to establish and manage the relationship of the parties to the client. The client's account handler can be seen in various configurations of contractual relationships to the client and other team players, but a focus on the relationship of the client and account handler is important for project success.

The client

Managing relationships across life cycle stages and providing continuity of service are dependent upon the type of client and the client–client representative or client–client handler relationship. Many authors have discussed the client and produced client classifications (Blismas *et al.*, 2004; Chappell, 1991; Chinyio *et al.*, 1998; Walker, 1998; Wilkinson & Scofield, 2003). They can be classified according to the types of project they are initiating – residential, commercial, industrial or civil – as well as the value of the projects. They can be grouped into categories of expertise and skill: those who are experienced and qualified in the construction industry, those who have a degree of experience but are not specifically qualified, and those who have no industry-specific qualification and may have project responsibility only once or twice in their lifetime. Clients can also be categorised as to size of their organisations. In some cases an individual will be the sole decision maker for the organisation. In many cases there will be a number of people involved in the decision process, making it advisable for the client to appoint a representative within the organisation to be the point of reference for the project:

> *'Project outcomes are directly influenced by construction clients, particularly with respect to their needs and responsibilities in the construction process.' (Kometa et al., 1995 p.135)*

Different clients mean that different relationships are likely to be formed, from strong personal relationships to distant business relationships. Jawaharnesan and Price (1997) report that having client participation improves the utilisation of people, material and other resources and achieves customer satisfaction. It is likely that the strongest project relationships are formed between the client and the client representative, or project account handler, with a second group of equally strong relationships between the client representative and the other project participants. Pryke states:

> *'Where clients have the skills and capacity to manage the supply chain, a very effective and efficient system of procurement is possible which avoids the need for elaborate systems of financial and progress management.' (2004 p.32)*

This is supported by Wilkinson and Scofield (2003) who point out that clients that attempt to facilitate a project without professional guidance usually have sufficient confidence, if not always the skill, to set up all the relationships in a project themselves. Self-managing clients may have an interest in saving money, may be suitably qualified to undertake the facilitation of a project, or may require detailed involvement in the project. They may even design the building, as well as let the contracts and supervise the site works. Wilkinson and Scofield (2003) suggest caution for the self-managing client, as there is a danger that these clients may operate outside the scope of their ability and therefore end up dissatisfied with the end result, losing money in the process.

However, clients face a problem: while the institution's budget is often insufficient to pay for an outside representative, there is not usually an in-house professional with the skills to do the job (Scofield & Wilkinson, 2001). For commercial and industrial projects some large client organisations do have the in-house skills and prefer not to hire an independent representative (Wilkinson, 2001). Government authorities tend to act this way when procuring civil engineering works. In many cases, the design and all the supervision will be conducted in house and only the construction of the project will be contracted out (Love *et al.*, 1998). This system works well when the client has experience in procurement and facilitation of projects, and in-house skills and experience to complete the project. In the residential market, using the constructor, builder, or main contractor as the project representative is a common method of project completion (Wilkinson & Scofield, 2003). Research by Wilkinson (2001) suggests that not all projects are suitable for using independent project management services, the reasons for not using these services being that companies can provide all the services in house or can use alternative procurement forms such as turnkey design-build. Clients, whatever the type of project, need to make decisions on how the project is to be managed. Choosing a client representative is a common first step.

Choosing a client representative

The choice of client representative is critical to the success of a project, as demonstrated by Kometa *et al.* (1995). Client representatives include independent consultants, project architects and in-house client representatives. Client handlers tend to work within construction organisations to handle an external client's business with that company. Both client handler and client representatives have an interest in building strong relationships with the client and they have common skills needs. The usefulness of a client representative in a project has been discussed by Wilkinson (2001) as providing the client with clear definitions of the roles of parties involved, ensuring accurate and adequate supervision of consultants, reducing in-house squabbles and fiddling, and providing clear leadership.

In addition, Walker (1998) produced advice for clients on the type of client representative who would be 'good' for construction time performance. Client representatives, he believed, should be sophisticated in terms of knowing what is involved with the project, its scope and complexity; able to offer and accept advice about both design and construction; have good communication skills; have good team-building and interpersonal skills, and clearly communicate the priorities of the client's objectives (Walker, 1998). Walker also studied the impact that a client representative (either internal or external to the client organisation) has on the construction time performance of a project. He suggested that there is a need for the client representative to concentrate on relationship-building skills as this has a major impact on the project success. It is highly desirable for the client representative not only to be well informed, but also to be well suited to the position. The representative is in the best position to facilitate the smooth execution of the project and resolve contentious issues before they become disputes (Walker, 1998).

Other authors have studied the client–client representative relationship and problems that may arise. For instance, Ma and Chan (1998) suggested that, from the viewpoint of the client representative, the main issue is getting the client to provide a clear brief in order for the project to be successful. Bowen *et al.* (1999) suggest that client representatives believe clients are inexperienced about the procurement methods available, that some clients are inflexible and reluctant to heed advice from professional consultants, and that effective briefing is blocked by a lack of clarity with regard to communication networks. Bowen *et al.* also claimed that client representatives believed clients were often unfamiliar with the design and construction process. That client representatives are used in the construction industry is not in dispute. For instance, Wilkinson (2001) reported that clients were using client representatives in the form of project managers for most of their projects and that these managers were having sole management responsibility, acting as an intermediary between the project team and client.

Wilkinson and Scofield (2001) showed that the most common type of person chosen to manage the project for the client and to act as the client representative throughout the project was an independent consultant, followed by the project architect and in-house representative equally. All these people act, for the duration of the project, as a type of account handler, facilitating the transition of the project through its various life cycle stages.

The client can face difficulties when choosing a client representative. Some of the disadvantages have been discussed by previous researchers, such as Chappell (1991), Hughes (1997), Kwayke (1997), Masterman (2002), Murdoch and Hughes (1996) and Spinner (1997). In particular, Chappell (1991) reported that a representative can worsen communication between the employer and the design team. Kwayke (1997) mentioned the lack of authority given to the representative, the lack of financial risk the representative takes and the increase in costs. On the other hand, Kwayke (1997) reports that there are advantages to using a client representative, such as having the brief prepared by a skilled professional, having a professional manage any confrontations, releasing the architect from management, and having an independent and sole point of contact for the client. An inexperienced person without the protection of an independent agent is unlikely to notice, for instance, low-quality work until it shows up as a defect.

Because of these issues, Wilkinson and Scofield (2001) reported that clients felt a client representative should be independent of the project. However, without professional expertise choosing an independent client representative was not considered easy. One employer representative noted that there had been poor performers passing themselves off as specialists and was consequently hesitant at making a choice without professional assistance (Wilkinson & Scofield, 2001). The research by Wilkinson and Scofield (2001) showed that client ratings for the construction professionals suggest a quantity surveyor is the preferred option for a project representative and an architect the least preferred.

The survey by Wilkinson and Scofield (2001) gave an indication of why a client would choose a representative to manage a project. The common reasons were previous successful use, lack of in-house resources, to save on fees and to control project costs. Protecting the client's interest by providing a single point of responsibility was also mentioned. A majority of clients would choose the same representative again if they had produced the expected results and had worked well with the client company. This indicates that repeat use of successful project representatives who have built up relationships with the client organisation is common.

Clients were likely to choose the person to handle their project account at the inception stage of the project. This is in order that the representative thoroughly understands the scope of the project and has time to become familiar with the characteristics of the project and the project teams.

Since there are different types of clients (Walker, 1998), there will be different preferences for the types of project representative chosen. However, as Wilkinson (2001) pointed out, using a representative to manage the project is particularly useful when there is a trusting relationship between the parties. This is irrespective of whether an internal client representative is used to manage the process, or an external company, or any other configuration.

The client–client representative relationship

Once appointed, the client and client representative need to establish roles and develop their relationship. Jawaharnesan and Price (1997) reported on the role the representative plays, suggesting that their role can vary from simply designing the project to its full monitoring. The varying roles thus lead to differing relationships being formed. Wilkinson and Scofield (2003) discussed the relationship of the client and the client representative. One of the questions raised was that of impartiality in cases where the client's representative needs to appraise the design objectively, or compare detailed design proposals by the contractor, and advise on life-cycle costing issues while at the same time maintaining programme and cost constraints. They suggested that if the client's representative is part of the client organisation this is not such an issue. However, if the representative is employed externally then this can be of significance. Building a relationship with the client in order to serve the needs of the client becomes the first priority of the external representative (Wilkinson & Scofield, 2003). This calls into question what the client representative should concentrate upon. In most cases this will be the service quality provided by others in the project, in particular the contractor and professional service providers – designers and quantity surveyors, for example.

If design professionals are to be successful representatives, it is likely that they will not be specifically design orientated but will have an in-depth knowledge of design and material alternatives, building code requirements and contract management. Whereas historically architects have led the design and construction process and carried out contract administration (Hughes, 2001), more recently other participants are taking this role (Wilkinson, 1998). This was affirmed by Pryke, who points out:

> *'The traditional roles of architect and quantity surveyor are being displaced to some degree by the role of cluster leader, which combines both the design co-ordination and financial management within one role.' (2004 p.29)*

Wilkinson and Scofield (2003) argued that if architects act as representatives they need to be careful not to focus too heavily on aesthetics. Similarly, if engineers take on this role they need to be careful not to focus on the technical design detail, and if quantity surveyors act as representatives they should not make decisions solely based on financial limitations. Wilkinson (2001) showed that co-ordinating professionals come from a variety of backgrounds, including those outside the construction profession, such as accounting or commerce. Clients need to be aware that the background of the representative can influence the decisions the representative makes and the project focus.

Clients can use their own staff as project managers. The role of this in-house project manager may not be so wide as to include all the tasks of the external representative, such as document preparation, design management, safety management and site supervision (Jawaharnesan & Price, 1997). However, the role should include developing the project definition, procurement, measuring performance, communications, in-house co-ordination and post-project evaluation.

An external client representative will usually be commissioned at the outset of the procurement process in order to determine how best to add value to the company. When projects go wrong, the client's representative is blamed. A survey by Wilkinson and Scofield (2001) into the client's view of representatives showed that negative traits experienced by clients were reported as:

- No ownership of project
- Lack of effort
- Too friendly to contractor
- Short-tempered
- Egocentric
- Doesn't listen

- Not strong enough in confrontational situations
- Lack of industrial experience
- Reliance on consultant
- Lack of site management experience

In choosing to employ a person to handle the project account, clients often devolve responsibility, hoping for a more successful project outcome. The client must therefore have built a trusting relationship with the representative. Being clear on divisions of management and the lines of communication and what each participant's role is expected to be will improve this relationship. This way, in principle, all the players can be clear about their role relative to the management of the project and to an anticipated reduction of confrontations. The representative needs to ensure, at the project outset, that clients who are unable to read the drawings and visualise long-term implications do not affect project success. Examples of the ways in which clients may do this are by failing to understand the cost requirements of changes, requiring changes outside the project scope, or expecting the project to be completed within the original timeframe despite having required time-related variations.

Relationships between the client and client representative or client handler can become poor for several reasons: for example, when clients are not kept informed about the project or when the client representative tries to force their opinions onto the client (Bowen *et al.*, 1999). Similarly, the client's perceived inability to align scope and budget has been viewed as a source of strain on the client–client representative relationship (Smith *et al.*, 1998). Often, the problems can be attributed to a client's decision to save money without adequate appreciation of the consequences (Scofield & Wilkinson, 2001). In order to ensure that the client–client representative relationship does not deteriorate, it is important for the client representative to ensure continuity of the project service throughout the project life cycle.

Ensuring continuity of service between client representative and client project at life cycle stages

This section develops research by Scofield and Wilkinson (2001) concerning the role of the client representative and the client at each stage in the life cycle, assessing how this relationship ensures continuity of service throughout the life cycle. It provides a new and extended interpretation of this data including the addition of further research by Wilkinson and Scofield (2003). Previous research (Scofield & Wilkinson, 2001), using interviews with clients, showed that there are some extra functions that can be performed by the representative to ease transition through the life cycle stages.

At inception stage, when the representative helps the client to establish needs, it was found that to ensure service continuity representatives should be selected to develop the project concept and its feasibility. Initial investigation should centre on what is actually needed, but the client representative should be guided by the client and not pressurise the client. Clients should be appointing their representatives at inception so that they identify the important issues, such as whether to build or relocate, and the relative costs of each. At initiation stage the important functions of the representative become those of information gatherer and relationship builder. Information gathering and advising continue to the feasibility stage and are joined by the key roles of information analyst and programmer.

Usually the transition from initiation to feasibility is smooth, especially if the client representative is already appointed and has been able to quickly establish a relationship with the client. However, if the client representatives are not appointed until after the funding application they sometimes have little idea as to the background of the proposal. This can present difficulties in handling the client's project. The client's representative assists the client to commission the players to undertake the design, contract documents and tendering. This is often done under time pressure, leaving insufficient time for thorough consultation with user groups. When this happens relationships are compromised, as tasks become a priority above team development.

Scofield and Wilkinson (2001) reported that where there is no clear line of responsibility set up for a single client representative very limited authority is available for the client representative. Thus the client at initiation and feasibility stages becomes reluctant to hand over full responsibility to the representative, and so the client's representative may feel unable to perform their role adequately. In order for there to be less chance of dissatisfaction between the team members and the client, the client representative should involve team members as early as possible in the life cycle process. Established lines of responsibility and authority, and better feasibility studies and site visits, will help the client representative to form a clearer idea of the client needs.

The research found that, in the client's view, staff movement away from a project had an adverse effect on project process particularly at design stage. Examples given were budgets being set that became difficult to understand by successors, or communications breaking down. Reducing these movements is important for the client so that developments of relationships are continuous and not fragmented, leading to improved continuity of service. The client's representative needs to coordinate all the different facets of information from the various team players and report these to the client. A coordinating role is very difficult to fulfil at design stage as there are many opinions and uncertainties about what is wanted. Skills the client most desires of a client representative are the need to be analytical and to communicate well. The client representative needs to focus on team building, listening to their team and coordinating the different activities and people.

Interestingly, the research found that, rather than the client representative making a selection, the method of appointing professionals was likely to be based on previous association. Using previous associations can lead to alternative ways of procuring construction projects, such as partnering or relational contracting. The research showed that clients are forming relationships with companies and using these companies on subsequent projects. Clients usually have preferences for particular companies that may or may not be the best for the project. These preferences are formed because either the client has worked with them before, or the consultants have past experience from similar projects. There was also a preference for local companies. This suggests that clients are strategic about their choice of companies and, it is reasonable to assume, about the representative chosen to handle the client's project needs. The client representative needs to be cautious at this stage – weighing the appointment of client-preferred companies with the best interests of the project.

At design stage, one of the biggest problems found was the failure of the consultants to coordinate their work with other professionals. Here the client representative needs to have an understanding of the construction process and be able to relay this information to the client. The way in which the consultants are appointed in a contract is an important factor in the types of relationships that will be formed. If consultants are separately engaged then overall responsibility and authority for their work needs to be established. As project issues at tendering stage can alter the outcomes of the project, the client representative needs to be confident in the selection of the company chosen.

A particular case reported by Scofield and Wilkinson (2001) was where tender prices were perceived to be too high, the consultants were replaced and the project was value-engineered, at a considerable cost in terms of time and fees. In another case, the quantity surveyor made a mistake in the estimate. In a third case, the electrician underestimated the cost and spent the rest of the contract trying to make up the difference with cheap labour.

Client representatives are often good at managing the project through the early stages. When a project reaches the tender stage, it is hard for the client representative to evaluate prices and answer construction queries unless they are sufficiently skilled at doing this. It is important for the client representative to receive good professional advice, thus bringing into the project more participants with whom to form relationships.

Once the project reaches construction stage, the project relationships become even more complex as the various specialist contractors are brought in. The research found that the client representative sometimes changes during the project: a professional consultant at the design development stage, followed by a project management firm at the construction stage, for example. This may have an impact on the continuity of service provided, where problems similar to those reported in Smyth's (2000) relay system would occur. In the interviews reported by Scofield and Wilkinson (2001), those client representatives who had delegated the task of representative at construction stage commented in hindsight that they had probably expended too much on costly and unnecessary details at the expense of the more important services and they regretted the lack of opportunity for client input.

There were problems in the projects that only became apparent in the construction stage, by which time it was costly to rectify them. For example, impractical odd-shaped rooms and wasted space were commonly reported problems. These faults were blamed on the client's lack of ability to read drawings accurately and the client representative's failure to communicate the reality of the project. Clients could not appreciate the future implications of decisions they were making in the design stage. They felt that it would have helped them to handle projects effectively if they had been involved with the client representative on site. Clients reported that good project management by a person involved throughout the process could have averted many of the problems. This person would see the project through all the life cycle stages and act for the client, as per the account handler proposed by Smyth (2000). The client representative needs to be able to organise teams and work, manage the project for the client and act as mediator if problems arise.

Previous research by Wilkinson (2001) reported that at construction stage the problems are, generally, relationship-based. Relationships with clients and relationships with the other professionals are the main sources of problems faced by these client representatives when trying to project manage construction projects. This research (Wilkinson, 2001) showed that a lack of contact with the client could be a source of grievance for the other professionals that is often transferred into frustration with the client representative. Similarly, research by Hughes (1997) suggested a communication problem at construction stage where contractors were unable to get clients to talk to them.

Following construction and commissioning, analysis of the interviews showed that at occupation clients felt the final move to the new premises was often disruptive and could have been better handled. There were often defects to rectify and in some cases the basic planning of the building occupation was a problem, showing a lack of continuity between practical completion and occupation. In most projects the client thought that the end result was reasonably successful and that the experience on the whole was good. However, evaluation is not always carried out and, in the cases where it is, all stakeholders are not always present.

Because projects are often only done once by any one project team, the client does not have an incentive to evaluate the project's success. Conversely, professionals involved in projects do have a vested interest as success may determine future commissions. The evaluation process should be organised by the client representative, as it is useful as a tool to measure quality assurance and thereby improve performance and reduce costs. Thus the client representative becomes an assessor of the success of the project. Clients reported that having a partnership relationship throughout the project between the client, client representative, the designers, contractors and the management would be essential.

Life cycle stages, critical relationship success factors and the changing client representative role

At each project life cycle stage the role of the client representative changes and the ways in which the client representative handles the client alters. The

Table 6.1 Relationship roles and success factors at project stages.

Stage	Critical relationship success factors	Development roles
Inception	Appointment of client representative. Accurate needs analysis Establishing client–client representative relationships	Initiating adviser Information gatherer Relationship builder
Feasibility	Continuity of representative Identification of users and project philosophy Professional guidance and advice to client Knowledge and research of project	Adviser Analyst Programmer Information gatherer
Design	Definition of requirements Clear authority for representative Clarification of roles Compatibility of teams	Team builder Listener Coordinator
Tender	Liaison and good technical assistance	Analyst
Construction	Delegation Clarification of team network Independent assessment and control	Organiser Mediator Manager
Commissioning	Early confirmation that project is operational Project evaluation organisation	Quality manager Assessor

model in Table 6.1 is proposed for increasing the likely success of a construction project by focusing on the relationships at different stages and the representative's changing role. This model was first proposed by Scofield and Wilkinson (2001) and has now been significantly developed and updated in light of the new analysis presented in this chapter.

From Table 6.1 it is clear that the client's representative has to be multiskilled over the project life cycle. Typically the client's representative may have skills for one aspect of the project life cycle but not all. By being aware of the critical success factors and the changing roles, clients and client representatives are in a better position to achieve project success. This may be through having a single client representative, or having multiple representatives at different stages with skills matched to the stage requirements.

Conclusion

There are many relationships that are formed during a construction project. The ability to handle the client's requirements and to facilitate the relationships between the various parties in the project is often the responsibility of a client representative. The configuration of this role is varied and impacts on the continuity of service at different life cycle stages. In this chapter, the analysis has focused on the different requirements of the client representative at different life cycle stages and the changes that are required to the role in order for the client representative to handle the project effectively.

The model developed in this chapter helps explain what these roles are and how the client representative can more effectively manage the client's needs throughout these project life cycle stages. It is recommended that clients and client representatives attend to the different roles they play at various life cycle stages, particularly ensuring that the skills required of the representative are present and attention is given to the critical relationship success factors. By doing this, a client representative should be better able to understand how to manage the client and the project, and when the role they play needs to change in order to ensure smooth continuity of service.

References

Blismas, N. G., Sher, W. D., Thorpe, A. & Baldwin, A. N. (2004) Factors influencing project delivery within construction clients' multi-project environments. *Engineering, Construction and Architectural Management*, **11** (2), 113–25.

Bowen, P.A., Pearl, R.G. & Edwards, P. J. (1999) Client briefing processes and procurement methods selection: a South African study. *Engineering Construction and Architectural Management*, **6** (2), 91–104.

Chappell, D. (1991) *Which Form of Building Contract?* Architecture Design and Technology Press, London.

Chinyio, E., Olomolaiye, P. & Kometa, S. (1998) A needs based methodology for classifying construction clients and selecting contractors. *Construction Management and Economics*, **16** (1), 91–8.

Hughes, W. (1997) Construction management contracts: law and practice. *Engineering Construction and Architectural Management*, **4** (1), 59–79.

Hughes, W. (2001) Evaluating plans of work. *Engineering Construction and Architectural Management*, **8** (4), 272–83.

Ireland, V. (1995) The role of managerial actions in cost, time and quality performance in high rise commercial building projects. *Construction Management and Economics*, **3** (1), 59–87.

Jawaharnesan, L. & Price, A. D. F. (1997) Assessment of the role of the client's representative for quality improvement. *Total Quality Management*, **8** (6), 375–89.

Kometa, S., Olomolaiye, P. & Harris, F. (1995) Quantifying client-generated risk by project consultants. *Construction Management and Economics*, **23** (2), 137.

Kometa, S., Olomolaiye, P. & Harris, F. (1997) Validation of the model for evaluating client-generated risk by project consultants. *Construction Management and Economics*, **14**, 131–45.

Kwakye, A. A. (1997) *Construction Project Administration in Practice*. Longman, London.

Love, P. E., Skitmore, M. & Earl, G. (1998) Selecting a suitable procurement method for a building project. *Construction Management and Economics*, **16** (2), 221–33.

Ma, T. & Chan, A. (1998) Fast track procurement path: a study of construction management in Australia. *Second International Conference on Construction Project Management*, Singapore.

Masterman, J. W. E. (2002) *Introduction to Building Procurement Systems*. Spon Press, London.

Meredith, J. & Mantel, S. (2000) *Project Management: A Managerial Approach*. John Wiley & Sons, New York.

Murdoch, J. & Hughes, W. (1996) *Construction Contracts Law and Management* (2nd edn). E & F N Spon, London.

PMBOK (2000) *Project Management Body of Knowledge*. Project Management Institute, USA.

Pryke, S. (2004) Twenty-first century procurement strategies: analyzing networks of inter-firm relationships. *RICS Foundation Research Paper Series*, October, 27, London.

Royal Institute of British Architects (2000) *Royal Institute of British Architects Plan of Work*. RIBA, London.

Scofield, R. & Wilkinson S. (2001) Acting in the client's best interests: an analysis of the role of the client's representative. *Proceedings of CIB World Building Congress*, Wellington.

Shtub, A., Bard. J. & Globerson, S. (1994), *Project Management: Engineering, Technology, and Implementation*. Prentice Hall, NJ.

Smith, J. M., Kenley, R. & Wyatt, R. (1998) Evaluating the client briefing problem: an exploratory study. *Engineering Construction and Architectural Management*, **5** (4), 387–98.

Smyth, H. J. (2000) *Marketing and Selling Construction Services*. Blackwell Science, Oxford.

Spinner, M. P. (1997) *Project Management Principles and Practices*. Prentice Hall, New Jersey.

Walker, D. (1998) The contribution of the client representative to the creation and maintenance of good project inter-team relationships. *Engineering Construction and Architectural Management*, **5** (1), 51–7.

Wilkinson, S. (1998) The growth of project management in the New Zealand construction industry. *Second International Conference of Construction Project Management*, Singapore.

Wilkinson, S. (2001) An analysis of the problems faced by project management companies managing construction projects. *Engineering Construction and Architectural Management*, **8** (3), 160–70.

Wilkinson, S. & Scofield, R. (2001) A skills profile for a successful project facilitator. *Proceedings of CIB World Building Congress*, Wellington.

Wilkinson, S. & Scofield, R. (2003) *Management for the New Zealand Construction Industry*. Prentice Hall, Auckland.

7 Applying teamworking models to projects

Mohan Kumaraswamy and Motiar Rahman

Teamworking models can draw both inspiration and substance from scenarios ranging from a pack of predator animals trapping its prey in a well coordinated manoeuvre, to a football team slicing through the opposing defence. The common goal-orientated focus of all team members suggests the advantages of:

(1) The value of team relationships.
(2) How such relationships can develop naturally when a group focuses on a common goal (as a football team does) in what may be hereafter termed a 'teamworking by objectives' (TBO) framework.
(3) A high degree of 'trust' in each other to adjust their approaches towards common objectives when encountering risks and changing conditions. The important issue of trust is dealt with elsewhere (see **Chapter 4** by Smyth).

In terms of the proposed focus on the above-mentioned TBO, the objectives of both the organisation and individual team members must be clearly aligned with overall project objectives. This could be achieved by well-structured incentivising mechanisms, by trust and through relationship-building strategies. However well these structures and mechanisms are defined, the contractual and 'transactionally binding' forces between organisations (and individuals) need to be significantly strengthened by 'relational bonding' forces (Kumaraswamy *et al.*, 2002a; Palaneeswaran *et al.*, 2003) if synergies are to be achieved across the multi-disciplinary and multi-cultural project teams of today. This is approached from the standpoint of previously proposed transactionally efficient 'relational contracting' (RC) models (Macneil, 1974; Rahman & Kumaraswamy, 2002a).

This chapter describes and develops the above ideas by following this introduction with a general overview section on what is expected from teams and teamwork. Thereafter the first focus is on the relationship-building phase. To demonstrate this, a summary of the main relevant findings from a structured set of Hong Kong-based surveys yields insights into how a 'relationally integrated supply chain' (RISC) can be assembled in the first instance, and how a 'relationally integrated project team' (RIPT) can then be drawn from it (Rahman & Kumaraswamy, 2004a). The RIPT can be developed in an already less opportunistic and longer-term relational scenario. Trust and relationships in the RIPT are then specifically boosted through strategies and operational mechanisms, such as for joint risk management (JRM) and pain-share gain-share arrangements. Outcomes from the survey exercises as well as the subsequent case study demonstrate the increasing importance of relational approaches, including JRM, in anticipating and minimising project risks and other problems. The ongoing case study on a risk-intensive railway extension project exemplifies the success of such a strategy and related mechanisms.

This case study also demonstrates the benefits of focusing on project 'value' through value engineering. It reinforces the recognition of focusing on 'best value' for all concerned. This is conceptualised as a convenient approach to formulating a unifying hierarchy of objectives. A corresponding conceptualisation of 'relationally integrated value networks' (RIVANs) is proposed as a useful progression from the RISC-RIPT model to an even more integrated and value-focused team. It is proposed to develop this concept from the specific viewpoint of a more value-focused and improved teamworking model for construction projects in the first instance.

The chapter discusses finally a few other relevant theories and teamworking models. In this context, a small but relevant cross section of existing models and approaches (for example, Belbin, 2004a, 2004b; Rippin, 2002) are examined in terms of their theoretical bases and practical applications. Identification, comparison and analysis of relevant theory together produce a valuable reminder of other elements that may be incorporated to extend the potential scope of teambuilding and teamworking theory and applications in general, as well as in construction specifically.

More precisely, relevant group theory, organisational dynamics (Handy, 1993; Mullins, 2002) and teamworking models (Belbin, 2004a; Rippin, 2002) are drawn upon, for example to assess the general interplay between existing organisational cultures, individual roles and personalities, informal groupings, dominant coalitions and other drivers in the development of a project culture and team dynamics – both of which can enhance or constrain project management, or indeed make or break a project.

Teams and teamwork

Teams are defined as:

> *'groups of people with complementary skills who are committed to a common purpose and hold themselves mutually accountable for its achievement. Ideally, they develop a distinct identity and work together in a co-ordinated and mutually supportive way to fulfil their goal or purpose.' (Constructing Excellence, 2004)*

Constructing Excellence (2004) also describes 'effective teamwork' as resulting from:

- A team whose membership and size matches the task
- Good leadership and attention to teambuilding
- Commitment by team members to understand and identify with one another's goals
- The development of team goals – a shared vision
- A sense of common ownership of the task at hand and joint responsibility for its achievement
- Coordinated effort and planned sharing of tasks evenly across the team
- Open exchange of information within the team
- Honesty, frankness and trust among team members

It is increasingly appreciated that:

(1) The brilliance of a few outstanding individuals may only provide sporadic and unsustainable results, and therefore 'while not ignoring, or neglecting the individual, we should devote far more thought to teams: to the selection, development and training of teams' (Jay, 2004).
(2) However excellent the working systems and however streamlined the organisational structures, excellent performance can only be delivered by capable, integrated and motivated teams. This is true in any area of collective performance, whether in the industrial–commercial or sports arenas.

This appreciation was confirmed, for example, in a Cambridge University-based study (Management Briefing, 2002), which demonstrated that construction meetings work better when they are not dominated by a few individuals. The researchers mapped interactions at different types of meetings, in search of ways to forge teamworking based on integration and support. They concluded that 'teamworking is not rocket science', but needs a major cultural shift and support tools such as for self-assessment. A tool was devised to help identify whether the pooling of resources and harnessing of combined expertise was truly effective.

The need for a major cultural shift away from traditional confrontational scenarios and adversarial standpoints has been central to many recent construction industry reports worldwide. Kumaraswamy *et al.* (2002b) illustrate a proactive approach to moulding project cultures towards desired objectives. This approach can be adapted and specifically directed towards developing the desired teamworking culture, bearing in mind the incentivisation and integration needed for synergising all the contributory cultural components as shown in Figure 7.1.

This adapted approach to developing the desired project culture can be 'operationalised' by 'feeding' into the conceptual model of 'the construction project as a network of inter-firm relationships' developed by Pryke (2004a, 2004b; see **Chapter 9** by Pryke). Pryke (2004a) highlighted three main groups of networks as 'contractual', 'information exchange' and 'performance incentives'. All three groups of networks can be taken as important for

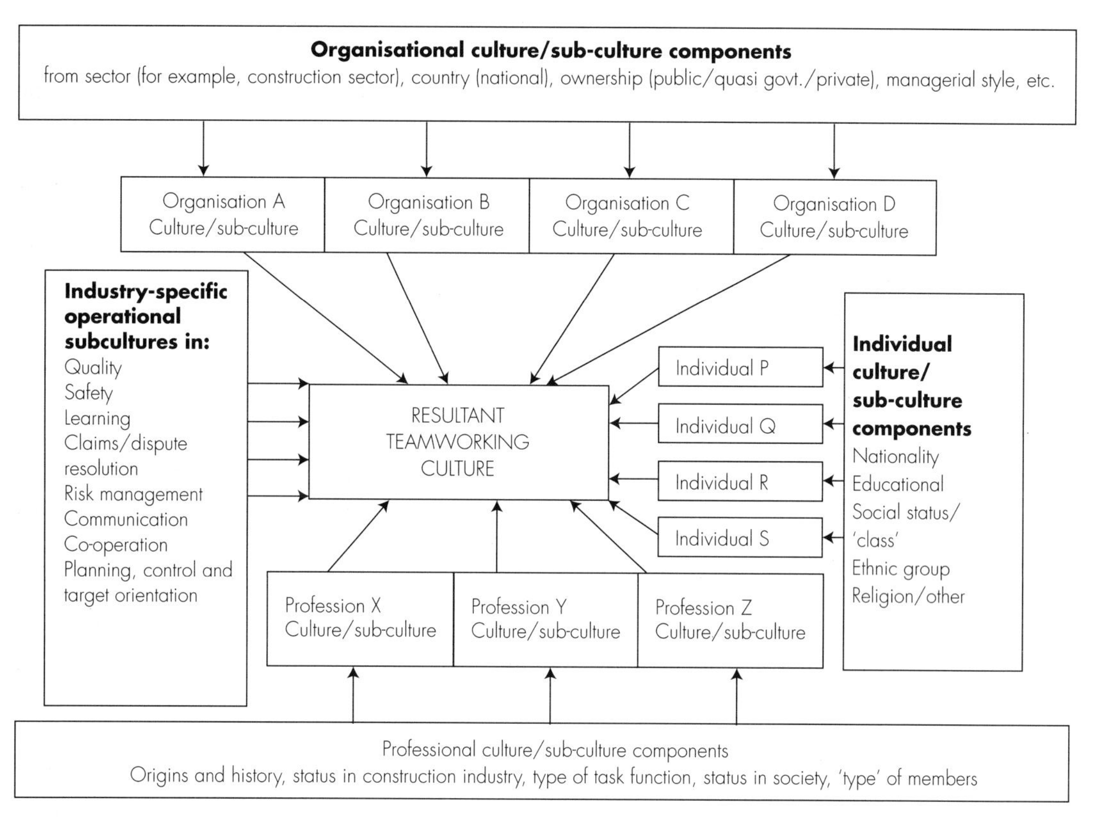

Fig. 7.1 Sources of a typical construction project team culture.

teamworking, since the first should provide an appropriate framework while suitably designed information exchange and incentive mechanisms can encourage, if not empower, teamworking.

A number of respected reports (for example, Emmerson, 1962; Latham, 1994) aimed at reforming the construction industry in the UK over the last few decades have recommended practical measures to reverse the much deplored industry fragmentation. In particular, the value of teamworking has been espoused in the *Accelerating Change* report by the Strategic Forum for Construction (2002), setting specific targets for 20% of construction projects by value to be undertaken by integrated teams and supply chains by end-2004, and for this to be increased to 50% by 2007.

Benefits from successful teamworking are expected to include a wider range of ideas, more effective responses to change and more efficient use of resources, for example (*Constructing Excellence*, 2004). The imperative for effective teamworking models and guidelines for their efficient implementation is therefore very evident.

Teamworking in construction organisations

Sources of knowledge

The construction industry is characterised not only by fragmented teams as mentioned above (Latham, 1994) but also by its multi-disciplinary nature. Of course, multi-disciplinary teamworking is not unique to construction. For example, Wilson and Pirrie (2000) deal with a number of 'indicators of good practice' for multi-disciplinary teams in a different context, but lessons learned in other scenarios can have a bearing on the construction industry as well.

In the construction industry itself, an initiative entitled *Teamwork – learning by doing* was launched following *Rethinking Construction* (Egan, 1998). *Teamwork 2001* was the second in a series of demonstration projects reportedly conveying evidence that collaborative working in design and construction can reduce waste, cut cost, rationalise processes and promote a working culture of trust and high performance. The next year brought the *Accelerating Change* initiative launched by the Construction Industry Council, intended to expedite the pace of reform (Strategic Forum for Construction, 2002).

The body of 'good practices' in construction project teamworking can also draw heavily on the vast experiential knowledge being developed in partnering and alliancing-type approaches. Partnering is extensively documented, yet would benefit from the development of a coherent and easy-to-use knowledge base as proposed by Kumaraswamy *et al.* (2003a). Alliancing moves beyond the various levels or generations of partnering to develop virtual, if not real, cross-organisational teams that approach project success together (see, for example, Hauck *et al.*, 2004).

Meanwhile, toolkits for effective partnering and alliancing are continuing to be developed at a broader level, for example by the European Construction Institute (1997, 2001). Furthermore, knowledge developed in generic supply chain management should also be drawn in to address shortcomings in present partnering arrangements in construction, for example where subcontractors, suppliers and even consultants are often sidelined in the envisaged cooperative teamworking (Sze *et al.*, 2003), thereby curtailing its efficacy.

A solid theoretical foundation is also useful in justifying the use of such collaborative approaches to those who could be apprehensive of the dangers of collusion and corruption that can possibly arise from 'too close' a collaborative scenario. *Relational contracting* approaches have been found to provide such a sound foundation (Rahman & Kumaraswamy, 2002a) to underpin and promote well-balanced good practices in partnering, alliancing and indeed in broader teamworking.

Relational contracting in principle

While there are many contributors to poor project performance, much of the blame is often attributed to the traditionally adversarial contracting systems. These adversarial dynamics waste resources and energies on:

- Inter-organisational interface management (for example, with many levels of supervision, some of which are duplicated by different organisations)
- Disputes on perceived responsibilities and contractual liabilities, between various contracting parties – across the construction project supply chain

The situation has become more complicated with the many sub-subcontractors, sub-suppliers and specialists now participating in construction supply chains. This necessarily involves more transactions, contracts, interacting interfaces and resultant complexities.

Relational contracting (RC) principles provide a framework for exploring the ways in which the friction and disputes associated with formal transactions can be reduced. These approaches might include, for example, a major re-alignment towards *common objectives* and more *collaborative approaches* as in partnering, and also injecting a healthy dose of joint risk management (JRM) over and above 'traditional' or *classical* contracting and risk allocation principles that involve clear demarcations of all possible risks (Rahman & Kumaraswamy, 2002a).

RC principles and JRM practices can empower proactive procurement and project delivery strategies, for example by providing suitable incentives for designers, contractors, subcontractors and suppliers, aiming at a longer-term vision, and a more holistic teamworking mindset.

Relational contracting in practice

RC is built on the recognition of win–win scenarios and mutual benefits derived from cooperative-collaborative relationships and teamworking between the parties to a contract. It provides principles and a framework for approaches such as partnering, alliancing and joint risk management in general.

Partnering and alliancing arrangements were introduced into many construction projects from the 1980s commencing in the USA. The foundations of such collaborative arrangements have been more recently traced to the win–win principles of RC, as against the 'zero-sum' transactional approaches generated in *classical contracting*.

In the above context, such contracts had been classified into *classical*, *neo-classical* and *relational* by Macneil (1974) and their basic differences were identified by Williamson (1985). Relevant issues have been summarised by Rahman and Kumaraswamy (2002a) in relation to recent trends in construction project management. RC has been shown to be a useful approach to reducing 'transaction costs' in the longer term, and promoting cooperative relationships that in turn facilitate joint risk management, and thereby improve productivity and reduce disputes.

The usefulness of RC has been confirmed by a series of Hong Kong-based surveys, as reported by Rahman and Kumaraswamy (2002b, 2004b). Collaborative working practices have also been developed in some key infrastructure projects in Hong Kong (Lighthouse Club, 2004). The conceptual framework, theoretical justifications and legal implications of such 'relationship-orientated' practices are undergoing further investigation. Meanwhile, the practical outputs are presented in the following models that can be applied to teambuilding and teamworking.

Relationally integrated supply chains (RISCs) and relationally integrated project teams (RIPTs) for teamworking

Structural integration of, for example, the 'design' and 'construction' functions in a 'design and build' project is not enough in itself to ensure effective teamworking. Relational integration throughout the supply chain is also needed, as has been demonstrated by Rahman *et al.* (2001), Kumaraswamy *et al.* (2002a) and Palaneeswaran *et al.* (2003) – cf. **Chapter 5** by Cox and Ireland. Kumaraswamy *et al.* (2000) also illustrated the importance of team member selection in optimising supply chain performance. Appropriate selection was also seen to be a useful tool in moulding project cultures (Kumaraswamy *et al.*, 2002b; Rahman *et al.*, 2002). A generic framework has been proposed (Rahman & Kumaraswamy, 2002b, 2004b) to construct 'relationally integrated supply chains' (RISCs) for the longer term, from which project-based 'relationally integrated project teams' (RIPTs) could be drawn together for specific projects (Rahman, 2003; Rahman *et al.*, 2003).

Collaborative working arrangements can be better integrated if different team 'partners' know each other's strengths and weaknesses before entering into a particular project team. This will help them to reach a collective understanding on various issues. Longer-term interactions of both a formal and an informal nature are expected to help in building trust and reliability. As a preliminary step in such relational integration, while targeting longer-term benefits, it is proposed that clients (as project initiators) should maintain relationships with other potential 'partners' as indicated in Figure 7.2.

It is also proposed that clients maintain databanks that contain performance records against both 'hard' and 'soft' factors of their current and potential 'partners' along the supply chain. 'Hard' factors include items such as 'pricing levels' and 'timely completion of previous projects', whereas 'soft' factors include items such as 'attitude towards teamworking-teambuilding' and 'approach to negotiation'. The scores obtained from the hard factors are to show the competency of specific partners, while those originating from soft factors are to justify the compatibility of partners to work with each other. The target is to:

- Select the most suitable team players in each category (for example, consultants, contractors, subcontractors and suppliers) for a particular project.
- Build the 'optimal' project team.
- Build the project team based on previous performances and working relationships.

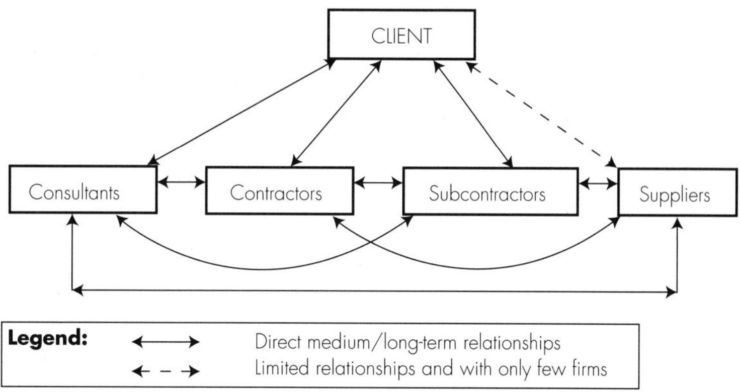

Notes:

1. Each contracting party will have its own supply chain linkages, which are not necessarily the same as others.
2. A particular category of contracting party may have relationships among themselves (for example, consultants with consultants, as in forming JVs), despite being potential 'competitors'.
3. Client may also have relationships with other clients for some projects.
4. In the long run, there can be a centralised databank maintained by the government/large clients where other interested parties may enter and search for their potential partners.

Fig. 7.2 Basic structure of a relationally integrated supply chain (RISC).

Consideration of the above points will also include an estimation of potential compatibilities for the relational integration and resulting synergies in the new project scenario (Rahman, 2003).

In the proposed RISC, other 'partners' would also maintain their own databanks. For example, contractors would maintain databanks of their subcontractors and suppliers and also of consultants with indicators of past and potential working relationships; the clients' databank would include all the parties, as some contractors work as subcontractors in some projects and vice versa, and some specialist suppliers' information is needed to nominate a particular supplier. Other contracting parties can also select partners for other teams, as in joint ventures (JVs) or alliances, from these databanks. Moreover, each party's databank would also include information about others in the same category: for example, other consultants' information may be kept and updated by a consultant for any future collaboration, as in forming a JV. Different parties from such RISCs will be able to work with common objectives in a particular project as one team. However, since they would probably compete on some other jobs some might also see a need to safeguard certain core strength 'business secrets'. Nevertheless, these approaches will generate opportunities between different contracting parties to practise long-term RC-based approaches, for example in strategic partnering and longer-term alliances.

The envisaged client-driven exercise of assembling a relationally integrated project team (RIPT) is graphically indicated in Figure 7.3. It begins after the feasibility study and along with the envisaged 'project analysis' indicated for strategy formulation at Stage 1. This is based on both the client's preferences

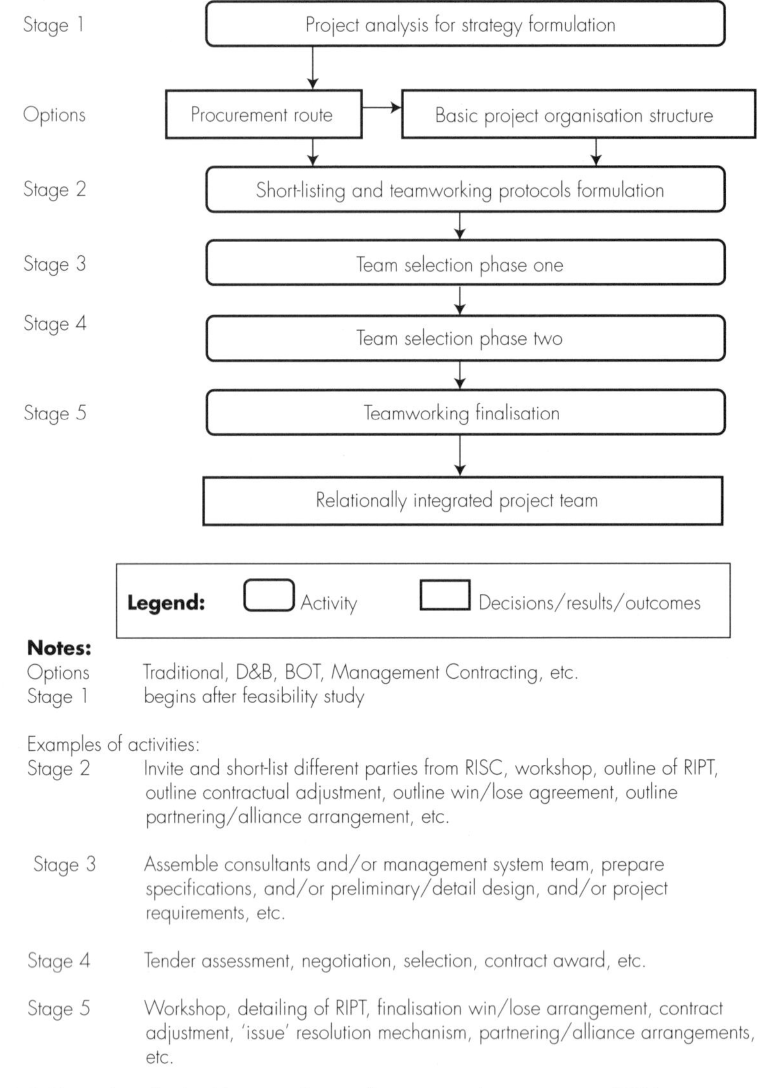

Fig. 7.3 Flow chart for building a relationally integrated project team (RIPT).

and project requirements. The outcomes of this analysis will lead the client to select a particular procurement system that may be familiar to the industry, or a hybrid of two or more different, but commonly used, procurement systems – with a project organisation that is best suited to the project and client preferences. Stage 1 will also guide the client to:

- Decide on the required 'quality' of different partners of the project team and the timing of their mobilisation.
- Prepare the contract documents incorporating 'proper' allocation of all foreseeable risks and required flexibility or adjustment mechanisms for handling unforeseen events.

- Arrange any additional teambuilding protocols for boosting relational integration, for example partnering workshops.

A few of each category of contracting party – for example, consultant, contractor – from the databank are proposed to be short-listed, based on their 'performance scores' against both hard (technical) and soft (relational) factors, and invited to bid for joining the project team. Since the previous performance levels against different soft factors of these potential team players are already known from the databank, they may be selected on the basis of hard factors at this stage. Selected team players will be mobilised depending on the project requirements and project organisation. Contractors will select their respective subcontractors and suppliers, but in consultation with the client. The client's consultation will be from the viewpoint of assembling the 'best' possible RIPT.

The main purpose of Figure 7.2 is to disseminate the outcomes of Figure 7.1 to all potential team members for any feedback from, and as a part of, the overall teambuilding exercise. Activities at Figures 7.3 and 7.4 will depend on the specific procurement arrangement. Some procurement arrangements (for example, design-bid-build or design-build) will result in more activities in either of these two stages. However, all required activities such as teambuilding protocols (for example, partnering workshops), contract award, selection and mobilisation of 'partners', issue resolution mechanisms, gain-pain share arrangements (if any), and any other sub-groupings or operational arrangements will be finalised by Figure 7.5, leading to the RIPT.

Irrespective of the procurement system selected, the RIPT will be expected to dynamically and proactively address all uncertainties and any changes during project progress, using the best available options to meet the project objectives. Benefits/losses experienced in addressing such events will be shared among RIPT partners according to agreed mechanisms (Rahman, 2003). However, some projects may not be suitable for such practices. Moreover, RIPTs are specifically assembled to suit the overall project requirements and meet clients' preferences, which may vary from project to project. Therefore, in appropriate cases the degree of such incentivised relational integration may essentially vary from post-contract partnering-type arrangements to alliancing, through to vertical integration – as if all project team members belong to a single organisation.

Relevant pointers from a case study

It is useful to assess the value and practicality of the framework of RISCs and RIPTs. Although not yet directly applied in practice as described above, a precursor case study of a live project using collaborative working arrangements is being carried out on a project for extensive remodelling and improvement of a busy underground railway station. The client maintains lists of registered

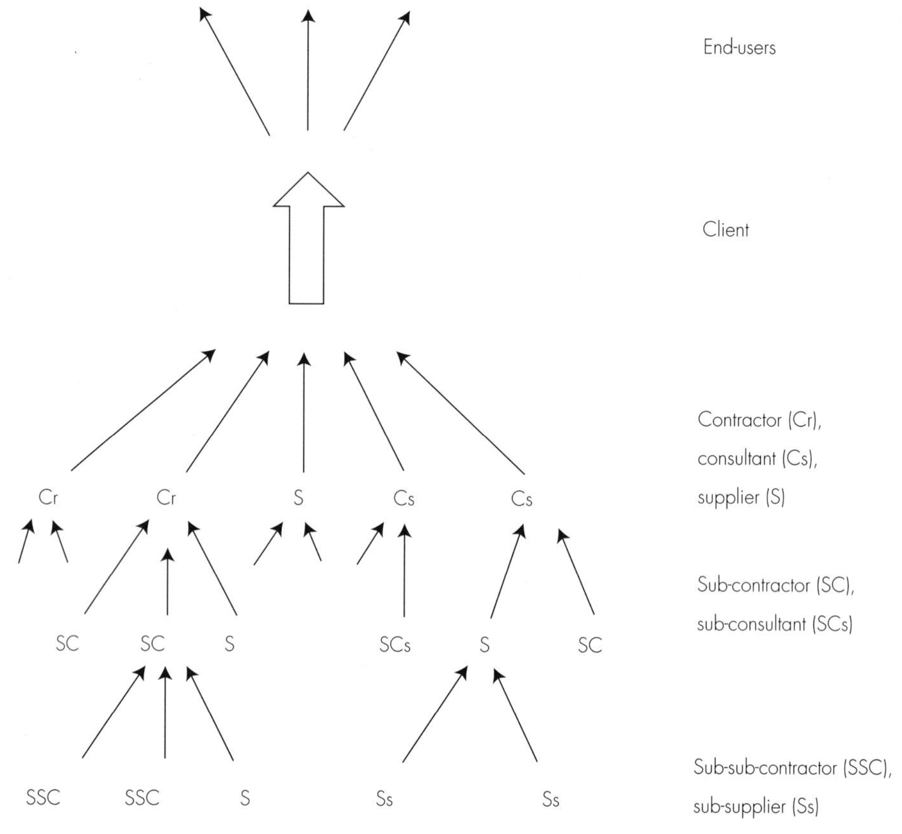

Notes:

1. Each arrow represents the principal 'added value' contributed by each project team member.
2. Only the principal value flow is shown. Complementary lateral flows that reinforce/add to the value of other team members are not shown here.

Fig. 7.4 Project value stream – a basic visualisation.

contractors, consultants and suppliers, and conducts periodic performance appraisals on them. Many of its smaller contractors and suppliers also work as subcontractors for other contractors. The case study also uses open prequalification for special projects, with its own set of criteria for registration and short-listing of different parties and various kinds of projects, for example greenfield, renovation and extensions. The client organisation usually uses consultants for design services and standard lump sum forms of contract for construction works, but manages projects by strong teams of in-house staff. A wealth of useful information has been collected from this ongoing project, but only the relevant extracts are presented here.

The project was conceived to improve facilities and extend, modify and alter the station, relieve the current and predicted congestion, and improve access to the station and passenger flow inside the station. It involves many

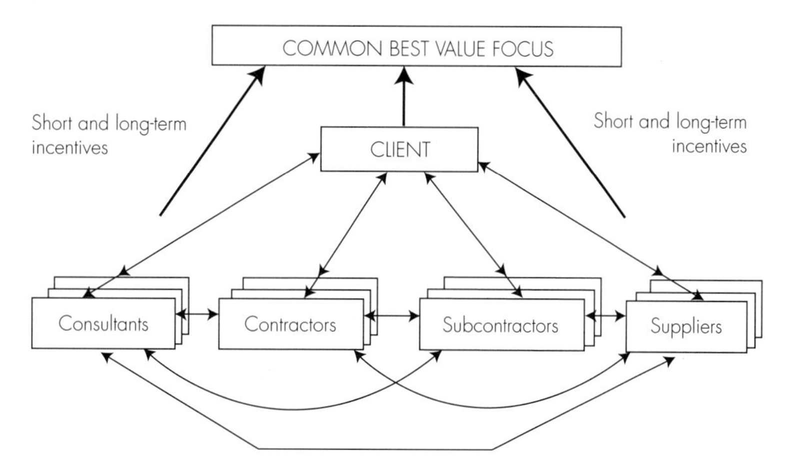

Notes:
1. Each contracting party (for example, a 'consultant') will have its own network, which is not necessarily the same as others.
2. A particular category of contracting party may have relationships among themselves (for example, consultants with consultants).
3. Client may also have relationships with other clients for some projects.
4. At a broader industry level: a centralised databank (with relevant information on all potential partners) can be maintained by a consortium of large clients (including government). Authorised parties may be allowed to access and search for potential partners.

Fig. 7.5 Proposed structure of a relationally integrated value network (RIVAN).

interfaces and critical risks, and the cost of temporary works exceeds that of permanent works. Based on the timing, duration and scope of works, it was envisaged that a single contract encompassing the majority of works should be awarded for the civil, building and mechanical and electrical services works. Specialist contracts were to be awarded separately, whereas various operational projects must proceed in parallel. Contractor cooperation was needed to minimise disruption to pedestrian movement, mitigate environmental issues, and to provide convenient access to ground-floor businesses, timely completion and controlled overall cost. The client's objective was to achieve the best overall solution without compromise to the safety of the railway, while offering a realistic balance between the programme and total project cost. This involved a unique approach, which began by changing the client's corporate policy towards enhanced collaboration, and outlining the basic features of an innovative contract, strategies for pre-qualifying the contractors and selection of the main contractor. It was decided to negotiate the content of parts of the contract documents jointly with the short-listed or selected contractors.

A target cost contract approach was adopted with information sharing and pain-gain share arrangements, early involvement of the contractor and targeted multi-stage value engineering. At the pre-contract stage, the client identified foreseeable and potential risks to the works. Risks were divided into three categories: client's risks, contractor's risks and shared risks. These formed a part of the bidding documents and provided a clear indication of the planned responsibilities to all bidders.

The screening or prequalification of contractors was accomplished by inviting expressions of interest from among a group of contractors who were working on other collaborative projects for the client. This approach can be compared with that suggested from the RISC above and many parallels noted. The expression of interest was invited in the format of a questionnaire booklet to assess contractors' ongoing performance on collaboration, to identify attitudes and abilities of the contractors toward a more collaborative working arrangement, and to estimate their relational compatibility to work in a coalescent team.

Apart from the above screening, the bidding process was arranged in two stages, to select two contractors after the Figure 7.1 exercise and the winning contractor at the end of Figure 7.2. Among the six pre-qualified contractors, five contractors submitted their bids in Stage 1. Along with provisional method statement and specific proposals on different technical issues, the contractors were asked to propose or quote a percentage for their maximum possible overrun against the client-supplied figure of 10% of the final target cost. They were also asked to quote fixed amounts for the costs of items such as direct labour and employees, site establishment, profit, head office overheads and contractor's risks, and a fixed fee for the service in the next stage, that is if selected for Stage 2. These formed the fixed cost element of the contract, which was carried forward to the tender target cost summary for Stage 2 and would not change during the contract period if the contract were awarded.

In addition, bidders were asked to identify any new or potential risks, including those originating from their methodology, and to allocate and price them accordingly. A budget estimate of the 'actual cost' of the works – temporary and permanent works – was calculated by the client's consultant engineer, and bidders were notified of this figure. Therefore, the contractors were not asked to bid on the 'actual cost' at this stage, but to quote profit and head office overheads based upon the estimate provided to them. The initial target cost was to comprise the sum of the client-provided estimate of actual cost and the contractor-quoted fixed cost. The bids were assessed against both technical and commercial criteria, and the two top ranked bidders were carried forward to Stage 2.

During the Stage 2 exercise, the two contractors from Stage 1 developed two 'comprehensive proposals' with many innovative solutions and extensive value engineering (VE) inputs. The client provided a 'referral panel' to explain the client's requirements and assist them in developing each comprehensive proposal; to review, assess and comment on various deliverables relating to that proposal; and to help the two contractors in several other agreed issues – while maintaining confidentiality and sincerity throughout the development of each comprehensive proposal. The contractors were informed that significant emphasis would be placed on a carefully assembled risk register that would consider method-related risks and suggest methods of avoiding or mitigating those risks. Although the client provided the anticipated date of completion for the works, the two contractors were invited to propose their own contract periods in order to strike a realistic balance between programme and total project cost. The lists of subcontractors, gain-pain share arrangements and several target obligations were agreed separately with the contractors. As a result of extensive VE, the estimate of the project was reduced by more than one-third of the client's consultant engineer's estimate by both the bidders. The bids were considered close, and the contract was awarded to the apparently very marginally more favourable bidder.

The project is ongoing as this chapter is being finalised. The contractor and the client's full project management team – along with their respective staff – are sharing the same office. A dedicated bank account in the name of the project is being maintained for an 'open book' arrangement. The contractor receives all the payments into this account and pays out to his subcontractors and suppliers, and meets other project-related costs from this account. The contractor maintains the account, but it is transparent to both parties, client and contractor. Joint decisions are being made on cash flow forecasting and adjustments. The contractor has a similar target cost and pain-gain share arrangement with his subcontractors.

The contract requires execution of all variations, changes and amendments to the contract through the process of VE proposals and any party to the target cost and gain-pain share arrangement can initiate a VE proposal at any time. These proposals are discussed monthly, but meetings can be and are arranged weekly depending on the nature, number and requirement of the VE proposals. There have already been over 200 accepted VE proposals. Although most of those arose from client's variations, both the contractor and the main subcontractor raised a considerable number of proposals themselves. In comparison to the total cost of the project the savings from these VE proposals is not considerable, but they provide useful additional value to the client from an already reasonably optimised scenario that had benefited from extensive VE at previous stages. Moreover, there is already a total saving of three weeks on the critical path and it is expected that the contract will be completed well ahead of schedule.

Relationally integrated value networks (RIVANs) for value-focused teamworking

Incentivised teamworking strategies and tools would be useful for empowering the implementation of the previously discussed RISC-RIPT framework and approaches as in the foregoing case study. In this context, RIVANs were first proposed by Kumaraswamy *et al.* (2003b).

The specific focus on *value* as a common objective drives the relational integration that was discussed as important in the previous section on RISCs and

RIPTs. Techniques and tools for mapping value, value chains and value streams have been proposed by others, for example Porter (1985), Kelly *et al.* (2002) and Hines *et al.* (2001). These are expanded upon below.

A firm's *value chain* was conceptualised by Porter (1985) as:

- A firm disaggregated into its strategically relevant activities
- Activities embedded in a larger stream of activities, which were then called a value system
- Activities integrated (through the above value system) in value chains of other organisations, for example the firm's suppliers (upstream value) and buyers (downstream value)
- Including primary activities (such as logistics, operations, marketing, sales and services) and support activities (such as firm infrastructure, human resource management, technology development and procurement)

Porter's *value chain* model was itself aimed at delivering competitive advantage to a commercial firm in general; Male (2001) applied the conceptualisation of a *project value chain* to construction scenarios, as did Kelly *et al.* (2002) who also illustrated the integration of various project value chains with the corporate value chain.

It is proposed to integrate the above approaches with the 'value stream mapping' for distribution organisations visualised by Hines *et al.* (2001). They identified a 'value stream' as a 'more useful unit of analysis than the supply chain or the individual firm'. Integrating their approaches with those discussed above, the 'added values' at each point of the construction value chain would contribute and converge into the ultimate value stream. This may be broadly visualised as in Figure 7.4, which shows typical project supply chain members at different levels, for convenience. The tributaries of added value should ideally merge synergistically to a confluence, integrating into a well-directed value stream that meets client requirements and user needs.

A well-formulated value stream focus is expected to empower the optimal integration of project teams and lead to resultant synergies through relationally integrated processes. Also, shifting from visualising less interconnected linear supply chains to recognising a cross-linked value network can help to map out and incentivise potentially divergent value streams towards convergence in consolidated high performance levels.

A research agenda has been proposed to develop and integrate these concepts into a visualisation of 'value networks' based on the flows in Figure 7.4, and with the basic framework in Figure 7.5 that has been developed from the RISC framework in Figure 7.2. This approach was outlined by Kumaraswamy *et al.* (2003b) and Kumaraswamy and Rahman (2003). Each of the twin-headed arrows represents a series of relationships, for example, between the many contractors and sub-contractors. Relational integration in each of these relationships needs to be structured and motivated by non-adversarial procurement frameworks, appropriate delivery systems, cultures and individual 'trust'.

Powerful tools such as *social network analysis* should be useful in analysing the networks of such inter-organisational relationships (Mead, 2001; Pryke, 2004a; see **Chapter 9** by Pryke), in the particular context of the value chains that would be mapped as above. Findings from such analysis could help to identify practical approaches to aligning the value focus of different sub-teams and team members. For example, these could include sub-team-specific incentive mechanisms as in the above case study and as described by both Bower *et al.* (2002) and the European Construction Institute (2003). In summary, it is seen that RIVANs would provide a sound framework and useful value-based focus for the envisaged TBO (teambuilding by objectives) approach that is proposed in this chapter.

Other relevant general teamworking models

Overview

Having developed in this chapter a basic proposal for value-focused teamworking in construction scenarios, it is worth comparing this with recent developments in other industries. For example, it should be worth tapping into experiential knowledge and the vast reservoir of literature on the dynamics of teamworking that has accumulated over the last few decades in other sectors. A relevant cross section is accessed here, with a view to providing some innovative applications for the construction industry. The strong individualistic and collective human elements that can drive or diminish teamworking processes point to the value of an initial overview of good practices and their theoretical foundations. Attempts are thus made to juxtapose the 'state of the science' with the 'state of the art', and to suggest ways forward to achieve 'higher states' in both dimensions. For example, Rippin (2002) has compiled a concise overview, starting with:

- *Team basics*: small numbers of people, complementary skills, common purpose, common set of specific-purpose goals, commonly agreed-upon working approach, mutual accountability
- *Types of team*: that 'do' or 'make' things; that 'recommend' things; and that 'manage' things
- *Characteristics of high performance teams*: high levels of communication, trust, optimism, high expectations of themselves, participation by all members, dedication to common goals
- *Differentiating teams from work groups*: groups also pool resources, knowledge and skills, but have a single leader; a team involves interdependent people with a clear collective vision, and while they may report to a manager they could function on their own for extended periods.

Drawing on group dynamics and organisational behaviour models

Given the commonalities, teamworking strategies can benefit from knowledge acquired on the workings of groups (Handy, 1993) and organisational dynamics (Mullins, 2002). For example, the development of 'informal groups' (such as those based on common background and interests), their interactions with formal groups and the emergence of 'dominant coalitions' may cut across organisational lines and communication channels, shortening and even 'short-circuiting' (and thereby disrupting) planned activities.

Also, progressive group development has been identified (Mullins, 2002) through phases of:

(a) Group forming – involves some uncertainties about what to do and how to interact
(b) Group storming – involves tensions, rivalries, conflict on interpersonal issues
(c) Group norming – some cohesion, establishing customs, rituals, jokes, values
(d) Group performing – focus on tasks

Mullins (2002) describes how such 'group norming' develops, thereby building up the group culture that will eventually contribute to the project culture, as previously projected in Figure 7.1. Informal groups and group dynamics may be used or abused, for example by applying group peer pressure for conformity.

Typical sets of group 'roles' that emerge in many groups have been identified as 'friend and helper', 'strong fighter' and 'logical thinker', while wider classifications include 'organiser', 'commentator', 'comedian' and 'deviant'. Other roles may emerge depending on the group type and activities, for example 'fixer' (trouble-shooter), and of course there may be some strong individual personalities.

Teamworking models

The development of teamworking requires that all members understand and subscribe to the common team goals and that their endeavours towards these goals are efficiently coordinated. The 'Belbin Teamworking Model' (Belbin, 2004a, 2004b) provides an interesting overview of theory and practice in moving beyond general group theory and organisational dynamics to specific teamworking needs and attributes. For example, Belbin (2004b) defines a team role simply as 'a tendency to behave, contribute and interrelate with others in a particular way'. An appreciation of such informal roles allows both team members and the collective team to benefit from each other, for example in responding rapidly to both internal needs and external pressures.

A nine-year UK-based study of the behaviour of managers worldwide (Belbin, 2004b), used psychometric tests and managerial exercises to track manager traits, styles and responses. The study identified clusters of

Table 7.1 Belbin team roles, contributions and weaknesses (Belbin, 2004b).

Belbin team-role type	Contributions	Allowable weaknesses
Plant	Creative, imaginative, unorthodox. Solves difficult problems.	Ignores incidentals. Too preoccupied to communicate effectively.
Coordinator	Mature, confident, a good chairperson. Clarifies goals, promotes decision-making, delegates well.	Can often be seen as manipulative. Offloads personal work.
Monitor evaluator	Sober, strategic and discerning. Sees all options. Judges accurately.	Lacks drive and ability to inspire others.
Implementer	Disciplined, reliable, conservative and efficient. Turns ideas into practical actions.	Somewhat inflexible. Slow to respond to new possibilities.
Completer finisher	Painstaking, conscientious, anxious. Searches out errors and omissions. Delivers on time.	Inclined to worry unduly. Reluctant to delegate.
Resource investigator	Extrovert, enthusiastic, communicative. Explores opportunities. Develops contacts.	Over-optimistic. Loses interest once initial enthusiasm has passed.
Shaper	Challenging, dynamic, thrives on pressure. Has drive and courage to overcome obstacles.	Prone to provocation. Offends people's feelings.
Teamworker	Cooperative, mild, perceptive and diplomatic. Listens, builds, averts friction.	Indecisive in crunch situations.
Specialist	Single-minded, self-starting, dedicated. Provides knowledge and skills in rare supply.	Contributes only on a narrow front. Dwells on technicalities.

behaviour, which it was found could be easily taken up by the commonly prevalent managerial personality types, and classified these types of behaviour into nine team roles (see Table 7.1).

Another exercise, based at the University of Nebraska–Lincoln (Varvel *et al.*, 2004), sought ways of developing more effective teams by identifying and using information on each member's psychological type. The 'Myers–Briggs Type Indicator' (MBTI) was applied to senior engineering design students in order to identify their various psychological types, based on Jung's classical classification (Jung, 1921). A team effectiveness questionnaire (TEQ) was developed for each team to rate its own effectiveness. In this exercise no significant correlation was found between psychological type dimensions and team effectiveness, but individuals were trained to appreciate the personality types of fellow team members and this training helped them to improve communication, trust and interdependence, which in fact are three attributes that had been previously identified as essential for effective teamworking. This confirms that while a knowledge of personality traits is important and can help slot members into appropriate roles, it needs to be complemented by over-arching strategies, for example formulating suitable goal-directed incentives that would drive the team towards higher performance.

While the Belbin model of team roles and the MBTI test are widely known, it may be noted that there are other classifications, such as those for mapping personality types. For example, the 'Enneagram' is reportedly used by big corporations, government bodies and universities in the USA for training leaders and staff (Knott, 2004). The corresponding 'enneatypes' are classified as:

- Perfectionist
- Helper
- Performer
- Artist
- Thinker
- Loyalist
- Epicure
- Leader
- Peacemaker

These personality types may be compared with the 16 MBTI types, from tests reportedly taken by over two million people each year (Knott, 2004).

Other approaches to measuring team performance are also worth exploring, for example as described by Telleria *et al.* (2004) in terms of general business and manufacturing processes. They reported on the ongoing development of a framework for the more systematic management of teams. This was targeted through a specific framework for performance measurement and management. The framework was in turn based on three central building blocks of business process, team and environment, and a 'deployment path' that guides the implementation, for example on how to align strategy, business processes and team performance.

Of course, applications in specific sectors such as construction, or indeed in sub-sectors or segments such as design organisations, would require adaptation and fine-tuning of the chosen model. In so doing, it would be valuable to also draw on and synergise with emerging concepts, best practices and experiential knowledge from those sectors, such as have been presented in relation to the construction industry in the previous sections of this chapter.

Concluding observations

Benefits can evidently be generated by effective and successful teams. Effectiveness is the extent to which a team is successful in achieving its task-related objectives. Successful teams are said to achieve:

- A wider range of ideas than individuals working in isolation, and better decisions
- Improvements in participants' confidence, attitudes, motivation, and personal satisfaction

- Greater clarity in expressing ideas through group discussion
- Greater optimism by focusing on positive outcomes and putting less weight on problems
- More effective responses to changes as improved trust and communication help a team to adapt
- Better understanding by team members of the nature of their individual contribution and of the needs of other team members
- More efficient use of resources, especially time (*Constructing Excellence*, 2004)

This chapter attempts to compare relevant elements of the 'state of the science' in teamworking models to the 'state of the art' in innovative good practices of teamworking, and thereby to develop and promote better approaches to applying teamworking models to construction projects for enhanced performance. It also attempts to unravel the intricacies of trust-based relationships, to model the formation of a 'project culture', to mobilise the benefits of *relational contracting* (RC) and to convert all of these into commercially incentivised and beneficial teamworking relationships that add significant value to the outcomes for each party.

Furthermore, the chapter introduces innovative concepts such as *team-building by objectives* (TBO), as well as expanding and integrating recently proposed models of *relationally integrated supply chains* (RISCs), *relationally integrated project teams* (RIPTs) and *relationally integrated value networks* (RIVANs), with a view to enhancing and integrating approaches to better teamworking and improved project performance.

References

Belbin, R. M. (2004a) *Management Teams: Why They Succeed or Fail* (2nd edn). Elsevier Butterworth-Heinemann, Amsterdam.

Belbin, R. M. (2004b) *Belbin Team-working Model*, http://www.belbin.com/meredith.html.

Bower, D., Ashby, G., Gerald, K. & Smyk, W. (2002) Incentive mechanisms for project success. *ASCE Journal of Management in Engineering*, **18** (1), 37–43.

Constructing Excellence (2004) http://www.constructingexcellence.org.uk.

Egan, J. (1998) *Rethinking Construction*. HMSO, London.

Emmerson Report (1962) *Survey of Problems before the Construction Industries*. Ministry of Works, HMSO, London.

European Construction Institute (1997) *Partnering in the Public Sector*. ECI, UK.

European Construction Institute (2001) *Partnering in Europe: incentive based alliancing for projects*. ECI, UK.

European Construction Institute (2003) *ECI ACTIVE Manual of Value Enhancing Practices for Small Projects*. ECI, UK.

Handy, C. B. (1993) *Understanding Organisations* (4th edn). Oxford University Press, New York.

Hauck, A. J., Walker, D. H. T., Hampson, K. D. & Peters, R. J. (2004) Project alliancing at National Museum of Australia – collaborative process. *ASCE Journal of Construction Engineering and Management*, **130** (1), 143–52.

Hines, P., Jones, D. & Rich, N. (2001) Lean logistics. In: A. M. Brewer, K. J. Button & D. A. Hensher (eds) *Handbook of Logistics and Supply Chain Management*. Pergamon, London.

Jay, A. (2004) Foreword. In: R.M. Belbin *Management Teams – Why They Succeed or Fail* (2nd edn). Elsevier Butterworth-Heinemann, Amsterdam.

Jung, C. G. (1921) Psychological types (trans H. Godwyn Baynes, 1923). In: *Classics in the History of Psychology*. Internet resource developed by C. D. Green, York University, Toronto. http://psychclassics.yorku.ca/Jung/types.htm.

Kelly, J., Morledge, R. & Wilkinson, S. (eds) (2002) *Best Value in Construction*. RICS and Blackwell, London.

Knott, K. (2004) Personality portraits. *South China Morning Post*, 9 Feb. 2004, C7, Hong Kong.

Kumaraswamy, M. M. & Rahman, M. M. (2003) Relational contracting, joint risk management and relationally integrated value networks. Workshop on *Integrated Risk Management*, Netherlands Ministry of Housing, Spatial Planning and the Environment and the Government Buildings Agency, Hague, 30 June, p. 19 of slides in handout.

Kumaraswamy, M. M., Palaneeswaran, E. & Humphreys, P. (2000) Selection matters – in construction supply chain optimisation. *International Journal of Physical Distribution and Logistics Management*, **30** (7/8), 661–80.

Kumaraswamy, M. M., Rahman, M. M., Palaneeswaran, E. & Ng, S.T. (2002a) Innovative initiatives in integrating construction supply chains. *Proceedings of 1st International Conference on Construction in the 21st Century*, Miami, April.

Kumaraswamy, M. M., Rahman, M. M., Palaneeswaran, E., Ng, S. T. & Ugwu, O. O. (2003b) Relationally integrated value networks, *Proceedings of 2nd International Conference on Innovation in Architecture, Engineering and Construction*, June, UK.

Kumaraswamy, M. M., Rowlinson, S. M., Rahman, M. M. & Phua, F. (2002b) Strategies for triggering the required 'cultural revolution' in the construction industry. In: R. Fellows & D. Seymour (eds), CIB TG29CIB publication no. 275.

Kumaraswamy, M. M., Yeung, N., Sze, E., Law, S. & Rahman, M. M. (2003a) Knowledge-building for successful partnering. *Proceedings of the Joint International Symposium of CIB Working Commissions*, 22–24, October 2003, Singapore, **1**, 468–80.

Latham, M. (1994) *Constructing the Team: Joint Review of Procurement and Contractual Arrangements in the United Kingdom*. HMSO, London.

Lighthouse Club (2004) *Collaboration not Confrontation in Executing Construction Contracts*, Proceedings, 21 May, Hong Kong.

Macneil, I. R. (1974) The many futures of contracts. *Southern California Law Review*, **47** (3), 691–816.

Male, S. (2001) A cost of value orientation? Aligning project cost management within the project value chain. Keynote address, *International Conference on Project Cost Management*, 25–27 May 2001, Beijing, HKIS.

Management Briefing (2002) Teamworking is not rocket science. *New Civil Engineer International*, April, 36–7.

Mead, S. P. (2001) Using social network analysis to visualise project teams. *Project Management Journal*, **32** (4), 32–8.

Mullins, L. J. (2002) *Management and Organisational Behaviour* (6th edn). Financial Times/Prentice Hall, Harlow.

Palaneeswaran, E., Kumaraswamy, M. M., Rahman, M. M. & Ng, S. T. (2003) Curing congenital construction industry concerns through relationally integrated supply chains. *Building and Environment Journal*, **38** (4), 571–82.

Porter, M. E. (1985) *Competitive Advantage: Creating and Sustaining Superior Performance*. Free Press, New York.

Pryke, S. D. (2004a) *Twenty First Century Procurement Strategies: Analysing Networks of Inter-firm Relationships*. RICS Research Paper Series, RICS Foundation, London, **4** (27), 38.

Pryke, S. D. (2004b) Analysing construction project coalitions: exploring the application of social network analysis. *Construction Management and Economics*, **22** (8), 787–97.

Rahman, M. M. (2003) *Revitalising construction project procurement through joint risk management*. PhD thesis, University of Hong Kong, Hong Kong.

Rahman, M. M. & Kumaraswamy, M. M. (2002a) Joint risk management through transactionally efficient relational contracting. *Construction Management and Economics*, **20** (1), 45–54.

Rahman, M. M. & Kumaraswamy, M. M. (2002b) Risk management trends in the construction industry. *Journal of Engineering, Construction and Architectural Management*, **9** (2), 131–51.

Rahman, M. M. & Kumaraswamy, M. M. (2004a) Contracting relationship trends and transitions. *ASCE Journal of Management in Engineering*, **20** (4), 147–61.

Rahman, M. M. & Kumaraswamy, M. M. (2004b) Potential for implementing relational contracting and joint risk management. *ASCE Journal of Management in Engineering*, **20** (4), 178–9.

Rahman, M. M., Kumaraswamy, M. M., Rowlinson, S. & Palaneeswaran, E. (2002) Transformed culture and enhanced procurement: through relational contracting and enlightened selection. In: T. M. Lewis (ed.) *Proceedings of the CIB Joint Symposium on W092, W063, TG36 and TG23*, January 14–17, Trinidad & Tobago.

Rahman, M. M., Kumaraswamy, M. M. & Ng, S. T. (2003) Re-engineering construction project teams. In: K. R. Molenaar & P. S. Chinowsky (eds) *Winds of Change: Integration and Innovation in Construction*. Proceedings of ASCE Construction Research Congress, 19–21 March, Hawaii, USA, CD Rom.

Rahman, M. M., Palaneeswaran, E. & Kumaraswamy, M. M. (2001) Applying transaction costing and relational contracting principles to improved risk management and contractor selection. *Proceedings of the International Conference on Project Cost Management*, May 25-27, Beijing, HKIS.

Rippin, A. (2002) *Teamworking*. Capstone Publishing, Oxford.

Strategic Forum for Construction (2002) *Accelerating Change*. Construction Industry Council, London (also at http://www.strategicforum.org.uk/pdf/report_sept02.

Sze, E., Kumaraswamy, M. M., Wong, T., Yeung, N. & Rahman, M. M. (2003) *Weak Links in Partnering Supply Chains? Consultants' and Subcontractors' Views on Project Partnering*. 2nd International Conference on Construction in the 21st Century, Hong Kong.

Telleria, K. M., MacBryde, J. C. & Bititci, U. S. (2004) *Facilitating Team Performance Management*, Centre for Strategic Manufacturing, University of Strathclyde, Glasgow, http://www.smartlink.net.au/library/teams.htm.

Varvel, T., Adams S. G., Pridie, S. J. & Ulloa, B. C. R. (2004) Team effectiveness and individual Myers-Briggs personality dimensions. *ASCE Journal of Management in Engineering*, **20** (4), 141–6.

Williamson, O. E. (1985) *The Economic Institutions of Capitalism*. Free Press, New York.

Wilson, V. & Pirrie, A. (2000) *Multidisciplinary Teamworking – Indicators of Good Practice*. September, Scottish Council for Research in Education (SCRE), http://www.scre.ac.uk and *Multidisciplinary Teamworking – Beyond the Barriers? A Review of the Issues*. September, SCRE, Glasgow.

8 Managing project risks

Martin Loosemore

The rather immature field of risk management is best described as a spectrum of competing doctrines. Although numerous schools of thought have emerged to describe how risks should be identified, measured and managed, it is possible to identify two broadly competing doctrines – *homeostatic* and *collibrationist* (Hood & Jones, 1996). The *homeostatic* view represents current orthodoxy in risk management practice and advocates a scientific approach to risk management. This emphasises rationality in decision-making, prevention rather than cure, anticipation rather than reaction, blame rather than forgiveness, objectivity and quantification rather than subjectivity and qualitative measures, independence rather than interdependence, elitism rather than collectivism, confinement rather than consultation, and structures and controls rather than people and processes. Herein lies the traditional actuarial world of the risk management industry with strong traditions in the mathematics of probability. It is an approach that also dominates the construction industry (Loosemore, 2000).

In contrast, those that have a *collibrationist* perspective argue that people are not necessarily rational in responding to risks, but perceive risks according to cultural and social networks in which they are embedded. Furthermore, they believe that reliable forecasting is impossible in many business areas and that the variety of stakeholders in organisations prevents aggregate goal setting that can be translated into precise technocratic decision rules. The collibrationist view emphasises consultation, collective responsibility and social responsibility in the management of risks. It also argues that in an increasingly uncertain world, organisational resilience – an ability to respond to problems and opportunities after they have arisen – is as necessary as prevention. While prevention is better than cure, in reality it is impossible to create a crisis-free environment, which means that an overemphasis on prevention is misguided, inefficient and even dangerous.

If the homeostatic approach relies heavily on the mathematics of probability, the collibrationist might seek to manage project outcomes in a context of human irrationality in relation to perceptions and responses to risk. The management of risk in a collaborationist world must place considerable emphasis on human relationships within the project environment.

The issue of rationality

Underlying the two broad positions is the oldest and most enduring controversy in risk management – the issue of *rationality*. More than any other development in business, it is the rise of *scientific rationalism* that has distinguished modern managers from their predecessors. However, Bernstein (1996) questioned whether we have gained from this increased reliance on science where probability analysis has supplanted hunches and intuition in many areas of business. It is indisputable that the development of scientific techniques has enhanced our quality of life by allowing businesses to take more risks. Yet there is a downside. In particular, the scientifically driven devices of modern business tend to underplay the importance of people in organisations (Richardson, 1996). Furthermore, businesses often fall victim to the mathematical regularity that logisticians portray, failing to plan for the unpredictability that often lies in wait in the real corporate world (Berry, 2000).

Consider, for example, the months preceding the stock market crash of early 2000, when unpredictable and vicious price swings in stocks followed many months of illogical investment in high-tech companies, which became grossly overvalued in comparison to their asset values and profit potential. It is impossible for scientists, computers or mathematical models to predict such irrational consumer behaviour with accuracy and it is foolish to believe that we can attribute reliable figures and numbers to it. These types of events remind us that people form their own subjective perceptions of risk, which often differ from the objective assessments made by experts and scientists. They illustrate that ordinary *perceptions* play an important part in the risk management process and cannot be discounted on the basis that they are uninformed, irrational, biased and subject to error. To the people that hold these perceptions, it is the objective assessments that are irrational and, ultimately, there is no other way for managers to interpret risks than in terms of human values and emotions.

Risk perceptions

Traditionally, society's explanations of risk events relied upon myth, religion, metaphor and ritual at a community level (Barnes, 2002). However, in modern times the evolution of science has elevated explanations beyond the realm of folklore and superstition to become rational and objective. The transfer of responsibility for risk from the community to scientists and experts has led to the emergence of risk management as a profession in its own right. This profession has grown rapidly as the state increasingly offloads to the private sector its traditional responsibility for risk, and the professionals who now manage it have become a technical elite that has developed complex and specialised vocabularies, concepts and techniques of management.

Nevertheless, while risk managers have become more sophisticated in their approach to managing and measuring risk, the majority of the public continue to rely on cultural and social explanations of risk events, leading to significant perceptual differences between the public and the private business sector. These differences lie at the core of the increasingly prominent and frequent conflicts between construction companies and an increasingly empowered public. In many companies there remain significant institutional 'blind spots', ignoring the contextual experience of risk and the perceptual issues that are relevant to public concern.

Furedi (2002) argues that the growing public interest in the concept of risk has been driven by the 'politics of fear', fuelled by the media's negative reporting of the risks to society posed by big business. Over the years, media coverage of badly managed public health crises such as the Exxon Valdez oil spill in the US, the Bhopal gas leak in India, the BSE crisis in the UK and SARS have heightened consciousness of business-related risk. In this environment of increased vulnerability and mistrust of the private and public institutions responsible for the control of risk, managers can expect a much more informed and empowered response to incidents than their predecessors. The management of people's behaviour will not be made easier by the fact that the response to an incident is likely to be shaped not so much by the objective facts surrounding a risk or incident but by the deeper consciousness that prevails in society at that point in time. For example, in the post-September 11 2001 era, managers can expect people to react far more irrationally to security risks than before.

Today's society is more risk averse and safety conscious than ever before and any organisation dealing in risk must understand that it must also deal with the public's perception of it. In particular, high-risk industries like construction, which have a relatively poor public image, are likely to have their capacity for growth and innovation severely curtailed if they do not.

Risk communication

The key to better managing the public perception of risk is fundamentally one of communication, consultation and involvement in decision-making. It is also about identifying and understanding the stakeholders in a business – who they are, what their needs are and what is likely to influence their perceptions of risk. It is particularly important to realise that objective, actuarial and technocratic expressions of risk, although easier to operationalise, have little meaning if they are separated from the social and behavioural context in which risk is experienced by the public.

The need to consider the social and cultural aspects of risk does not mean that managers should abandon technical risk analysis. If reliable data is available on risks, which can be meaningfully and accurately measured, then it is right to use it. The point is that risk management should not be confined to objective outcomes and measures, but should also take account of people's perceptions of risk. Individuals respond to decisions according to their perception of risk and not according to the objective levels associated with it. In other words, community consultation must be recognised as a crucial part of effective risk management and decision-making and managers should appreciate that expert assessments of risk are relevant only to the extent that they are integrated into individual perceptions. For example, contrary to what most construction companies would believe, the establishment of safer projects is not dependent entirely on technical advances, but also on understanding what safety means to the stakeholders and communities involved in those projects. This means that the emphasis of management strategies should not solely be on the development of technical solutions to safety risks, but also on a better understanding of how project stakeholders live their lives and perceive and cope with the uncertainties created by construction activities.

This need for meaningful community consultation about risk has been an aspect of risk management literature for over a decade. We have begun to develop an understanding of how lay people think about risk, what shapes their perceptions and how best to manage them. For example, we have learnt that when communicating with stakeholders it is important to ensure that the language used assists communication and is not a medium of dominance disguised by indecipherable jargon (Bowden *et al.*, 2001; Edkins & Millan, 2003). Furthermore, we know that any communication needs to be based on a mutual sense of respect for the positions and perspectives of all parties (Kasperson & Kasperson, 1996). Finally, we have learnt that risk management should be a multi-way process that is designed to make risk management accessible to everyone and to promote mutual understanding, if not consensus (Barnes, 2002). Following these three simple principles does not guarantee the elimination of conflict between organisations and their stakeholders, but it does offer an opportunity for people to become part of the solution rather than part of the problem.

Understanding risk perceptions

The chasm of mistrust that has emerged between the general public and certain parts of the business sector in response to past approaches to managing risk is one of the greatest challenges facing managers today (Barnes, 2002). In contrast to those of the past, where managers focused on seeking explanations for the public's misunderstanding of 'real' risks, contemporary risk management strategies must seek to understand why the public perceives risks in the ways it does in order to accommodate those perceptions. It is increasingly being recognised that, in a democratic society, all stakeholders affected by a development have a right to be considered in the decision-making processes affecting their interests. It has also become evident that failure to gain a comprehensive appreciation of the nature of public perception and concern with risk renders attempts to manage risks at best ineffective, and at worst magnifies risks and becomes counterproductive (Renn, 1996). The project management team needs to understand and manage risk as perceived by project stakeholders. There is also a need to recognise that project risk perceptions affect those comprising the project team. Project team responses to risk represent an aggregation of individual project actors' risk perceptions and motivations towards those perceived risks. The following section reviews chronologically the various perspectives that have been taken on this issue.

The economic perspective

Initially, the economic perspective assumed that people respond rationally to hazards on the basis of comparisons of costs and benefits, reacting best to the path of maximum potential benefit (utility) to them personally (von Neumann & Morgenstern, 1953; Willett, 1951). This implied that managers should consider the perceived impact of business decisions on the personal objectives of the different stakeholders affected by their decisions. Practically, it also meant reducing reliance on traditional risk assessment methods. These methods aggregate data over large segments of a population affected by a decision, focusing more on strategies tailored to individual stakeholder groups.

The psychological perspective

The psychological perspective sought to explain the observation that people do not always behave rationally by basing their risk judgements on expected values (Argyris, 1990; Tversky & Kahneman, 1981). This research revealed a whole range of biases that cause people to attenuate or amplify a risk. For example, it was found that information about a risk that challenges current beliefs tends to be downplayed, a process called *cognitive dissonance*. It was also found that recent events, which come to people's minds easily, tend to be rated as more probable than distant memories, and this is *availability bias*. Higher probabilities tend to be attributed to events for which information is available or when the source is perceived to be significant, an *anchoring bias*. The value of the psychological perspective was the revelation that these biases were often the underlying cause for seemingly irrational public responses to business decisions. Furthermore, it emphasised for the first time the importance of giving as much attention to stakeholder beliefs as to the data relating to risks.

This is critically important, because evidence suggests that people who believe they have been exposed to risk, even if they have not, tend to exhibit some kind of stress response. Using an historical example to illustrate this point, the effects of stress among residents living near Three-Mile Island nuclear reactor persisted for a number of years after the reactor accident, even though the scientific evidence and data indicated that the risk of radiation was no greater there than anywhere else in the US. So it is not just the perceived level of danger that is important but the perceived degree of control over that danger. A lack of control over a situation is enough in itself to initiate a seemingly irrational behavioural response.

The sociological perspective

Risk is an interactive phenomenon and people are influenced by the perceptions of others around them, making sense of the world in collaboration with other people in social situations. This reliance on community as a source of perceptions is amplified when there is not much information or knowledge about a risk or when there is mistrust in external regulators who have the responsibility to provide that information. The implication for managers is that one cannot understand and manage the reaction to a decision without understanding how a community functions, and that these reactions must be engaged and considered by decision makers if perceptions are to be managed effectively. For example, in many developing countries poverty, a lack of education and corruption often suppress the behavioural response to development activity, which is not the case in Australia or the UK, where most people have satisfied their basic needs. For this reason, social debate about a development is often much more intense outside immediately affected communities than it is within them.

Balfour Beatty's experiences on the ill-fated $1.4 billion Llisu Dam project in Turkey provide a vivid insight into the potential for large projects to have enormous ripple effects that reverberate around the world. As a result of this project, the company attracted considerable criticism from environmental groups and was targeted by Friends of the Earth, working with Kurdish groups who claimed that the dam would have made 30,000 Kurdish people homeless and destroyed historic cities. Protestors also dug into the firm's other worldwide activities, threatening to provide substantial negative publicity for the company and forcing 40% of its institutional investors to withdraw backing for the project (Richards, 2002). Other companies involved in the project received similar attention. For example, once Friends of the Earth noticed Amec's involvement in the Llisu Dam project they went on to campaign against their involvement in many other projects, including road-building projects in the UK, another dam in Belize and paper mills in Indonesia. This made Amec a risky stakeholder to have involved in any large project – a reputation no company can afford to have!

The sociological perspective also reiterates the critical importance of risk communication to the risk management process (Kasperson & Kasperson, 1996). It recognises that many people do not learn about risks through direct experience, but through an array of information systems such as the media, social institutions and organisations responsible for managing risk, pressure groups and informal personal networks of friends, neighbours and colleagues. Particularly important in shaping perceptions is the extent of media coverage, the volume of information provided, the way in which risk is framed and the symbols, metaphors and discourse used in depicting and characterising a risk. The rhetoric, which can result from the attempts of each stakeholder to grab the decision maker's attention, can be overwhelmingly complex and difficult to align, and this is one of the great challenges facing managers today.

The cultural perspective

Finally, and most recently, cultural approaches to risk have helped managers understand how different stakeholder groups develop specific cultures (that is, shared beliefs, values and ways of seeing the world). Cultural theory was developed by Douglas (1966), expanded by Douglas and Wildavsky (1983) and further theorised by Thompson and Wildavsky (1990). The essence of cultural theory is that people form into groups of common objectives and perceptions, assigning particular meanings to risk events. That is, people rely on patterns of habit and socialised reinforcement of their values and behaviours in order to make sense of the world. In this way, risks are perceived and responded to according to principles that are embedded in particular forms of social organisation. For example, from a cultural perspective, arguments about a construction project would not just be concerned with choosing a safer technology, design or production process, but would be linked to fundamental questions about the social and political meaning of technologies to societies and in their broader societal implications.

Case study – Multiplex Facilities Management's approach to community consultation

This is a case study of Multiplex Asset Management's relationship approach to risk management. It documents the process by which Multiplex Asset Management (a construction company) developed a new risk management system based on a critical analysis of traditional technocratic approaches to risk management in the construction industry. The philosophy underpinning Multiplex Asset Management's system is that risk management should consider risk in its full social and cultural complexity. This requires greater community consultation and mechanisms to consider the wide variety of different interests that have a legitimate role to play in the development process.

The Multiplex Group is a listed international construction company, based in Australia, with an annual turnover of over AU$2 billion. It is a major force in construction across South East Asia, the Middle East and more recently the UK. Multiplex's culture revolves around its willingness and ability to successfully undertake high-risk projects and it has been responsible for the construction and management of some of the world's most famous landmark buildings. For example, it built the Olympic Stadium for the Sydney 2000 Olympic Games and was contracted to build the new Wembley Soccer Stadium in the UK. The way in which the Wembley project has unfolded could be analysed in terms of the need to extend risk management in the form of engagement into its UK supply chains in terms of culture and modes of operation.

Multiplex Facilities Management (MFM) is a subsidiary of the Multiplex Group and was established in 1998 to offer developers, building owners and investors an integrated range of facilities and asset management services. MFM extends the services offered by the Multiplex Group to the entire building life cycle – from inception, through design and construction, to the ongoing use and management of the built facility. MFM's business is based on the increasing recognition by many property owners that the land, buildings and support infrastructure owned by an organisation is a vital resource that makes an important contribution to customer core business objectives. The role of MFM is to ensure that it effectively manages its customers' facility-related risks to ensure that their real estate assets support and contribute to core business objectives.

The start of MFM's journey

The development of a new approach to risk management is a journey, not a destination. This case study is about an ongoing process that started over two years ago. At that time, MFM had many good reasons to review and improve the effectiveness of its risk management procedures. For example, MFM existed in a high-risk environment and was managing some of Australia's most prominent landmark buildings. The company had also grown rapidly since its formation in 1998 and the directors were aware that they needed to keep updating their existing management systems to keep pace with this growth. In addition, MFM existed in an increasingly legislative and competitive business environment where the penalties for non-compliance were becoming more severe. This was particularly so in occupational health and safety and environment management. A reputation for non-compliance could very easily destroy good customer relations and public relations, leading to a loss of business, increased insurance premiums and increased finance costs. In parallel with this trend towards increasing legislation, the world of corporate responsibility was also emerging fast, as the public became more sensitive to the impact of the construction industry's practices on their lives, and more empowered to do something about it.

This trend towards greater moral, ethical and social responsibility in business has continued and it is now estimated that over 35% of all decisions made on the UK stock exchange are based on a company's non-financial performance. Similarly, the US Dow Jones Index has an in-built sustainability index to guide investors, and in Australia the recently published ASX Corporate Governance Guidelines (the system by which Australian listed companies should be managed and directed) promote accountability and control systems to manage the social risks involved in achieving corporate objectives. While such guidelines only apply to listed companies at the moment, it is likely that they will be increasingly used and adapted by financiers in assessing the reputational risks associated with non-listed companies in the future.

In addition to the above, MFM was seeing the character of its customer base changing, its clients becoming more informed, more sophisticated, more demanding and more risk averse. In particular, those large and potentially lucrative clients who themselves had sophisticated risk management systems were expecting the same standards of their business partners and were insisting on risk management audits as part of the pre-qualification process. MFM's future involvement in public private partnership (PPP) projects presented a particular challenge with significantly greater levels of new risk being accepted over extremely long periods of time. Being an organisation that considered itself customer focused, MFM inevitably found itself facing the prospect of taking new risks, which were challenging to manage and control.

One of the major problems facing MFM, peculiar to the emerging profession of facility management, was the need to manage risks that had been passed 'down the line' by contractors and designers, who misunderstand the facility management business and over whom there had traditionally been little control. Unlike Multiplex Construction, MFM were also exposed directly to the general public who used their buildings or were affected by them. The MFM directors realised that eventually, if risks were going to be effectively and fully managed, they would also have to change practices along the entire MFM supply chain and involve stakeholders in the decisions made. This meant that any new risk and opportunity management system – opportunity linked to risk on the basis that they are often indistinguishable, particularly in the early stages of identification – would have to be suitable for use in all stages of the construction process from inception, through feasibility, design, planning, construction, facilities management and redevelopment, and that extensive consultation with people involved in these stages would be necessary.

In summary, MFM saw risk management as a *core business function*, a crucial capability that would provide it with confidence to move forward into an increasingly competitive business environment, taking risks and opportunities that its counterparts did not have the confidence to manage. MFM also recognised that risk and opportunity management should be an integral part of good management practice and that the benefits could be enormous in terms of improved efficiency, better performance, increased competitiveness and higher profitability. Indeed, research at the time indicated that these opportunities were huge. For example, a survey of the UK's top 75 companies in the construction industry indicated that 57% regularly declined tenders on the grounds that projects were too risky (Smee, 2002).

Finally, MFM also knew that companies prepared to take a close and critical look at their risk management practices would be seen more favourably by insurers and financiers, potential and existing clients (particularly sophisticated clients), shareholders, the general public, pressure groups and the courts, if something did go wrong.

MFM's consultative approach to risk management

MFM adopted the classic definition of a stakeholder as being any person or organisation who can affect or be affected by the outcome of a decision

(Freeman, 1984). The organisation identified that the variety of potential stakeholders associated with an MFM project could be enormous, and could vary from decision to decision. For example, stakeholders may include:

- The client and its staff
- Any company with a stakeholder interest, its owners, shareholders and staff
- Customers of the products or services provided by the project's end product
- Users and tenants of the final end product
- Operators of the final end product
- Contractors, sub-contractors and suppliers
- Business partners
- Business counterparts
- Union groups
- The media
- Financial and insurance institutions
- Legal advisers
- Specialist consultants
- Regulatory, licensing and approval authorities and inspectors
- People who may be living near the project and affected by its activities
- The general public
- The government, politicians and community leaders who may be affected by associated employment opportunities
- Special lobby/interest groups relating to issues associated with the construction or operation of a facility, such as native rights, heritage, the environment, pacifists, animal rights
- Professional bodies

Collectively, the stakeholders listed above represent the construction supply chain, along which there is typically little sense of collective responsibility for the management of risks. In the construction industry, the dominant culture is one of self-preservation and survival of the fittest – or more precisely, the most powerful. MFM realised that this selfish approach to the management of risk and opportunities is in no one's interests. A chain is only as strong as its weakest link and in a construction project interdependencies are so strong that everyone's interests are ultimately linked. There was little point in MFM managing its own risks if it could not manage those that arose along its supply chain – from customers to suppliers.

MFM decided that the best way to achieve collective responsibility for risk management was to insist on the same standards and principles of risk management from their supply chain. They involved key stakeholders and business partners in the development of their new risk management system, so that they understood what MFM wanted to achieve and what type of behaviour was expected from them. Also, the risk management system would need to incorporate extensive consultation arrangements, a requirement that everyone was expected to look out for and notify everyone else's risk and opportunities – as well as their own. Furthermore, MFM would require every party to an MFM contract to demonstrate compliance with the risk management system. In this way the risk management system would also drive supply chain reform. There was a belief that if MFM's business partners could not live up to MFM's high standards of risk management, then perhaps MFM should not be working with them at all.

MFM also consciously decided to measure its success against its customers' objectives rather than against its own objectives. Few companies do this and the idea was that if MFM could keep its customers happy, then its business would automatically be successful, profitable and healthy. MFM also broadened its idea of who its customers were and ensured that it had extensive stakeholder consultation guidelines and requirements in the risk management system. The risk management system includes a simple stakeholder management tool, which allows managers to identify and categorise stakeholders at the start of the risk management process and consult them in an appropriate manner according to their interest in a decision outcome and their ability to influence it. MFM recognised that not only was it an increasingly stringent legislative imperative to consult stakeholders, but that badly managed stakeholders could be an enormous source of disruption and costs to its business. In contrast, well-managed stakeholders could be an enormous source of support, of expertise and information that would help MFM manage its risks and opportunities more effectively.

One of the biggest barriers to effective stakeholder management in the construction industry is the tendency for many companies to rely on insurance and back-to-back contracts as a substitute for good risk management. There is a habit of passing responsibility for risk management down the procurement chain until it reaches the point of least resistance. By this time, the problem is usually a lot larger and more difficult to resolve. Many companies do not realise that transferring a risk inappropriately to a party that does not have the knowledge, capacity or capability to manage it does not offload a risk. It merely gives the illusion of control, and when the company that is given the risk is unable to deal with it, the seemingly offloaded risk will simply default back to the originator as a bigger problem that is much more difficult to manage.

MFM addressed this problem in a number of ways. First, the risk management system would incorporate guidance on good risk and opportunity management practice and deter managers from inappropriately passing manageable risks (and therefore potential opportunities) to their business partners. Second, staff would be trained in risk and opportunity management so that they would better appreciate the dangers of arbitrary risk transfer and be more confident to deal with potential risks internally, turning them to advantage for MFM. Finally, MFM decided it would review all its risk transfer mechanisms (contracts, insurance strategies and so on) to ensure that they reflected this more confident and equitable approach to risk management.

Stakeholder consultation processes in MFM

To be successful, MFM managers needed to make the right choices relating to the involvement of stakeholders in decision-making processes. The MFM risk management system identified a number of key considerations that needed to be made in the stakeholder management process. These related to:

- Objectives
- Timing
- Participants
- Techniques
- Information provision

Objectives

Many stakeholder consultations are ineffective because objectives are not clearly formulated. These objectives can be numerous and include:

- Compliance with regulatory requirements
- Giving stakeholders opportunities to voice opinions
- Educating and informing stakeholders about hazards and opportunities and of strategies to manage them
- Tapping stakeholder knowledge and experience of hazards as a supplement to technical data
- Learning about stakeholder perceptions of hazards and preferences for dealing with hazards
- Building a collaborative culture to mobilise an active constituency of stakeholders to support development activity
- Securing the participation and trust of stakeholders in the risk management process

Given the increasing prominence of regulatory control in the construction industry and the increasing penalties for non-compliance, MFM recognised the danger that many technically minded managers would choose compliance as the motive for consultation. Burby (2001) argues that such a narrow approach would be disastrous and result in significantly reduced benefits from the risk management process. For example, it could result in potential risks and opportunities going undiscovered, fewer ideas for dealing with those that are discovered and reduced commitment to management control strategies. Indeed, Burby found that when decision makers pursued three or more of the above objectives, their constituents adopted 55% more mitigation strategies than was the case in constituencies where none or only one of the objectives was pursued. MFM realised that the greatest increase in stakeholder adoption was when decision makers emphasised 'fostering citizen influence in hazard mitigation' as their main reason for consultation. Here the research showed a 76% increase in adoption of mitigation strategies compared with decision makers who did not have this objective. 'Learning about citizen preferences' was also important, with a 70% increase in adoption.

Timing

Effective consultation requires effective planning and dedicated time. Ad hoc meetings attached to the end of other meetings send the wrong message to stakeholders, implying that the process is not meaningful and that their contributions are undervalued. The result is an adversarial atmosphere, which is not conducive to constituency building. Clearly, decisions about timing are linked to the objectives underlying the consultation process. If the objective is simply compliance, then meetings will be strictly dictated by laws and regulations. Similarly, if the objective is to inform stakeholders of a development rather than meaningfully to gain their views, the consultation process might be limited to public hearings at the end of the development process. However, since the objective in MFM was to tap stakeholder knowledge to assist in the identification of risks and opportunities and in the development of more effective control strategies, the interaction had to be far more frequent, occurring throughout the decision-making process. Indeed, Burby found that large dividends in adoption of risk mitigation measures resulted from early stakeholder participation in decision-making processes (85% compared to decision makers who initiated stakeholder consultation in later stages).

Participants

One problem in managing the consultation process is that, over time, the mix of stakeholders who need to be consulted may change, meaning that consultative arrangements should be regularly monitored and reviewed. Furthermore, organisations cannot attend to all potential claims on their decisions and it is impossible and irrational, within the time and resource constraints of a project, to consult all stakeholders. Indeed, Burby (2001) found that decision makers indiscriminately embarking upon widespread consultations are only slightly more successful in seeing mitigation strategies acted upon. For these reasons, it is useful to employ a stakeholder management strategy, which will disentangle the important stakeholders from the others.

However, the literature on stakeholder management is enormous and bewildering with numerous models that are often difficult to operationalise (Mitchell *et al.*, 1998). For simplicity's sake, MFM used a model based on Freeman's classic definition (1984) of a stakeholder. This is illustrated in Figure 8.1 and classifies stakeholders as *key*, *important* or *minor* according to their capacity to affect or be affected by a decision outcome. Key stakeholders should be a top priority and be intimately involved and consulted in the decision-making process. Important stakeholders should be a medium priority and involved to a lesser extent, being kept informed of and happy with the decision-making process. Minor stakeholders are a low priority and justify only minimal involvement in the decision-making process.

Techniques

MFM recognised that the appropriate consultation technique depends on the objectives of the risk management process. A variety of consultation techniques would have to be used in seeking opinions about any one decision. For example, if the intention is to simply inform the public of development

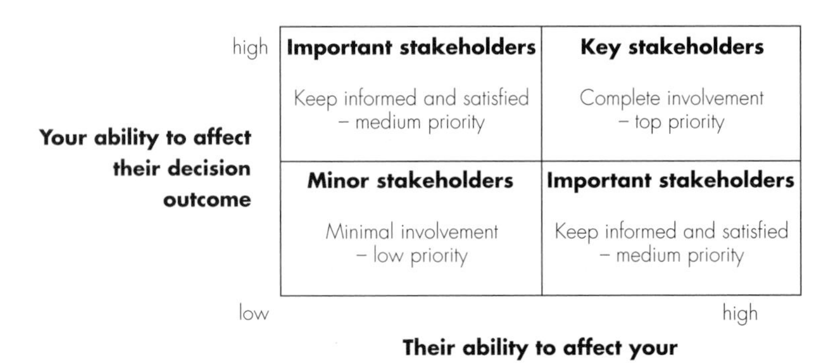

Fig. 8.1 MFM's simple stakeholder management model. Adapted from Freeman (1984).

activity, then public hearings are a widely used technique. Other common techniques include educational workshops, talks to community groups, brochures, newsletters, stakeholder advisory committees or working groups, expert panels, interviews, surveys and telephone hotlines. These techniques are used for different purposes, resulting in different coverage of stakeholders and producing different types of information. For example, some stakeholders will not, or are not able to, attend public meetings. Surveys are able to solicit the views of a large number of people over a wide geographical area, yet are unable to generate the depth of understanding, insight and trust that interviews can.

Regardless of the consultation technique used, stakeholders can be consulted individually or in groups, MFM realising that this should be done at all stages of the risk management process – identification, analysis and control. Some consultations would be at regular intervals and consider broad business issues, while others would be dictated by specific project decisions such as whether or not to have public spaces in a residential development. Ongoing discussions can often reveal new risks and opportunities as they arise, whereas consultations focused on particular decisions are much more specific in their boundaries of scope and timing. At a project level, the extent and method of consultation depends on the significance of the decision being analysed and the practicalities of contacting stakeholders.

While individual consultations may be appropriate for smaller decisions, MFM decided that the most effective consultation mechanism for decisions that involve significant risks and opportunities is to construct an expert panel (see also Bowden *et al.*, 2001). Since the opinions of an expert panel will form the basis of the risk management process (actuarial data not being available for many risks and opportunities), the combination of skills and knowledge levels within the panel is an important consideration, as is the quality of facilitation. A well-conceived panel uses people that are trusted and respected, that have the right mix of skills, experience and expertise, that are well networked in their field and that have the support of considerable resources and expertise from their own organisations. In many ways, it is the resources available to a panel rather than the individuals on it that are of primary interest. A panel with the right mix of qualifications is unlikely to omit major events from consideration and people are more likely to act on their recommendations. To this end, an expert panel should consist of people of high credibility and experience who genuinely understand the business and the range of potential risks deriving from its activities. When decisions can involve major risks, panels should consist of a combination of nationally and internationally recognised experts.

Information provision

Access to adequate and appropriate information is essential to empower stakeholders and enable them to contribute effectively to the risk management process. For example, Burby (2001) found that the more types of information were provided by planners in the development process, the more likely it was that recommendations proposed would be accepted by the community. Of particular importance in securing stakeholder acceptance was information relating to the goals, hazards, alternative designs and risk mitigation strategies being considered.

However, there are a number of potential problems that can starve the process of essential information and make the whole process of managing a vexing one. For example, many stakeholders involved in a consultation process may not have any risk-related information to contribute. This is highly likely in the construction industry where traditionally there is very little objective risk-related data collected and analysed – even by the largest organisations. Much risk information therefore remains anecdotal. Another issue relates to the inevitability that many stakeholders are likely to have conflicting interests in a decision's outcome, adding a strong political/bargaining dimension to the decision as to how much information to provide. There may also be confidential issues that one party may not wish to divulge. Indeed, even if this information is revealed, some stakeholders may lack the ability to make sense of it. Therefore information provision is an issue not just of access and quantity, but of content.

Since different stakeholders are likely to understand information in different forms, MFM required their managers to disseminate risk-related information in different ways and forms. They also ensured that any stakeholder meetings were facilitated by an independent consultant who was trusted by all parties, who could reconcile their different interests, and who could be trusted with confidential information. Finally, MFM began to build a database of risk-related information in its main areas of risk to contribute to stakeholder meetings and encouraged its stakeholders to do the same.

Barriers to stakeholder management

Unfortunately, the construction industry has a poor record of stakeholder consultation in its decisions, which is one of the reasons that it has a relatively poor public image compared with other industries (Moodley & Preece, 1996). Although the reasons for this lack of consultation are unclear, it is likely to be related in part to the vast number of stakeholders that can be involved in a construction project. This makes widespread and meaningful consultation administratively unworkable and a potential cause of heightened conflict rather than consensus. It may also be related to the difficulties that some stakeholders have in participating effectively because of physical distance or apathy, or to a lack of resources and time on the part of decision makers for engaging in meaningful consultation or a lack of knowledge about how to do so effectively. Given these constraints, it is not surprising that many managers see little benefit in involving stakeholders in their decisions. This is particularly so when constituents are not supportive of construction activity. Understandably, without positive signals from stakeholders, decision makers only see problems in involving them.

MFM realised that the above problems often create a significant gap between what is claimed to be 'best practice' and what actually occurs. However, they believed that many of the obstacles to effective consultation could be overcome if MFM could develop active constituencies. To develop such constituencies, they had to ensure that their project stakeholders were aware of their potential exposure to hazards. This required a collaborative planning process on each project where various stakeholders could become well informed about their relative needs, objectives and perspectives, and where they could suggest and comment on alternative approaches to manage them. MFM believed that if a degree of consensus could be achieved between a project's stakeholders about ways to manage risk then the political support needed to sustain efficient development could be secured.

Conclusion

The study of human perception tells us that making effective decisions about risky activities is difficult. The risk manager who relies wholly on scientific expertise and who ignores the human dimension of risk management is likely to create more risks than he or she solves, even in the most technical situations. Our technical skills have come far, but the future challenge of risk management in the construction industry is to rise above the limitations of individual minds, reconciling the interests of different stakeholders to reach a consensus about the risks that face a project and ways of dealing with them.

This chapter has presented a case study of an organisation that has sought to achieve this through the development of a new risk and opportunity management system that ensures that:

- Stakeholders feel involved
- Information for making decisions is maximised
- Perceptions of risk and opportunities are equalised among all those who can affect the outcome of a project
- Everyone understands the basis upon which a decision is made
- People feel involved and valued in the process

- People feel that their interests are being considered
- People understand their responsibilities and role in the risk management process

As one of MFM's employees commented after using the system:

> *'The major benefit is that the system facilitates the need to communicate with the key stakeholders to ensure that we have identified the decision objectives and KPIs together. In other words we are not just analysing the decision requirements and solutions from our perceived key stakeholder requirements, we are confirming the same in a structured way and working on the decision as a team.'*

References

Argyris, C. (1990) *Overcoming Organisational Defenses*. Allyn and Bacon, London.

Barnes, P. (2002) Approaches to community safety: risk perception and social meaning. *Australian Journal of Emergency Management*, **15** (3), 15–23.

Bernstein, P. L. (1996) The new religion of risk management. *Harvard Business Review*, March–April, **74** (2), 47–52.

Berry, A. J. (2000) Leadership in a new millennium: the challenge of the 'risk society'. *The Leadership and Organisation Development Journal*, **21** (1), 5–12.

Bowden, A. R., Lane, M. R. & Martin, J. H. (2003) *Triple Bottom Line Risk Management*. John Wiley and Sons, Sydney.

Burby, R. J. (2001) Involving citizens in hazard mitigation planning: making the right choices. *Australian Journal of Emergency Management*, Spring, 45–58.

Douglas, M. (1996) *Purity and Danger: An Analysis of the Conceptions of Pollution and Taboo*. Routledge and Kegan Paul, London.

Douglas, M. & Wildavsky, A. (1983) *Risk and Culture: An Essay on the Selection of Technological and Environmental Dangers*. University of California Press, Berkeley and London.

Edkins, A. & Millan, G. (2003) *Construction Risk Identification: A Review of the Literature on Cognitive Understanding of Risk Perception*. The Bartlett School of Architecture, Building, Environmental Design and Planning, Paper no. 16, University College London, London.

Freeman, R. E. (1984) *Strategic Management: A Stakeholder Approach*. Pitman, Boston, MA.

Furedi, F. (2002) Paranoid and proud of it. *The Sydney Morning Herald*, 4 May, 4–6.

Hood, C. & Jones, D. K. C. (eds) (1996) *Accident and Design: Contemporary Debates in Risk Management*. UCL Press, London.

Kasperson, R. & Kasperson, J. (1996) The social amplification and attenuation of risk. *Journal of Environmental Risk Planning and Management*, May, 64–77.

Loosemore, M. (2000) *Crisis Management in Construction Projects*. American Society of Civil Engineers Press, New York.

Mitchell, R. T., Agle, B. R. & Wood, D. J. (1998) Toward a theory of stakeholder identification and salience: defining the principle of who and what really counts. *Academy of Management Review*, **22** (4), 853–86.

Moodley, K. & Preece, C. N. (1996) Implementing community policies in the construction industry. In: D. A. Langford & A. Retik (eds) *The Organization and Management of Construction*. E and F N Spon, London.

Renn, O. (1996) *Three Decades of Risk Research*. Center of Technology Assessment, Stuttgart, Germany, June 3.

Richards, M. (2002) Conscientious objectors. *Building*, 28 March, 20–2.

Richardson, W. (1996) Modern management's role in the demise of a sustainable society. *Journal of Contingencies and Crisis Management*, **4** (1), 20–31.

Smee, R. (2002) *Construction Risk*. Presentation Services, London.

Thompson, M. E. & Wildavsky, A. (1990) *Cultural Theory*. Westview Press, Boulder, CO.

Tversky, A. & Kahneman, D. (1981) The framing of decisions and the psychology of choice. *Science*, **211**, 453–8.

Von Neumann, J. & Morgenstern, O. (1953) *Theory of Games and Economic Behaviour*. Princeton University Press, Princeton, NJ.

Willett, A. (1951) *The Economic Theory of Risk and Insurance*. University of Pennsylvania Press, Philadelphia.

Note from the author

The author would like to thank the following people for contributing to the principles that underpin the MFM's Risk and Opportunity Management System:

Dave Higgon, Industrial Relations Manager, Multiplex Construction NSW Ltd.
Charlie Reilly, Deputy Managing Director, Multiplex Facilities Management
Melissa Teo, University of New South Wales

Section IV

Relationships across Project Clusters and Supply Chains

Context

Albert Einstein said: 'Not everything that can be counted counts, and not everything that counts can be counted.' Many of the most important advances in science, whether in quantum physics, psychology or evolution, no longer depend upon direct observation. Indirect observation, surrogates and measures of inference are frequently used. This is widely accepted even though it challenges many of the parameters of what is considered scientific method. In social sciences measurement is used, but again direct observation is not always possible, and social science methodologies embrace values and interpretation with little challenge today. And maybe this is a good thing in practice when creativity is becoming of increasing value, for replication and emulation have short 'half lives' in competitive advantage and profitability.

We need to embrace the new and examine the constraints, but in a considered way as not everything new is good and not everything that seems good is. In one sense our whole endeavour of linking theory to practice is action research for the benefit of society. We are able to socially reconstruct our circumstances (cf. Berger & Luckmann, 1967). Social reconstruction depends upon what sense we make of current and emerging situations. A modern approach to management emphasises goal achievement the most, whereas the post-modern one emphasises process. Perhaps the sense we make of our context is more post-modern: concepts of partnering, supply chain management and agile production being used yet without precise definitions, and thus without precise goals; and accountability, benchmarking and key performance indicators being stressed rather than the actual outcomes in practice.

This section takes a critical look at the ways in which such social reconstruction has faired to date in networks, in supply chain management, in relationship management and for PFI/PPP projects. This section is more critical and evaluative of a relationship approach both in positive pursuit and in questioning the premise.

An overview of the contribution made by each chapter is set out below and then the chapters are located in the framework.

Contributions of chapters in this section

Pryke adopts a social network analysis methodology in **Chapter 9**. In *Projects as networks of relationships*, Pryke takes the view that the construction industry is in a state of transition as it tries to deal with the unfamiliar business environment imposed by post-Latham relational contracting strategies. The move away from a contractually focused approach to project governance has forced construction industry project actors to adopt supply chain management, among other strategies, to prevent deteriorating project performance over multiple-project, strategic-partnering-related timescales.

The non-linear, complex, iterative and interactive project environment typical of construction projects is difficult to analyse using traditional task dependence, structural analysis and process-mapping-based approaches. Pryke argues that the solution is to examine the contractual relationships, financial incentives and information exchange networks associated with the principal functions of the project coalition. In this way a comprehensive understanding of the systems of which the construction project is comprised is possible, and it becomes feasible to begin to articulate a social network theory of project coalition activity and effectiveness. A summary of the findings of current research in this area, involving a number of international construction projects, is provided.

Green in **Chapter 10** is critical of many of the process management trends, including supply chain management. He is not advocating the status quo, yet challenges the form and implementation of the new practices. He argues that the concepts are neither implemented comprehensively, nor can they map out in practice as theory would expect. The lack of clear definition of many of the concepts, such as supply chain management, while in theory giving room for competitive advantage, in practice becomes a 'sleight of hand', whereby it serves to act as a means to induce flexible working practices designed to push responsibilities and risk along the supply chain in one direction and require lower prices to be achieved in the other. In essence, these practices are a reconfiguration of the *machine metaphor* of production, which fails to respect or harness relationships in ways that other metaphors of interaction in production do, such as the brain analogy of the organisation, which implies learning and knowledge management, or the orchestral conductor whose relationship to the orchestra is to draw out the best from each section.

Cox and Ireland in **Chapter 11** focus upon *Relationship management theories and tools in project procurement*. They assert that market forces and the players' positions of power in the supply chain render a relationship approach in general and relationship management in particular inappropriate and inoperable in almost all circumstances. Their focus is largely concerned with *what is* rather than what *can be*. Many may be sceptical about the long-term ability to sustain new procurement approaches and to adopt relational contracting and relationship management in the long run; however, governments and clients have consorted to engineer this. The challenge becomes to address what the limits and constraints are in practice, Cox and Ireland clearly arguing that they are reached sooner rather than later. Perhaps this is typified by the difference between a procurement emphasis for intervention, stressing market management, and a marketing emphasis, stressing market management and market creation, which changes the nature of competition and the balance of power, and can change transaction costs, for example by raising switching costs as more service value is added.

Mutuality is a term frequently used – mutual trust, mutual benefits, gain-pain share and so on. Mutuality is a term of *equality*, where the benefits must be equal. Cox and Ireland advocate recognition of the unequal balance of power, and thus their analysis invites retention of the notion of *equity*; that is, benefits are proportional to the relative power of the parties. Therefore, 'win–win' does not necessarily imply an equal share of the gains.

Chapter 12 is by Ive and Rintala who consider *The economics of relationships* and are more positive about the ability to reconstitute relationships in positive ways. They provide an analysis of the shifting balance of power during the project life cycle, thus stating that the balance during procurement is not retained throughout. They use a case study approach in the Private Finance Initiative (PFI) and Public Private Partnership (PPP) market to illustrate their argument. Their analysis is that the PFI/PPP market does reconstitute relationships at the procurement stage too, one conclusion being that the design-build-operate option under PPP is likely to lead to closer relationships and greater innovation. This is especially true at the procurement and design stages because the design-build-finance-operate option may transfer financial risk, but the banks are risk averse and tend to shun innovation in technology and management practices.

Table IV.I Contributions within market relationships. Developed from Gummesson (2001).

Relationships	Chapter 9 Pryke	Chapter 10 Green	Chapter 11 Cox and Ireland	Chapter 12 Ive and Rintala
Classic market relationships				
R1. The dyad (supplier-customer)	Direct	Direct	Direct	Direct
R3. The network	Direct			Direct
Special market relationships				
R4. Relationships via full-time and part-time marketers		Very indirect	Very indirect	
R5. The service encounter	Direct	Direct	Direct	Direct
R6. The many-headed customer and many-headed supplier	Direct			Direct
R7. The relationship to the customer's customer	Direct	Direct	Indirect	Direct
R8. Close versus distant relations	Direct	Direct	Direct	Direct
R9. The dissatisfied customer	Indirect			Indirect
R10. The monopoly relationship				Direct
R16. The law-based relationship	Direct			Indirect
Mega relationships				
R19. Mega marketing	Very indirect			Direct
R20. Alliances that change market mechanisms	Indirect			Direct
R21. The knowledge relationship		Indirect		
R22. Mega alliances that change the market structure				Direct
Nano relationships				
R24. The internal market				Very indirect
R25. The internal customer relationship				Very indirect
R26. Quality and customer orientation				Indirect
R30. The owner and financier relationship.				Direct

Locating the chapters within a relationship framework

Each chapter for this final section is located within the framework outlined in the **Introduction** and developed in **Chapter 1**. It has been stated that the framework anchors this body of work. However, discourse will move understanding onwards; hence the framework itself becomes changed or perhaps discarded.

This introduction to **Section IV** has focused upon the context and framework of the four chapters and their contribution to a relationship approach with particular emphasis on relationships across project clusters and supply chains.

Section IV is to be located within the framework outlined in **Section I** in order to anchor the work, and to improve our understanding and help stimulate a discourse.

The *objectives* provided in **Section I** are addressed in this section as follows. All the authors focus upon interpersonal and inter-organisational relationships involved in the project, especially in supply chains and clusters. Both Green, and Cox and Ireland look at concepts for understanding and for the inception of relationships in critical ways. Green in particular analyses the gap between the concepts and practice in terms of how we define project

Table IV.II Contributions for internal relationships.

Relationships	Chapter 9 Pryke	Chapter 10 Green	Chapter 11 Cox and Ireland	Chapter 12 Ive and Rintala
Internal structural relationships				
R33. Peer relations	Indirect			
R34. System relationships and communication	Indirect	Indirect	Indirect	Indirect
Internal functional relationships				
R36. Immediate job function relations			Indirect	
R39. Circumstantial relations	Direct			
Strategic relationships				
R40. Formal relationships	Direct		Indirect	Indirect
R41. Career and organisational politics	Very indirect			
R42. Embedded relationships	Direct			
R43. Cultural relations	Direct			
R45. Strategy identification		Direct	Direct	Direct
R44. Strategy implementation-tactical response	Direct	Indirect		Indirect
Personal relationships				
R49. The affirming-orientated relationships	Indirect			
R50. The serving-orientated relationships	Indirect	Indirect		
R51. The performance-orientated relationships			Indirect	
R50. The appearance-orientated relationships		Indirect		
R51. The blame-orientated relationships			Indirect	

success, while Cox and Ireland are more concerned with raising expectations of success that cannot be achieved. Cox and Ireland are therefore sceptical of the value of managing relationships in other ways than those that have prevailed in the market. Green is open to alternative possibilities to achieve successful outcomes, but critical of current trends as practised. Both Pryke, and Ive and Rintala see potential for improving delivery and project success. Pryke introduces social network analysis as a tool to aid understanding, but also as a practical tool to help managers improve the effectiveness of relationships, and hence project efficiency. Ive and Rintala see potential for successful project outcomes, including enhanced innovation, in the way in which they are set up to redistribute the balance of power in the transaction, and ultimately in the market. The formulation of a project strategy and the effective management of the 'front-end' of the project are seen as critical to successful projects.

The relationship context is business-to-business or organisation-to-organisation in this section. This is the focus of Pryke and key in all the other chapters, where power and leadership issues are to the fore of management. **Chapter 9** by Pryke, **Chapter 10** by Green and **Chapter 11** by Cox and Ireland are primarily concerned with *classic market relationships*, whereas Ive and Rintala in **Chapter 12** are primarily concerned with *special market relationships* and, to an extent, *mega market relationships*. Both Green, and Cox and Ireland

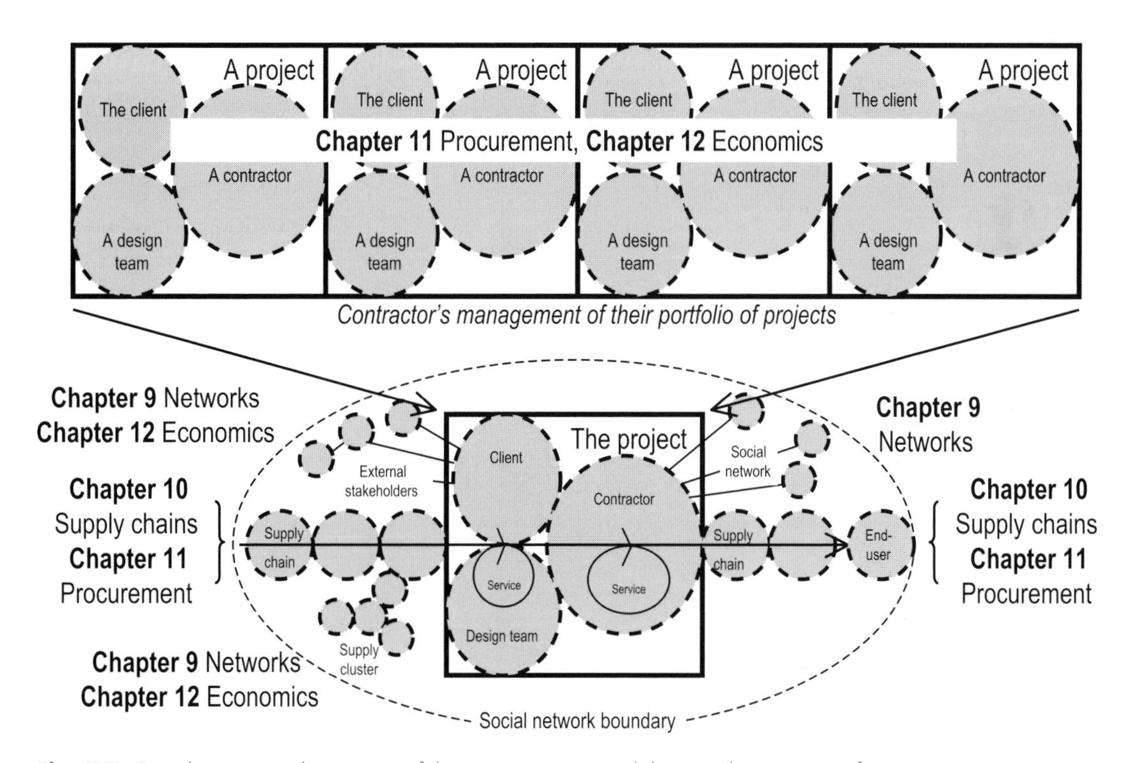

Fig. IV.I Contributions in relation to portfolios, programmes and the social environment for projects.

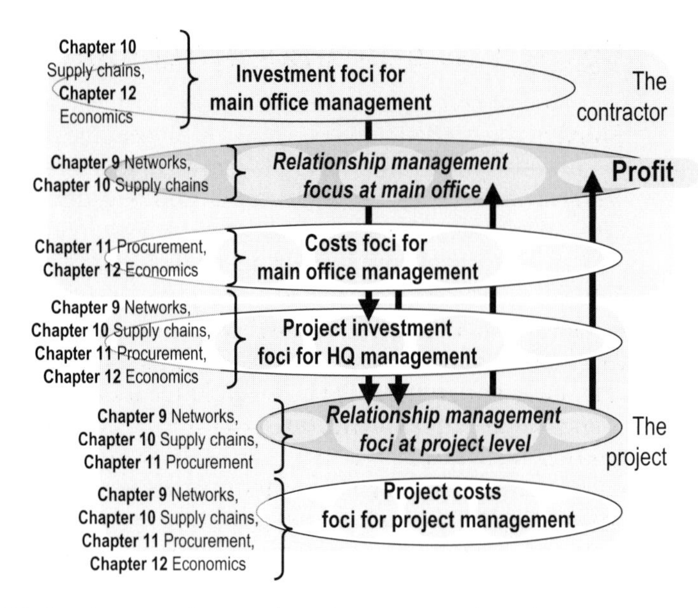

Fig. IV.II Contributions to the interrelationship between relationship management for projects and the main office.

are also addressing *special market relationships*. Table IV.I addresses these categories further. The allocation and weighting given to each relationship should be considered indicative as they are arrived at subjectively.

In **Chapter 1** a wider range of *nano* or internal relationships was recognised. Table IV.II addresses these internal relations for this section. This analysis can then be located within the context of portfolios, programmes and the social environment for projects (Figure IV.I).

Figure IV.I shows at the top the portfolio of projects that a contractor has underway and the strategic input management that managers can address in relation to the chapters for this section. The players immediately involved with project delivery, especially the supply chain and clusters, are related to the chapters too. Figure IV.II is based on Figure 1.10 under the relationship management paradigm, and pictures the connection between the corporate level of activity and the project level for the chapters.

This introduction to **Section IV** has focused on the context, contributions and framework of the four chapters to a relationship approach with particular emphasis on relationships located in supply chains and clusters.

Reference

Berger, P. L. & Luckmann, T. (1967) *The Social Construction of Reality*. Anchor Press, Garden City, NY.

9 Projects as networks of relationships

Stephen Pryke

Construction industries throughout the world have been driven, largely through the efforts of some of the biggest and most influential clients, to institute reform in procurement approaches and the governance of projects. Some might argue that these industries are still in transition as a result of these innovations. We observe a move away from governance based upon solely contractual relationships and the introduction of trust; the use of strategic partnering, framework agreements and other long-term, programme-related rather than project-related, relationships; the interest in supply chain management, partly as a means of maintaining value received by clients in the absence of competitive financial bidding; increasingly cross-disciplinary approaches to allocation of project roles and the management of those roles through effective project relationships. These emergent factors, arising in the context of a *non-linear, complex, iterative and interactive project system environment* typical of the construction project, require a radically different approach to analysing the processes that are involved in the delivery of a project in a way that provides client delight and minimises turbulence in relationships through adverse stakeholder reactions. We review the traditional methods of analysis adopted within construction and justify the need for innovative analytical approaches in the context of transition and reform.

It is proposed that the construction project can be viewed as a *network of relationships* between the firms that make up the project coalition. These relationships are many and varied but we settle upon a number of sets of relationships that represent the main project activities. Finally, findings from a programme of research projects are summarised, principally in relation to the legal, contractual, communications and financial incentive aspects of project governance.

Context

During most of the nineteenth and twentieth centuries, construction was procured through the use of separate contracts for design and production. A main contractor took responsibility for producing a building to the design and specification defined by the professional team (Winch, 2000). Management forms of procurement, where the construction management role became a discrete fee-earning activity rather than being rolled into a rather broader, risk-taking entrepreneurial role, became more popular during the middle of the twentieth century. Design and build became increasingly important during this period also, although contractors tended to subcontract design work. It is argued that the procurement reforms occurring during the latter part of the twentieth century (and reviewed in more detail by Masterman (2001) and Winch (2000) among others) have essentially comprised the shifting of roles and responsibilities within the project coalition members. Initially, we saw the role of the lead designer moving from being an independent consultant's role under traditional procurement, to being a more integral part of the production process under design and build procurement routes. Later, more radical changes in procurement saw the shift of design coordination away from a single designer to a group of cluster leaders (Holti *et al.*, 2000).

Construction industries throughout the world and their clients, as well as those engaged in research, have begun to understand more clearly the complex systems that constitute the activities of the project coalition. There follows a brief review of the methods that have traditionally been used to monitor and analyse the activities of the construction coalition, in order to provide a context for the discussion of a more innovative approach – *the analysis of the network of project relationships.*

How have we traditionally tried to analyse the activities of the construction coalition?

Much of the analytical output generated by, and aimed at, the practitioner has been qualitative: Simon (1944), Emmerson (1962) and Banwell (1964) and more recently Latham (1994) and Egan (1998) have all eschewed discussion of methodology and avoided the need to demonstrate how their findings were established. Analytical methods available to those seeking more quantitative methods can be classified broadly into three groups:

- *Task dependence* – focuses upon the time-related implications of dealing with an interdependent group of work packages, which critical path analysis exemplifies (see for example Higgin & Jessop, 1965).
- *Structural analysis* – refers to the representation of formal, usually authority-related relationships, in the form of a hierarchical 'family tree' type of diagram, which might be regarded as only loosely quantitative (see, for example, Masterman, 2001; Davis & Newstrom, 1989).
- *Process mapping* – a slightly more eclectic classification encompassing flow charts and cognitive mapping as well as linear responsibility analysis. For examples of process mapping refer to Tavistock Institute (1966) and Curtis, 1989; and see Winch and Carr (2000) for the proposed use of process protocols, modelling anticipated flows of information between project actors.

For a detailed and critical review of these methods refer to Pryke (2004).

In summary, the shortcomings of traditional methods of modelling construction projects have been:

- Dealing with interdependence – flow charts and critical path analyses tend not to cope with the non-linear, iterative and interactive characteristics of the construction process. In particular, the nature of such interdependence is difficult to define or classify.
- Appropriateness of level of detail – traditional forms of analysis are typically focused upon isolated micro-level decisions and these do not facilitate the study of complex processes such as construction, where a large number of unique decisions are made during each day of the design and production programme.
- Assumptions about hierarchy – when we represent the activities of any organisation, project or otherwise, using a 'family tree', we inevitably have to make certain assumptions about hierarchy. When representing project relationships, there is a similar argument applying to the representation of contractual relationships, which construction is evolving away from – note the increasing interest in lean construction methods and supply chain management.
- Traditional structural analysis is dyadic – discussions about contract and intra-coalition relationships have traditionally made an assumption that relationships essentially involve two parties, that is they are *dyadic*, and so much of our traditional thinking on contract administration in construction has been orientated towards managing these dyadic contractual relationships between firms. Proponents of supply chain management encourage us to look beyond immediate contractual relationships, and the introduction of the PPC 2000 form of partnering construction contract has challenged the dominant position of the Joint Contracts Tribunal (JCT) forms.

Conceptualising the construction coalition as a network of relationships

It is argued that the production of a building might be conceptualised as a nexus of contracts (Aoki, 1990), the governance of which can be analysed using observation of sets of transactions associated with each of the main components of project activity. The classification of these sets is:

- Briefing and client communications
- Design and specification
- Management of progress of production
- Management of budgets and costs

The governance of the project can be observed through inspection of the following:

- Contractual relationships between coalition member firms
- Financial incentives in place between these actors
- Information exchange networks

Why projects as networks?

Some of the traditional means of modelling organisational activity originated in an era of scientific management theory. Therefore, what these methods lack above all else is a sense of relating to the human being. Humans interact with one another for a wide range of social and business purposes. If we accept that all commercial activity comprises, at one level, social networks, then it follows that analysis of these networks might help us to understand how project coalitions operate and perhaps to distinguish effective, or perhaps profitable, network characteristics from those that are less effective.

At this point it is worth making the distinction between informal and formal relationships. Much has been made of the importance of 'contacts' – personal relationships that provide benefits in one's own work environment, perhaps on the basis of trading favours (see also **Chapter 5** by Cova & Salle). Indeed anecdotal evidence suggests that some work activities are closely linked with specific social activities. Although the area of informal relationships is undoubtedly relevant, this chapter focuses on the human being as a node within a project-related network, the purpose of which is to deliver a construction project; we are interested in understanding exchanges related to *project actor roles*.

Although exchanges in the project context involve human interaction, the project coalition is essentially a network of *inter-firm* relationships. The project actors are essentially firms, even if a number of them are very small. Hence, although we might personalise the role of architect, for example, the project actor is the practice employing the individual acting as architect; the individuals fulfilling these functions are not executing contracts between themselves personally. It is therefore appropriate that we regard relationships in the project coalition as between *firms*; if we deal with communication we can aggregate communications between individuals so that the communications, contract and financial incentive networks identified above are expressed in a comparative way.

The environment in which a construction project actor operates is a function of the networks with which he or she is connected. We can be associated with a number of networks simultaneously; indeed, this is an almost inevitable consequence of human activity. Our activity and the fulfilment of our role within one network and the manner in which we interact with other actors will be affected by the environment of other parallel and concurrent network activities. Individuals are in this way suspended within multiple, complex, overlapping networks of relationships (Nohria & Eccles, 1992) and we could not possibly fully understand the nature of an individual's role by reference to one network of relationships alone. Hence, to understand a person's role we need to look beyond their employer's office environment and study the project networks with which they are connected.

Behaviour and the effectiveness of such behaviour of a given project actor in the context of a single project environment can be explored and much more fully understood by relating the project actor to the position that they hold within the project networks. At a very simplistic level, if we look at a network of contractual relationships we may understand actors' roles and behaviour by the prominence or *centrality* that the actor might have within the contractual network. Conversely, we might learn that there is some divergence between the prominence conferred to an actor by the contract conditions and the prominence of that actor within another network type. How does the prominence conferred by the contract compare with the prominence observed when referring to communication exchange networks, for example? If there is a difference in these comparative prominences, what are the reasons for such differences?

Finally, before we leave this justification for the idea that the project might be conceptualised as a number of different but related networks, let us consider the relative crudeness with which we currently compare project organisations. Traditional structural analysis of organisations has little regard for project/organisation size or nature of activity. We typically represent an organisation as a *hierarchy of formal authority relationships* – who reports to whom, which director is responsible for a particular project function, and so on. It is argued that this is of relatively little use in understanding the effectiveness of either project or any other form of organisation. It is suggested that we are interested in project roles and functions; in the same way that the human brain will create neural networks to respond to particular types of function, the project coalition forms networks to respond to a given task. While the project coalition must remain constant throughout the project for contractual reasons, the project coalition actors will form into one configuration for the purpose of managing progress, and a different or modified configuration for dealing with design variations, or for managing costs. It is suggested that our understanding of the effectiveness of project teams will be advanced if we focus upon networks dealing with a particular project function and use these networks to understand the prominence or lack of prominence of project actors within those networks and the way in which financial incentives and information exchange govern these transactions within the network context.

How do we analyse these networks?

Social network analysis (SNA), as a technique, is widely used in the USA and had its origins among researchers at Harvard University during the 1930s, sociometric analysts and a group of anthropologists in Manchester, UK (Scott, 2000). It is based upon graph theory and its pioneers came from the areas of sociology, social psychology and anthropology (Wasserman & Faust, 1994). The application of SNA to construction projects is rare – an exception is the work of Loosemore (1996, 1998, 1999), which focused upon *interpersonal* networks in construction, however, as distinct from the *inter-firm* networks dealt with in this chapter.

Social network analysis is a concept and a methodology that enables any organisation or group to be represented as a system of nodes connected by links, where the links represent relations between the actors (or nodes). A group of actors can have a number of relationships. The attributes of each node (for example, the number of other nodes to which it is connected) can be represented mathematically; the configuration of the links between the nodes can also be represented mathematically, for example in terms of values for density of links. The networks can be represented graphically in a number of formats, including in the form of sociograms (see Figures 9.1 and 9.2 as examples).

Individual actors can be categorised as transmitters or receivers (of, for example, information) and a large number of specialist SNA terms exist to describe both the network as a whole and individual actors and their role within the network.

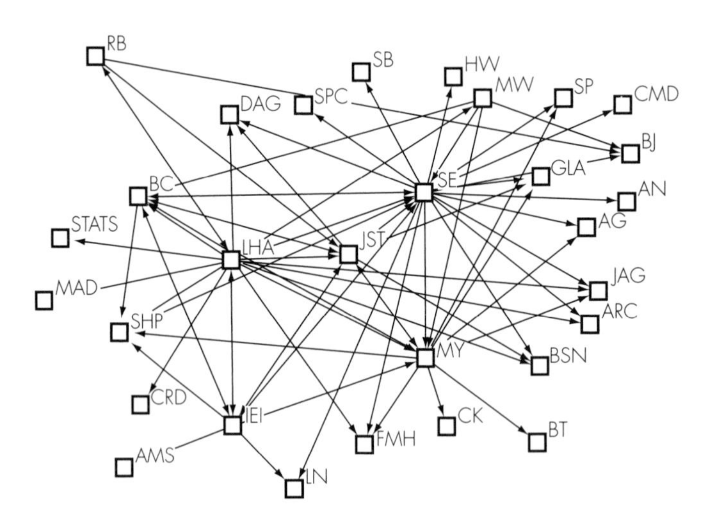

Key

Ref.	**Role**	**Ref.**	**Role**
AG	Atrium cleaning equipment	JAG	Architectural metalwork
AMS	Site electrics	GLA	Curtain walling subcontractor
AN	Window design consultant	JST	Structural engineer
ARC	Curtain walling subcontractor	LHA	Architect
BC	Contract manager	LN	Tenant/end user
BJ	Cladding QS for client	MAD	Bricklaying subcontractor
BSN	Precast concrete floors subcontractor	MW	Cladding supervisor
BT	Telephone systems	MY	Site manager; subcontractor
CK	Client QS	RB	Cladding draughtsman
CMD	Landscape consultant	SB	Landscape procurement consultant
CRD	Cradle equipment	SE	Client
DAG	Structural steel subcontractor	SHP	Local power provider
FMH	Groundworks subcontractor	SP	Piling subcontractor
HW	Atrium roof glazing	SPC	Fire protection subcontractor
IEI	M&E services consultant	STATS	LA and stat. undertakers

Fig. 9.1 Private sector partnering/SCM project: information exchange networks – design and specification.

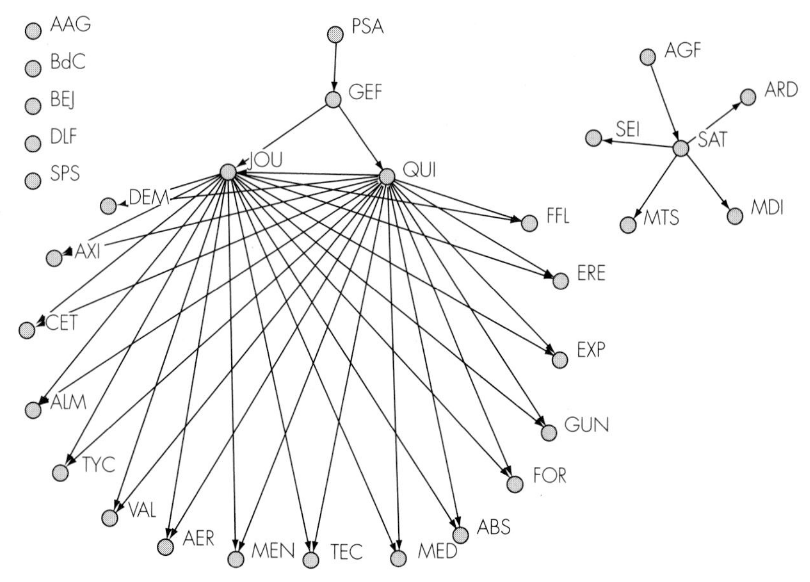

Key

Ref.	Role	Ref.	Role
AAG	Architect for permit only	GEF	Project managers for construction*
ABS	Tiling subcontractors	GUN	Steel doors subcontractors
AER	Plumbing subcontractors	JOU	External works subcontractors [+ main contractor 1of 2]
AGF	Fire insurers		
ALM	Lifts subcontractors	MDI	Specialist subcontractors
ARD	Lagging subcontractors	MEN	Structural steelwork subcontractors
AXI	Mechanical services subcontractors	MTS	Aerials subcontractors
BdC	Building Control Office*	PSA	Client
BEJ	Design consultancy	QUI	Main contractor [2 of 2]
CET	Security installation subcontractors	SAT	Sprinkler subcontractors
DEM	Electrical subcontractors	SEI	Underground services subcontractors
DLF	Project managers to client	SPS	Planning supervisor*
ERE	Ironmongery supplier	TEC	Internal partitions subcontractors
EXP	Specialist loading ramps subcontractors	TYC	Fire doors subcontractors
FFL	Rainwater installation subcontractors	VAL	Internal joinery subcontractors
FOR	Bricklaying subcontractors		

Nearest equivalent term for UK industry used here

Fig. 9.2 French private sector project: financial incentive networks.

It is possible to apply SNA to the following types of relationship (Wasserman & Faust, 1994):

- Evaluation of one person by another (expressed friendship, liking or respect)
- *Transfer of material resources* (transactions, lending or borrowing things)
- Association or affiliation (jointly attending a social event, belonging to the same social club)
- *Behavioural interaction* (talking together, sending messages)
- Movement between places or statuses (migration, social or physical mobility)
- Physical connection (a road, river, or bridge connecting two points)

- *Formal relations* (for example, authority)
- Biological relationship (kinship or descent)

The italicised headings indicate the types of networks that were analysed in the construction case studies used in this chapter. For further information on the subject of social network analysis, see the introductory material by Scott (2000) and Berkowitz (1982); those who become enthusiasts will want to read Wassermann and Faust (1994) and Wasserman and Galaskiewicz (1993). A more detailed discussion of the application of SNA to construction projects is contained within Pryke (2001). Figures 9.1 and 9.2, and Table 9.1 give exam-

Table 9.1 Comparable network centralities with normalised values and added notes of analysis.

	Project A	Project B	Project C	Project D
Client				
Contract	25	25	50	3
Programme	9	10	18	5
Design	14	0	3	3
Performance incentive	0	8	2	6
Consultants				
Contract	6	3	2	3
Cost	41	23	12	9
Instructions	18	5	18	3
Programme	41	15	2	8
Design	55	45	5	38
Performance incentive	0	0	0	6
Main contractor				
Contract	27	20	50	40
Cost	23	35	12	64
Instructions	14	23	18	41
Programme	32	63	18	71
Design	36	40	41	55
Performance incentive	27	20	2	56
Cluster leaders or subcontractors				
Contract	5	3	16	3
Cost	17	3	11	18
Instructions	20	3	4	3
Programme	36	6	18	21
Design	54	33	20	35
Performance incentive	8	4	Isolate	5

[1] importance of progress to private sector developer

[2] contractual centrality of LA client is reflected in other networks except F.I.

[4] low levels of cont. power for all consultants Project A max.

[9] Project C uses standardisation through partnering; avoids need for systems

[5] regardless of procurement system – this is the response to GMP contracts; contractor controls

[12] no evidence that clusters increase input of specialist subcontractors into design; traditional procurement seems to promote more subcontractor involvement

[6] experienced client

[3] inadequate control of design

[7] Project D client weak in all networks

[8A] powerful managem't of costs needed in traditional procurement

[8B] low power financial man, but good control achieved

[10] prime contractor prominent in contract, information exchange and performance incentives, essential where repetition and standardisation not possible (as in Project C)

[11] prominent position of cluster leaders in pilot project not yet supported by appropriate contract and performance incentives.

ples of the presentation of data in graphical form, and the following provides an overview of the type of analysis derived from a series of project case studies.

Case studies: applying SNA to the construction coalition

Private sector partnering and supply chain management project case

Figure 9.1 shows an information exchange network associated with design and specification activities occurring during the production phase of a project. This diagram shows a fairly dense network (moderately high levels of connectivity between the nodes). Inspection of the diagram reveals that actors LHA, JST, SE and MY are the most prominent, having the highest *centrality*, with IEI having slightly less *centrality*. We should not be too surprised that the main design consultants – architect and structural engineer – are most prominent, and that the client (SE) is so prominent, illustrating accurately the role played by one of the UK's most innovative property development organisations in performing its role as supply chain manager. The relatively low level of centrality of the services engineer (IEI) is related to the building type, commercial offices, where the heating and air conditioning and electrical services installations are of a relatively simple type.

We have considered *density* of the network and used *centrality* to understand the relative prominence of a number of key project actors in our project network. On this relatively small project, inspection of the network represented as a sociogram (Figure 9.1) is appropriate and accessible, and it provides easily assimilated information. On projects with very large numbers of actors and particularly where the networks are dense (large numbers of links to each node), visual inspection of the sociogram is less informative and we would use mathematical analysis. The research covered in this chapter used UCINET 6 social network analysis software, which provides a very wide range of routines, considerably beyond the scope of this chapter but covered extensively by Wasserman and Faust (1994). The software also converts the mathematical values into sociograms using a Windows-based 'draw' facility.

Another network characteristic that we might want to consider when looking at a sociogram like Figure 9.1 would include the existence of *isolates*, which are actors unconnected to another actor within the network. We presume that the isolate actors are connected to other team members in other project networks. Hence we might observe the financial manager or quantity surveyor as an isolate in the progress management information exchange network. We would, however, expect our financial manager to have a high level of centrality within the financial management information exchange network. Figure 9.1 has no isolates.

A further important network characteristic is the extent and location of *bridges*, which are nodes that link two other nodes that would otherwise not be linked. An example of this would be AMS (bottom left of Figure 9.1), which is only connected to the remainder of the network by IEI. Any information coming to AMS has to come through IEI. If IEI is removed or becomes dysfunctional for whatever reason, AMS effectively becomes an isolate. If IEI is particularly ineffective at processing information, through incompetence, or through the pursuit of vested interests, communication efficiency will be dramatically affected.

In summary, SNA data analysis can be presented in a wide variety of ways; academics and practitioners are continually evolving and inventing new approaches. In this case we have dealt with sociograms (Figures 9.1 and 9.2) and the next case presents tabulated comparative values for centrality measures (Table 9.1).

Comparable network centralities for four project cases

Table 9.1 presents a comparative analysis of network centrality values for four case studies. The findings are presented in more detail later in the chapter; we deal here with some examples of the kind of analysis that is possible by examination of *comparative centrality* values.

The four cases studies comprise construction projects executed in the south-eastern region of England; the projects were of approximately the same value, low rise with some office accommodation content, of similar levels of complexity and construction type. Other variables between projects are minimal, allowing us to study contractual relationships, performance incentives and information exchange variables. Project A was a public sector project for an English county council and employed traditional procurement methods; Project B was procured by one of the UK's largest private sector property developers, again using a traditional procurement strategy. Project C was a case study involving a private sector property developer actively involved in the use of partnering, supply chain management and work clusters in their projects. The management of this organisation was involved in innovation in procurement and contributed to the Latham Report (1994). Project D was a project carried out by the largest public sector procurement agency in the UK and a demonstration project, offered by its team as an example of innovative practice in procurement; this project also involved a research team, which advised the construction team and documented the findings of the collaborative team (see also Holti *et al.*, 2000).

Table 9.1 presents *centrality values* for the selected group of project actors. Centrality values give a mathematical value for the prominence of the actor with the network – measured by the number of connections that the actor has with other actors in the network. The figures in the table are normalised – multiplied by 100 and rounded to the nearest whole number – as we are interested in the relative values rather than the absolute values. Data is presented for the client actor, the private sector service providers or consultants, the main contractor or the nearest equivalent, the cluster leaders and the subcontractors, all of which may include several individuals. Within each of these actor groups we looked at the centrality of the actor within certain specific networks: contractual relationship networks, progress and design/

specification information exchange networks and financial performance incentive relationship networks. It is intended here simply to give an overview of the way in which analysis might be carried out by looking at a series of analytical comments (the numbers below correspond to the comment box numbers in Table 9.1).

- *Comment [1]* Looking at the client actor's centrality in the information exchange network for progress management, we see a marked difference between the centrality figure for public and private sector projects: figures of 9 and 5 for Projects A and D; 10 and 18 for Projects B and C. It would be incorrect, given that the figures shown are normalised and derived from four different networks, to state that the centrality for private sector client involvement in progress management, in broad terms, is double the equivalent figure for the public sector projects. The figures do give an indication of the order of magnitude of the difference, however. This would reflect a client organisation that is very 'hands-on' – managing the supply chain and taking a prominent role in the production process.
- *Comment [2]* The relatively high level of centrality of the client department in contractual networks for Project A is reflected in a similarly high level of centrality, relative to other projects, of the Project A client in progress and design information exchange networks. This tends to indicate that the contractual conditions reflect the behaviour of the project actors in relation to design activity and progress management. This might tend to indicate a certain *maturity* of contractual relationships on this traditionally procured, public sector project, where contractual conditions have been refined over time and relate closely to the behaviour of project actors.
- *Comment [3]* A centrality value of zero for the client on Project B in the design and specification information exchange network indicates a 'hands-off' approach by the client to the area of design, although we can see that the centrality of the client in the contractual network (a figure of 25) is relatively high. This indicates a client that deals with design process governance in a contractual manner, rather than taking a more managerially orientated approach. This reflects the approach of a client that has a very well-developed and formalised set of written documents that forms the basis of their contracts with consultants and contractors; having written down their requirements and signed an appropriate contract, they devote relatively little further attention to the matter during the course of construction. This might be regarded as reflecting a pre-Latham approach to construction. Both private sector projects display similar characteristics in this respect.
- *Comment [4]* Relatively low levels of centrality for consultants in (main contract) contractual conditions simply reflect a lack of prominence by the consultants in main contract documents. Consultants do not act as employers and are essentially in dyadic relationships with clients that do not (contractually) involve other actors engaged on the project. There are echoes of the Tavistock Institute's work highlighting the interdependence,

yet independence, of consultants within project teams (Higgin & Jessop, 1965; Tavistock Institute, 1966).

- *Comment [5]* Two of the projects involved GMP (guaranteed maximum price) supplements to the contractual arrangements, whereby the constructor was made responsible for the out-turn cost, or final account cost, of the project. Paradoxically, under procurement without GMP the client's quantity surveyor takes responsibility for the out-turn cost and predictions of such costs on behalf of the client. The process is complex but there is much anecdotal, as well as documented, evidence that this traditional (non-GMP) approach is unreliable on many projects. Our table shows the dramatic effect that incentivising the constructor to take responsibility for out-turn costs has. The contractor becomes central within the financial management information exchange network as well as the progress management networks; progress and costs are intimately linked from the constructor's point of view.
- *Comment [6]* Here we see the relatively high level of centrality of a very experienced client. This figure also reflects the type of network for contractual relationships that is of a 'star' sociogram configuration, with short, direct linkages to a large number of relatively small specialist subcontractors. This point highlights the need to avoid relying solely on SNA data when looking at complex projects; other qualitative material is required to make sense of, and provide a context for, the network data.
- *Comment [7]* The relatively low levels of centrality of the public sector client on Project D (public sector, innovative procurement approach) in all information exchange networks reflects a 'hands-off' approach to construction project management, associated with a lack of resources, knowledge, expertise and will from the centre to manage the supply chain from a central position.
- *Comments [8A] and [8B]* – the high centrality figure for consultants in the cost management information exchange network on Project A is perhaps, an indication of the need for intensive, independent financial management on public sector, traditionally procured projects. This is a situation where 'less is more' and the project with the most effective financial control (reflected in the variance between pre-contract estimate and out-turn cost) had the least prominent independent financial management activity. Effective financial management was an integrated part of a procurement approach that managed both design and progress and in this way avoided the need to put resources into independent financial management.
- *Comment [9]* The client in Project C has very low centrality in progress and design information exchange networks and does not employ performance incentives at all. This reflects a client that works hard to fulfil the function of supply chain manager, using in-house staff. Knowledge exchange between client staff and constructors is facilitated by short links (direct contracts) with specialist contractors, avoiding the contractual hierarchy seen on traditionally procured projects. Client C used a very high level of standardisation in design and component selection.

- *Comment [10]* The contractor on Project D (public sector, innovative procurement approach) has high centrality in the contractual network, performance incentive networks and all information exchange networks. This reflects a procurement route, prime contracting, where the client has set out to achieve a lower level of personal involvement and transferred the responsibility for management of the supply chain to the *constructor*. These are the actions of a lay client and are in stark contrast to the approach adopted by the experienced, 'hands-on' client for Project C. The high centrality in the performance incentive network indicates the need to use effective performance incentives where the client attempts to get the constructor to act as its agent in matters that are the province of consultants under traditional procurement routes (management of design and project costs).
- *Comment [11]* High centrality and hence prominence of cluster leaders for Project D indicate effective transfer of responsibility to this new role within the project coalition. The low centrality value in the contract network indicates that the contractual conditions are not yet in place to reflect this emergent role. The legal aspects of this are discussed in Pryke (2005b).
- *Comment [12]* There is evidence here that the cluster leaders on Project D (no other projects involved cluster leaders) might have had a more effective involvement in design management. The cluster leader is meant to manage a sub-budget for the cluster design and production on site. The fact that involvement in design was high but could have been even higher might be attributed to the choice of cluster leader; the cluster leaders were originally from a production rather than design background on this project.

In summary, these two forms of presentation, first the sociogram and then the table, are the most accessible for our purposes. We may also seek a higher level of details through the use of mathematical values against social network measures.

The analysis of project coalitions using SNA generates a very large volume of data, not least because the characteristics of project activity are represented by a large number of separate and discrete network activities. The classification of project coalition activity into the principle project functions and the further classification of the governance constituents enable a very detailed examination of the function and effectiveness of the project coalition. We have looked at two examples of representation of SNA data – the sociogram and mathematical data relating to centrality values for the main project actors. The intention has been to illustrate how analysis of construction project coalitions is possible using SNA.

Network study findings

At the time of writing four UK case studies were completed and a French case study involving Peugeot in Paris had been completed. A further case study in Beijing had started, data having been collected but not fully analysed. The following section presents findings based upon the first five (four UK and one French) studies in this programme. The findings are classified into five groups of issues. These are:

- Governance of the project
- Contractual
- Legal
- Effectiveness of communications
- Methodological

Project A comprised an archive building for a UK county council and employed a traditional procurement route without any post-Latham innovations in procurement or project management methods. Project B comprised a headquarters for an office equipment manufacturer built as a lease-back deal on land owned by the tenant for the complete building. This was chosen as an example of traditional procurement in the private sector. Project C was an office building for one of the UK's most prominent and innovative developers, involving partnered supplier relationships and hands-on supply chain management from the client actor team. This project was selected as an example of innovative practices in procurement and project management. Amongst the management team performing the client actor rôle were prominent policy makers within the UK construction industry. Project D was a sports facility, with office accommodation, for the UK armed forces. It was an example of innovative procurement and project management strategies within the public sector and was chosen as a demonstration project and a pilot study for prime contracting.

Projects were all of approximately similar value, corresponding to around £10 million at 2005 prices. Projects had a similar level of complexity in design and construction techniques employed. Data was gathered during the second half of the site construction phase of the projects.

Four case studies were featured in Table 9.1 and described briefly under the heading 'Comparable network centralities for four project cases' above. In the following section we draw some conclusions from the comparison of the four previous case studies with a fifth, French, case study. The French project was a logistics facility for Peugeot cars in Paris, comprising offices and warehouse accommodation using an example of *enterprises générales*, a relatively recent form of French procurement, similar to the UK 'with contractor's design' approach. Most importantly, this was seen as an opportunity to observe the use of performance incentives applied in a manner that would be unusual in the UK.

Governance of the project issues

Projects carried out on a 'with contractor's design' basis achieved single point responsibility and effective financial incentives to ensure that clients did not find the out-turn cost unacceptably surprising. Some might say that this is achieved at the expense of design quality. Where a 'with contractor's design' approach is not desirable, the use of a guaranteed maximum price supplement to a standard form of contract effectively changes the nature of the relationship between the client and the contractor, ensuring that the contractor manages the design development and vetoes any design features that have the effect of causing the project to exceed its budget. We found pain share-gain share (PSGS) types of arrangement to be less effective in changing coalition relationships simply because these arrangements tended to be informal and not applied promptly by clients.

The French case study used a tariff system of standard financial penalties for failure in relation to a wide range of specific areas of project performance. These included failure to provide design information in a timely fashion, failure to attend site meetings and other failures relating to noise and security matters, as well as more familiar failures relating to site production progress. These simple penalties were applied to all project actors on the French project and this helped to redress the balance of power in favour of the constructor. UK forms of procurement reflect the contractual hierarchy embedded in standard forms of building contract and do not help the constructor to apply pressure to other project actors in the pursuit of project timeliness and quality. Under the French system the constructor has more responsibility for design under traditional procurement approaches; the system of simple penalties applied *by the constructor* to other project actors, enables the constructor to apply leverage to the supply chain in a manner that would be welcomed by most UK constructors. The reluctance to employ this system in the UK might be related to the desire to preserve some sort of hierarchy within project coalition. Decisions about procurement are often made by, or with, professional service providers (PSPs) and it is not surprising that the resultant procurement and management systems tend to avoid applying penalties to PSPs as a group.

The use of PSGS, although limited by its administration by the client or their representative in the project coalition, does enable the client to effectively delegate the role of supply chain manager to a constructor, where the client lacks the knowledge, capacity or ambition to act in this capacity. Project D was procured by a client providing national defence and security. It was undesirable and impractical to manage the supply chain in the way that the client on Project C had, for example. Project D used work clusters and a PSGS arrangement to engage the constructor in the client-orientated tasks of managing design development and through this to manage contract out-turn costs.

Financial performance incentives seem to have been quite effective in all the case studies in changing the nature of relationships within the project coalition. The imposition of penalties upon professional service providers in order to support the constructor in attempting to achieve timeliness, quality and financial certainty for the client is important, not least in challenging the class system that seems to be maintained within the UK construction industry. The transfer of responsibility for design development, financial management and site production to cluster leaders must be regarded as unresolved and problematic, based on the evidence available so far. The role of cluster leader is new, informally defined and essentially unpaid, and there are some important issues relating to liability for design by what many might regard as unqualified project actors. This latter point has legal implications, which are explored in more detail under the heading of legal issues below.

The French case study utilised a large number of activity-related financial penalties, discussed above. Inspection of the network diagram in Figure 9.2 reveals a problem of *incentive contiguity*. Here we see that the client (PSA in Figure 9.2) is linked to the project manager (GEF) and the incentives cascade through the 'main contractors' (JOU and QUI) – there being two joint main contractors, a common device used in the French procurement system to increase the financial and skills base of contractors and thus the size of projects undertaken by those contractors. The incentives are then passed on to the various specialist subcontractors. The lack of contiguity is associated with the network isolates (AAG, BdC, BEJ, DLF and SPS) and a small cluster centred around SAT. None of these actors are connected to the main project network and are either not incentivised at all, in the case of the isolates, or only incentivised through another subcontract (and therefore not benefiting from the main incentive system initiated by the client). Further discussion of financial incentives in the governance of construction projects is contained in Pryke (forthcoming).

Networks clearly have an important part to play in the implementation of reforms because of their ability to identify the relationships and their significance in project delivery. The aggregation of the relationships helps identify the governance for projects. This is the case whether a traditional 'top-down' or a 'bottom-up' approach is chosen (Grantham, 2001). As implementation of reform is so often the most difficult part of project delivery (cf. Latham, 1994), the more that we can understand about relationships in the context of changing systems, the more likely we are to achieve effectiveness in implementation.

Contractual issues

Standing back from the case details, one can see a number of broader contractual issues emerge from the network analysis. The study of contractual relationships has revealed that the traditional JCT (Joint Contracts Tribunal) forms of contracts, including those based upon the JCT format, are structured around an early nineteenth century view of hierarchy in construction project teams. The architect or supervising officer has a high level of centrality within the coalition and is structurally superior to the main contractor, who in turn has superiority over the specialist subcontractors. The traditional forms of contract do not help the project coalition to operate as a construction supply chain and the supply chain manager's role is not identified within these traditional forms. The two-party or dyadic relationships that are implicit within traditional contract forms do not help to govern the activities of the project coalition; neither do they generate a system of financial incentives that functions sensibly and in the interests of the client and other stakeholders. An important departure from this inhibiting dyadic straitjacket came in the form of PPC 2000 in the UK, which was a brave attempt to redefine relationships, if not roles, within construction coalitions.

Clients with the knowledge, capacity and inclination to manage the supply chain from the position of client actor dramatically increase the client organisation's centrality in contractual as well as information exchange networks. Clients engaging with the design and production actors in construction find that they are able to achieve very short linkages with those holding detailed and highly specialist knowledge within the supply chain. This has several benefits. First, the contract can be quite simple in form – typically brief statements about time, costs, transfer of ownership of materials and insurance. Second, short linkages with specialist knowledge-holders in the coalition provide a knowledge transfer network, enabling the supply chain managing client to build a very high level of knowledge in-house. Finally, although simple contractual agreements do not necessarily reduce disputes over matters of content, reducing the number of contractual relationships *must* reduce the potential for contractual disputes.

Clients fall into two groups. There are those that do indeed have the knowledge, capacity and inclination to be placed in a position of prominence within the design and production supply chain. This type of client improves as a client and a supply chain manager over time provided that the staff they employ can be retained. There is an inevitable sense of self-perpetuation about this type of high-centrality client. Conversely, there are clients that, often for very good reasons, lack the knowledge, capacity or inclination and seek an agent to effectively manage the supply chain on their behalf. In this case the relationships can more easily reflect the traditional hierarchy of contractual relationship and the challenge for the client is to install an effective system of financial incentives to ensure that the project coalition functions effectively and in the client's interests at all times.

As far as the UK system is concerned, the design of effective financial incentives needs further work. At present consultants are, in effect, incentivised – through their conditions of employment – to reduce their engagement and time input into a project; contractors are equally not incentivised to innovate and reduce clients' out-turn costs. In some ways the use of strategic framework agreements, partnering and other forms of long-term relationships make matters even worse here. Unless the client can engage with the coalition and effectively manage the supply chain to improve efficiency and drive down costs, the abandonment of competitive tendering implicit within the partnering culture simply enables more opportunistic behaviour on the part of those who serve the construction client.

Legal issues

Projects C and D involved innovative procurement methods in the private and public sectors respectively, both of these projects entailing a radical change in project roles and relationships. Most fundamentally, both projects used work clusters or technology clusters (Gray, 1996). The use of work or technology clusters alters intra-coalition relationships in the following ways:

- Management of a cluster sub-budget becomes the responsibility of the cluster leader, effectively devolving this responsibility from the client's financial management consultant, which in the UK and some other countries is the quantity surveyor.
- Management of design within a given cluster (which might comprise the external envelope of the building, for example) is also carried out by the cluster leader to ensure that the trades within the cluster are coordinated and buildability and economy are optimised. There is a significant *disaggregation of design responsibility* here, and one that is not usually entirely welcomed by the design professionals, perhaps with good reason. Concerns about integrity of design are frequently raised.
- Finally, the cluster leader is responsible for the coordination of the work of the cluster during the production phase of the project.

Cluster leaders are individuals taken from one of the consultants or subcontractors employed on the site, and on our case study, Project C, were appointed informally and without separate compensation for the cluster leader role. Project D was a pilot study (see Holti *et al.*, 1998) undertaken by the one of the UK's largest public sector clients, and individuals acting as cluster leaders and their employers were motivated to cooperate by the prospect of a substantial future workload to be let to a small group of construction partners. Herein lies the most significant of the legal implications associated with the five case studies upon which this chapter is based.

Given the current structure of the UK industry, finding a suitably qualified cluster leader is problematic. Essentially the UK system of education and training provides constructors, financial managers and designers but not individuals with a good knowledge of all three areas. The legal liabilities for the cluster leader are substantial and dealt with in some detail in Pryke (2006). Perhaps the most important of these liabilities is that associated with design. While no standard conditions of employment appear to exist for cluster leaders, a recent judgment – *George Fischer Holdings* v *Multi Design Consultants* (*Construction News*, 2002) has established that the courts are likely to imply the existence of a contract of collateral warranty where exchanges of correspondence have indicated an intention to extend a duty of care to third parties. Having cooperated willingly to act in the role of cluster leader, our intrepid multi-disciplinary type might find that a project stakeholder, whom he or she has never met, is taking legal action relating to design defects some time after the project and the honorary role have been forgotten.

The case of *Pozzolanic Lytag* v *Bryan Hobson* [1998] BLR267 established an obligation for a professional whose contractual obligations go beyond his professional area of expertise to inform the client, so that the appropriate advice can be obtained. The courts are not really helping the industry in its endeavours to 'build down barriers' here and this is a powerful argument for the maintenance of the current status quo in relation to construction professions. It does have the effect of deterring the right-minded construction manager, for example, from providing design advice, or the designer from providing financial advice.

We have looked at one aspect of the legal issues associated with evolving roles and relationships involving innovative procurement and have drawn upon the supply chain management model for construction procurement. At the time of writing, legal problems arising out of the multi-disciplinary role of the cluster leader have not reached the courts. Clearly some unacceptably high risks have been transferred to cluster leaders in the interests of breaking away from traditional procurement methods.

Issues relating to effectiveness of communications

Some have argued that the construction coalition is a system of information management (Winch, 2002). If we accept this position, then understanding the operation and effectiveness of communications among coalition members is of fundamental importance. We looked at information exchange networks associated with financial management, progress management and design management. Analysis of these networks revealed the following:

- Procurement strategies involving long-term relationships coupled with effective management of the supply chain, either by the client organisation or by a professional service provider acting as an *agent* (the term agent deliberately used in a broader, economics-related, sense rather than the contractual definition here) for the client, appear to exhibit relatively low density values. In simple terms these communication networks appear to be less well connected than those found in traditional procurement environments. Evidence suggests that rather than this indicating badly organised and ineffective communications, it is indicative of a *lower level of need for extensive communications* (Pryke, 2005).
- Lower densities were found in both progress and design management communication networks under supply chain management environments. Linking this to the last point, it is suggested that when we manage the supply chain intensively from a central (possibly client role) position, we effectively manage the *causes* of both financial variations and problems associated with poor progress on site (timeliness of design information and buildability factors).
- Where the integration of the process of construction is taken seriously and is effective, the need for separate highly specialised management functions is reduced. In properly integrated systems (Holti *et al.* (1998) deal with these issues in relation to Project D above) external professional service providers (PSPs) were not in evidence – value management and risk management were neither desirable nor necessary as *stand-alone*

activities. The role of quantity surveyor was similarly of very low, or non-existent, prominence.

- The analysis of coalition communication relationships indicated that good clients make very effective creators and managers of integrated supply chains. Contractors, armed with the appropriate contractual authority and financial incentives by virtue of the highly central position created within the project networks, coupled with their knowledge and experience, also make effective managers of supply chains. Evidence suggests that the use of PSPs is generally less effective in the operation of integrated construction supply chains. It is suggested that this is partly because PSPs, certainly in the UK, tend not to be brought into financial incentive arrangements. The close link between fees and construction costs (even though 'lump sum' fees are almost exclusively applied) and the direct link between time allocated to task and profitability, tends to cause behaviour that minimises engagement by the PSP actors with each other and the remainder of the project coalition. The research found that communication was poor (evidenced by poorly connected, low density communication networks and high levels of isolates in these networks) and that as a result decisions relating to design and financial management appear to have been made on behalf of the client based upon insufficient information. In some instances PSPs appear to operate too independently of their coalition colleagues and do not therefore engage sufficiently with the coalition to effectively deal with their roles. In a very complex project environment the risky strategy adopted by individual PSPs may yield high profitability within each PSP; it may also lead to failure in delivering client satisfaction, or unacceptable levels of what Winch (2002) refers to as 'client surprise'.
- Although experienced clients appear to be most effective in the role of supply chain managers, even in projects where the client was very experienced and high levels of repetition in design detail and standardisation of specification were found, approximately 50% of all white collar management activity during the post-contract or construction phase of the projects involved design or specification matters. This indicates that the UK industry, at least, needs to achieve a step change in attitudes relating to bespoke design and standardisation. Significantly higher levels of standardisation will deliver smaller, less dense (cheaper, as a result of lower transaction costs) communication networks, improving profitability along the supply chain.

Network analysis issues

Social network analysis (SNA) is not yet commonplace in general management in the UK and is rarely applied to construction. Yet the *non-linear, complex, iterative and interactive project system environment* found in construction calls for a much higher level of sophistication than has hitherto been applied. SNA provides the tools to understand and analyse the relationships, and perhaps to move towards the formulation of a network theory of construction relationships. One reason for the lack of use of SNA in construction is that the barriers to entry are high because of the complexity of the language; indeed there is a lack of standardisation of key terms. The effort is, however, well worthwhile and this chapter has given some indication of the level of analysis that is possible through the application of SNA.

The use of SNA has required the acquisition of a certain level of knowledge in the social sciences and the need to interact with a group of individuals (SNA specialists) who have little or no knowledge or experience of construction. This in itself has been instructive and has provided an insight into how communities of practice work in non-construction spheres. Social network analysts are not members of professional bodies and organise themselves, perhaps not surprisingly, into networks for the purpose of sharing knowledge and educating newcomers to the discipline. As a group, social network analysts could provide considerable assistance to the traditional construction disciplines.

Summary

This chapter has presented an overview of SNA, an indication of how it might be used and the way in which analysis could be achieved. The chapter then moved on to the edited highlights of findings from five construction case studies. Space has not permitted that each individual point is proven here, but further details relating to such analysis have been referenced in the chapter.

Construction is not unique in its complexity or the nature of the systems employed to achieve completion. It is, however, an industry that has been satisfied with relatively unsophisticated methods of analysis, given the size of the industry and the size of its larger projects. We have presented some findings related to the way in which relationships within construction coalitions operate, with some emphasis on the use of financial incentives. We have also looked at the role that contractual relationships play in the effectiveness of the construction coalition. Innovation in any process brings inevitable legal problems and we looked at some of the key findings, particularly in relation to the transfer of responsibility of design and financial management responsibilities to those holding neither a relevant qualification nor professional indemnity insurance cover. It is hoped that the discussion in this chapter and research journals might herald the beginning of a debate about the exploitation of SNA in pursuit of excellence in construction systems.

References

Aoki, M., Gustafson, B. & Williamson, O. E. (1990) *The Firm as a Nexus of Treaties*. Sage, London.

Banwell, H. (1964) *The Placing and Management of Contracts for Building and Civil Engineering Works* (Banwell Report). Ministry of Public Building and Works, HMSO, London.

Berkowitz, S. D. (1982) *An Introduction to Structural Analysis: The Network Approach to Social Research*. Butterworth and Co. (Canada) Ltd.

Curtis, C., Ward, S. and Chapman, C. (1991) *Roles, Responsibilities and Risks in Management Contracting*. Construction Industry Research and Information Association, London.

Davis, K. & Newstrom, J. W. (1989) *Human Behaviour at Work*. McGraw-Hill, Singapore.

Egan (1998) *Rethinking Construction: The Report of the Construction Task Force to the Deputy Prime Minister, John Prescott, on the Scope for Improving the Quality and Efficiencies of UK Construction* (Egan Report). DETR at www.construction.detr.gov.uk/vis/rethink

Emmerson Report (1962) *Survey of Problems before the Construction Industries*. Ministry of Works, HMSO, London.

Grantham, A. (2001) How networks explain policy implementation outcomes: the case of UK rail privatization. *Public Administration*, Winter, **79** (4), 851–70.

Gray, C. (1996) *Value for Money*. Reading Construction Forum and The Reading Production Engineering Group, Berkshire.

Higgin, G. & Jessop, N. (1965) *Communications in the Building Industry: The Report of a Pilot Study*. Tavistock Publications, London.

Holti, R., Nicolini, D. & Smalley, M. (1998) *Prime Contractor's Handbook of Supply Chain Management*. Tavistock Institute, London.

Latham, M. (1994) *Constructing the Team: Joint Review of Procurement and Contractual Arrangements in the United Kingdom Construction Industry*. HMSO, London.

Loosemore, M. (1996) *Crisis management in building projects: a longitudinal investigation of communication and behaviour patterns within a grounded theory framework*. Ph.D. thesis, University of Reading.

Loosemore, M. (1998) Social network analysis using a quantitative tool within an interpretative context to explore the management of construction crises. *Engineering, Construction and Architectural Management*, **5** (4), 315–26.

Loosemore, M. (1999) Responsibility, power and construction conflict. *Construction Management and Economics*, **17**, 699–709.

Masterman, J. W. E (1992) *An Introduction to Building Procurement Systems*. E and F N Spon, London.

Nohria, N. & Eccles, R. G. (eds) (1992) *Networks and Organizations*. Harvard Business School Press, Boston, MA.

Pryke, S. D. (2001) *UK construction in transition: developing a social network approach to the evaluation of new procurement and management strategies*. PhD thesis, the Bartlett School, University College London.

Pryke, S. D. (2004) Analytical methods in construction procurement and management: a critical review. *Journal of Construction Procurement*, **10** (1), 49–67.

Pryke, S. D. (2005) Towards a social network theory of project governance. *Construction Economics and Management*, **23** (9), 927–39.

Pryke, S. D. (2006) Legal issues associated with emergent actor roles in innovative UK procurement: a prime contractor case study. Special issue on Legal Issues Associated with Relational Contracting. *ASCE Journal of the American Society of Civil Engineers*, **132** (1), 67–76.

Pryke, S. D. (forthcoming) Project governance: European case studies on financial incentives. *Building Research and Information*.

Scott, J. (2000) *Social Network Analysis: A Handbook*. Sage Publications, London.

Simon (1944) *The Placing and Management of Building Contracts*. The Central Council for Works and Buildings, HMSO, London.

Tavistock Institute (1966) *Interdependence and Uncertainty: A Study of the Building Industry*. Tavistock Publications, London.

Wasserman, S. & Faust, K. (1997) *Social Network Analysis: Methods and Applications*. Cambridge University Press, Cambridge.

Wasserman, S. & Galaskiewicz, J. (eds) (1993) *Advances in Social Network Analysis: Research in the Social and Behavioural Sciences*. Sage Publications, London.

Winch, G. M. (2002) *Managing the Construction Project*. Blackwell Science, Oxford.

Winch, G. M. & Carr, B. (2001) Processes, maps and protocols: understanding the shape of the construction process. *Construction Management and Economics*, **19**, 519–31.

10 Discourse and fashion in supply chain management

Stuart D. Green

This chapter offers a critical view of supply chain management (SCM) as a means of improving relations in the construction supply chain. SCM, partnering and collaborative working are all promoted as means of improving relations in the construction supply chain. A common theme is the need to overcome the construction sector's alleged adversarial culture. It is frequently claimed that construction would be much improved if only it would adopt the modern management practices utilised in other industries. Unfortunately, such exhortations tend to conceal more they reveal. Practitioners in the construction sector are too easily presented as being out of date and uninformed. Much of the current debate rests on an over-simplistic view of the extent to which organisations are capable of developing 'relationships'. The advocates of relational contracting tend to base their arguments on 'black box' models of the organisations involved. Attention is rarely given to the complex social processes that shape the diffusion of improvement recipes such as SCM.

This chapter delves within the 'black box' and develops an alternative view of organisations as pluralistic arenas where ideas for improvement are continuously contested. Espoused recipes for improving supply chain relations such as SCM can usefully be understood as management fashions. Of central importance is the concept of 'interpretive flexibility' and the way that improvement recipes frequently take on manifestations different from those envisaged. Developing reified models about the way relational contracting might operate in utopia is one thing; how such ideas are diffused and enacted in practice is another.

The argument unfolds as follows. Initially, the core concepts of relational contracting are rehearsed together with their supposed theoretical roots in transactional cost economics. The distinction is made between analytical concepts used for understanding and the more prescriptive improvement recipes that are mobilised for the purposes of promoting change. It is argued that relational contracting variants such as partnering, SCM and collaborative working fall within the latter category, and as such are best understood as management fashions. The basic characteristics of management fashions are described, with a particular focus on the diffusion process. If the micro-complexities of the diffusion process are to be understood, then organisations must be conceptualised as pluralistic arenas within which different improvement initiatives are contested, reinterpreted and enacted. Such a perspective challenges many existing assumptions about promoting 'better relations' in the construction supply chain. Attention is also given to the thorny issues of culture changes and trust. While the discussion focuses primarily on SCM, a similar analysis could also be applied to partnering, collaborative working, or any other fashionable variant of relational contracting.

Relational contracting

Core concepts

Relational contracting purports to offer an alternative to 'traditional', adversarial modes of engagement between contracting parties. The concept is based on the notion that relationships between contracting parties contain a significant social component (Macneil, 1980). Any enactment of relational contracting implies a long-term relationship with mutuality of objectives between the contracting parties. Relationships based on the recognition of mutual benefits supposedly lead to 'win–win' scenarios for the parties involved (Palaneeswaran *et al.*, 2003). These are the same guiding principles that underpin a range of 'enlightened' contractual practices, including SCM, partnering and collaborative working. The underlying script that unites these concepts is the alleged need to move away from the adversarial working practices of the past. Trust frequently figures large in the exhortations to adopt 'new ways of working', coupled with the requirement for a fundamental 'culture change'.

Transactional cost economics

The advocates of relational contracting frequently claim allegiance to the tradition of transactional cost economics (TCE), the basic principles of which have been well described in previous chapters. TCE claims to provide a framework for considering not only the costs of production, but also the costs of transactions between contracting parties (Williamson, 1985; Winch, 1989). Transactions are said to occur when a 'good or service is transferred across a technologically separable interface' (Williamson, 1985). Contracts are seen to provide governance structures that curtail the 'natural' tendency of economic agents to engage in opportunistic behaviours.

Much is made of the necessity for trust between contracting parties as a means of overcoming the inevitable deficiencies of legal contracts drafted on the basis of 'bounded rationality'. Trust is invariably seen as a property of organisations, with little recognition of the distinction between micro and macro levels of analysis. A recurring tendency is to assume that organisations display the same behavioural characteristics as individuals (Zaheer *et al.*, 1998). Rousseau (1985) refers to this confusion as the 'cross-level fallacy'. While it is easy to accept that there is a significant 'social component' between *people* engaging in business transactions, it is more contentious to extend this to business transactions between *firms*. This confusion between different units of analysis characterises much of the literature on SCM, partnering and collaborative working.

It is also important to recognise that TCE is essentially explanatory rather than prescriptive. The literature on TCE does not provide a recipe for 'improving' inter-organisational relationships. Such prescriptions are left to management gurus and the promoters of organisational change. Within these quarters, objective appraisal is too often abandoned in favour of the heady rhetoric of fashionable change recipes. Prior to progressing the argument further, it is necessary to address what is meant by 'management fashions' and the means by which they are diffused.

Management fashions

Interpretive flexibility

Management fashion has an important role in shaping the reality of modern organisations (Abrahamson, 1996; Clark & Salaman, 1998; Keiser, 1997). Previous exemplars include total quality management (TQM), business process re-engineering (BPR) and lean production (cf. Legge, 2002; Fincham, 1995; Benders & van Bijsterveld, 2000). Each of these has been marketed as a panacea for improved performance. They all share the characteristic of 'interpretive flexibility' in that they are subject to multiple interpretations. This vagueness of definition directly aids diffusion by enabling different storylines to be mobilised in different contexts. Such a diagnosis raises questions about what the 'implementation' of these recipes actually means. The relationship between managerial discourse and action is rarely straightforward.

While the notion of fashion is undoubtedly pejorative, this does not mean it can be dismissed as irrelevant to 'real world' issues. The way managerial discourses are mobilised and diffused impacts directly upon the 'reality' of modern organisations. However, the legitimacy of a management fashion depends upon the extent to which it echoes the discourse and ideology of the 'enterprise culture' (cf. Legge, 1999). Recurring themes include the iconic status ascribed to the 'customer' and an economic imperative to adopt 'new ways of working' in the face of irresistible global competition. Favoured discourses include the setting of performance targets together with an ongoing emphasis on measurement and continuous improvement. Such storylines invariably co-exist with exhortations for trust and improved inter-firm relations.

Consultants as intermediaries

Consultants frequently play an important role in the diffusion of management fashions by mediating between management gurus and end consumers (Alvarez *et al.*, 1998; Scarbrough, 2003). Of particular note is the ability of management consultants to legitimise new ideas in the eyes of practising managers. The mediating role is further supported through their involvement in inter-organisational networks dedicated to promoting change. Legge (2002) argues that consultants use such networks to build demand for their products. Consultants are especially adept at promoting a succession of management improvement recipes to ensure that there is a continuous market for their services. For example, the fact that BPR never lived up to its promise did not prevent them from presenting lean production as the next panacea for the industry's problems. These are the same consultants who are now advocating SCM.

Part of the skill of any successful consultant is the development of a range of alternative storylines under the same brand label. This enables them to maximise perceived relevance across a range of contexts. A common feature is the cross-colonisation of storylines. For example, numerous sources on SCM re-cycle previous ideas from BPR and lean production (for example, Childerhouse *et al.*, 2003; Christopher & Towill, 2000). Cross-colonisation frequently takes place to such an extent that there is little difference between management ideas other than the rhetoric within which they are presented (cf. De Cock & Hipkin, 1997).

Notwithstanding the above, practising managers are rarely passive and gullible consumers of management fashion. Management groups within receiving organisations mobilise the storylines that accord best with their own political agendas. Consultants and users therefore have a shared vested interest in the 'interpretive flexibility' of management fashions (Benders & van Veen, 2001). Consultants benefit through the continuous creation of markets for their services, and internal management groups have a continuous need for persuasive scripts against which they can act out the role of improvement champions. Promoters and users of management fashion therefore engage in a collusive interaction that serves the interests of both parties (Scarbrough, 2003).

Understanding organisations

Beyond the machine metaphor

The notion that practitioners mobilise management fashions as part of an internal power game invokes a political perspective on organisations whereby different interest groups compete for power and influence. The political complexities of organisations are invariably suppressed by the discourse of popularised management improvement recipes. The management fashions that are most readily accepted within the construction industry are those that conceptualise organisations as goal-seeking machines. The machine metaphor reflects the strong engineering orientation among practitioners and researchers in the construction domain (Bresnen & Marshall, 2001).

When viewed through the machine lens, the dominant task of management is to optimise efficiency of operation through the elimination of waste. Such storylines are based on an implicit assumption that organisations are unitary entities with all parties working towards a predetermined set of objectives. The rhetoric is frequently sweetened by lip service to teamwork and claims that 'people are our best asset'. But the implied model of a 'good team player' too often accords with the machine metaphor. The dominant view is that *a good team player is somebody who fits in. A good team player is somebody who tows the line.* It follows that too many 'good team players' are a liability in that the organisation is likely to become sterile. Homogeneous organisations comprising only 'good team players' are unlikely to possess the capacity to innovate, or to respond to changing circumstances. Senior management frequently espouses teamworking rhetoric while embarking on downsizing initiatives, usually without any obvious signs of irony.

The tendency to view organisations as goal-seeking machines is deep-seated within the field of micro economics, whose disciples frequently treat organisations as 'black boxes' in pursuit of profit maximisation. Economic analysis based on this perspective may well provide valuable insights into the functioning of the marketplace, but offers little in terms of understanding the way relational contracting is conceptualised and enacted within localised contexts.

Contested arenas

As described above, the machine metaphor sees individuals as mindless cogwheels operating in support of the organisation's goals. In contrast, political models of organisation focus attention on what happens inside the black box. Particular emphasis is given to the roles of power and conflict and the way that individuals use organisations as vehicles to achieve their own aspirations. Groups of individuals are seen to form temporary coalitions for the purpose of mutual interest. Such alliances are continuously changing as new power groups emerge and individuals transfer their allegiance from one group to another (Green, 1998). Management fashions are directly implicated in that different improvement recipes are mobilised by different interest groups. Ideas are continuously contested within pluralistic arenas. Some will be in the ascendancy, others will be in decline. Given the interpretive flexibility of management fashions, there is likely to be much substantive overlap between the competing storylines.

Such a model of organisations is by no means dysfunctional. Individuals can be highly motivated by such internal competitions, resulting in a less complacent and more innovative workforce. There is also a possibility that ongoing debate and reflection will trigger the creative adaptation of management ideas to suit localised contexts. The promoters of any given management fashion will thereby try different storylines until they find the one that 'sticks' the best. Accommodations would be sought with other improvement recipes as a means of 'buying off' sources of resistance. However, any deal would be transient and subject to re-negotiation across a series of contested arenas. The processes of innovation and diffusion are therefore not linear and sequential, but iterative and inter-dependent (see also **Chapter 9** on this point). Models of 'relational contracting' such as SCM or partnering will manifest themselves differently in different contexts. Any generalisations regarding the likely outcomes of such improvement initiatives are likely to conceal more than they reveal.

Dramaturgical metaphor

The political perspective developed above sees organisational life as being played out across pluralistic arenas where differences are contested, and to some extent reconciled. The outcomes of these engagements include a partial reconfiguration of the pre-existing factions and alliances. Additional insights can be gained by invoking the dramaturgical metaphor to shed light on how individuals act out 'roles' and utilise 'scripts' (Mangham, 1990; Clark & Salaman, 1996). From this perspective, the events acted out within organisational arenas are elements of performance. For example, any appointed (or self-appointed) 'champion' of SCM, acts out the role in accordance with the accepted script. Published 'best practice' guidelines and prescriptive textbooks provide the scripts against which they improvise.

The lexicon of 'supply chain management' therefore provides the language of the performance: 'customer responsiveness must be maximised', 'lead times must be reduced' and 'culture must be changed'. Others within the arena will dispute the meaning of such 'jargon' and will mobilise other resources as a means of resistance. Successful careers are built on the skills of performance. The implementation of any such initiative will provide a further series of acts in the unfolding drama. The closing scene will inevitably involve the search for the guilty when the initiative fails and is no longer judged fashionable. If the champions are wise they will have moved on by this stage to a lucrative appointment elsewhere.

Culture change

Behavioural compliance

The rhetoric of any fashionable improvement recipe tends to be hugely optimistic about the extent to which the construction sector's 'culture' can be changed. The associated assumption that organisational performance depends upon an alignment between employee values and managerial strategy primarily dates from Peters and Waterman (1982). However, others have questioned the extent to which culture can be manipulated and controlled (for example, Antony, 1994; Legge, 1994; Willmott, 1993). Of particular interest is the possibility that managerial action may promote unforeseen countercultures that are dysfunctional. Although the literature on organisational culture is huge, there are relatively few research studies of sector level ('macro-culture') change in construction. However, Ogbonna and Harris's review (2002) of culture change in UK food retailing provides a useful point of reference.

Ogbonna and Harris observe that managerial ambitions for culture change are invariably overly optimistic. In the first of two case studies, they describe how for the vast majority of employees the culture change programme resulted in 'instrumental behavioural compliance'. In the second, they found *some* evidence of limited and unpredictable value change. In others words, even when value change occurs it is not necessarily in the intended direction. Ogbonna and Harris further observe that employees are likely to be less persuaded by the discourse of culture change when the espoused storyline is in direct conflict with experienced reality. It should also be noted that in both reported cases the culture change programmes were initiated by the companies themselves, rather than by outside parties as tends to be the case in construction. Resigned behavioural compliance would therefore seem to be the most likely outcome in the face of the 'relentless pressure' applied by clients as part of partnering initiatives (cf. Bennett & Jayes, 1998). Indeed, it may be that the continued imposition of simplistic improvement recipes directly contributes to the 'bad attitudes' and 'adversarial culture' that industry leaders repeatedly decry.

Trust and power

Culture change and trust are inextricably linked in the discourse of SCM and other variants of relational contracting. 'Trust' between contracting partners is seemingly an essential ingredient of the desired cultural recipe. If culture is slippery territory, trust provides another metaphorical skating rink. Sabel (1992) defines trust as the *'mutual confidence that no party to an exchange will exploit the other's vulnerability'*. The debate is once again characterised by an assumption that firms are synonymous with individuals. Some sources even suggest that partnering relationships between firms are comparable to marriages between individuals. While trust between organisations is generally seen as an essential prerequisite of relational contracting, others have questioned whether economic cooperation is the product of trust, or whether trust results from economic cooperation (for example, Gambetta, 1988). Rarely is any attention given to the way that trust is shaped by imbalances in economic power between firms in the supply chain and the broader dynamics of the marketplace (cf. Korczynski, 2000).

Cox (1999a) argues that the nature of any relationship between organisations is inevitably mediated by power differentials between the contracting parties. Yet the more common assumption is that trust relies entirely on the interactions between individual actors without any reference to the constraints and pressures imposed by the broader context. In situations where there is a power imbalance between firms, cooperation may be enforced through power rather than trust. In such circumstances, the weaker party will feel exploited and, by definition, will not trust the stronger party. In these circumstances, adversarial relationships may well be suppressed by enforced cooperation. But this has little to do with 'culture change' and is likely to erode whatever trust previously existed. Indeed, such instances are likely to generate patterns of resistance and initiate distinctive countercultures that may well precondition the response of individuals to any future advocated 'enlightened practice'. For example, the Construction Clients' Forum's attempt (1998) to impose partnering on its members' supply chains directly undermined its stated commitment to relationships based on trust. Likewise, the infamous appeals of the Construction Industry Board (1997) to adopt partnering on the basis of 'fundamental belief, faith and stamina' are surely anathema to the concept of the 'learning organisation'.

Here lies a recurring theme. Proponents of industry improvement tend to adopt a top-down approach to workplace change. Even more bizarrely, the Egan Report (DETR, 1998) bypassed the construction sector's management to promote a change programme disproportionately biased towards the interests of large repeat clients. The legitimacy of large clients such as BAA and Tesco to impose ideas on a construction sector comprising 1.8 million people was blithely accepted in the cause of 'customer responsiveness'. The only seemingly acceptable role for managers and employees within construction companies was behavioural compliance.

Supply chain management: a model of imprecision

Relational contracting *per se* is unlikely to achieve the status of a management fashion, but SCM displays all the fundamental characteristics described above. Indeed, SCM has been promoted as an essential stepping stone towards the Holy Grail of relational contracting. While it is easy to accept SCM as a 'good idea', defining what it means is by no means straightforward. In common with many other supposedly new business improvement strategies, SCM suffers from definitional vagueness (cf. Bresnen & Marshall, 2001).

Definitional vagueness

The essential storyline of SCM is that it is no longer sufficient for managers to confine attention to their own organisation. They are now required to manage the 'integrated supply chain' (cf. Strategic Forum, 2002). Within this expanded domain of attention, most definitions accord with Cooper *et al.* (1997) in construing SCM as the 'integration of business processes from end user through original suppliers that provide products, services and information that add value for customers'. Such definitions are laden with managerial rhetoric and replete with associated assumptions. Of particular note is the unquestioned allegiance to the 'cult of the customer' (cf. du Gay & Salaman, 1992). The suggestion that activities are only worth doing if they add value to the customer is also suggestive of the 'slash-and-burn' strategies of business process re-engineering (BPR) (cf. Hammer & Champy, 1993).

Such approaches draw freely from simplistic machine metaphors with the implication that all parties within a complex social system will strive to achieve common objectives. Researchers also frequently portray organisations as homogeneous entities with little variation between the interests of individuals (Marchington & Vincent, 2004). The concept of SCM extends this dubious assumption even further. Firms, and even more contentiously individuals within firms, are assumed to be willing to subjugate their own interests for the common good of the supply chain. The rhetoric of SCM reduces complex and dynamic multi-organisational structures to black box entities in pursuit of process efficiency and customer responsiveness.

Strategic positioning

Several literature sources distinguish between strategic and operational SCM (for example, Cox, 1999b; Fuchs, 1998). The strategic view is primarily concerned with competitive positioning in the marketplace. Of central importance for firms is to decide which capabilities to develop in-house and which to outsource. This ongoing process of capability assessment is likely to be informed by a continuous appraisal of market opportunities. Such considerations are usually accompanied by an ongoing process of risk assessment. Strategic decision-making is conditioned by the wish to avoid the risk of key competencies falling under the control of competitors. The appeal of SCM is that it purportedly provides the means of retaining some degree of control over and coordination of outsourced activities. In this respect, the relevance of SCM is determined by the extent to which capabilities are being retained in-house or outsourced. The greater the focus on the latter, the more relevance is likely to be attached to the discourse of SCM. Such issues are likely to be continuously debated and contested at board level. Given that SCM is enacted by middle managers, any commitment to 'relational contracting' would always be subject to disruption from strategic decisions relating to the organisation's core capabilities.

But the 'strategy' storyline unfolded above is guilty once again of conceptualising organisations as single entities that engage in rational behaviour on the basis of perfect information. Reality is always much messier, and strategies are much more likely to be reactive and emergent. Decision makers are also prone to follow the pack, as this is perceived as being less risky. Spender's notion (1989) of industry recipes is relevant here, referring to the 'set of assumptions held in common within an industry about organisational purposes and a shared wisdom on how to manage organisations'. Changes attributed to strategic decision-making may therefore have more to do with isomorphism (commonality of forms) at the sector level (cf. DiMaggio & Powell, 1983). The discourse of SCM is therefore interlinked with the accepted industry recipe.

Enhancing efficiency

Leaving aside the heady world of strategy, the vast majority of the SCM literature is concerned with improving efficiency through the implementation of logistics (for example, Christopher 1998; Handfield & Nichols, 1999). Efficiency of operation is likely to be of central concern to middle managers who are themselves threatened by ongoing processes of restructuring and outsourcing. The discourse of SCM therefore provides them with a sense-making narrative in the context of an uncertain and changing world. Furthermore, it accords them with the legitimacy to act out the role of SCM champion. The espoused challenge is to implement more efficient ways of managing the flow of goods, services and information across the whole supply chain. The dominant metaphor is once again that of the machine. The quest is to make the machine more efficient, but the 'machine' now comprises the supply chain rather than the firm.

As described above, the enactment of SCM takes on different manifestations in different contexts. The discourse of SCM inevitably mixes with that of other improvement recipes in unique and transient combinations. Notions of 'integration', 'sharing the benefits' and 'continuous improvement' are essential parts of the script. The extent to which suppliers trust the advocates of SCM depends in part on their previous experience with the parties concerned. Cox and Ireland (2002) argue that the SCM cannot be understood in isolation from the power differentials between contracting firms. Firms are seen to enter into relationships for the purposes of appropriating value. Any espoused rhetoric of collaborative working may therefore serve to camouflage the fact that the dominant party continues to behave opportunistically. This tendency has been especially prevalent in 'partnering' arrangements where there is a significant power differential between the contracting firms (cf. Green, 1999).

Supply chain management in construction

If the generic literature on SCM lacks definitional certainty, this is even more true of the construction-specific literature. A dominant theme is the extent to which the 'generic theory of supply chain management' can be applied to construction (cf. Fisher & Morledge, 2002; Vrijhoef & Koskela, 2000). The main problem with this argument is that there is no single generic theory of SCM. Nevertheless, the dominant view is that the construction sector needs to follow the example of other industries and adopt the 'philosophy of SCM'. Sources tend to adopt an instrumentalist view of SCM and its supposed benefits. All that is required is to overcome outdated workplace culture and lack of management commitment (Akintoye *et al.*, 2000).

Vrijhoef and Koskela (2000) provide evidence of the cross-colonisation of storylines by arguing that SCM is based on similar 'theoretical concepts' as just-in-time (JIT) and lean production. All such concepts serve to reflect and reinforce the ideological construct of the enterprise culture. Once again, the machine metaphor dominates – as evidenced by their fixation with 'waste and problems'. The overriding assumption is that managers in the construction sector are in some way misinformed. What is seemingly absent is any recognition that SCM will be conceptualised and enacted differently in accordance with localised needs.

Other sources on SCM in the construction sector draw inspiration from the supposed lean and agile paradigms of production (cf. Towill & Christopher, 2002). The popularity of lean production is allegedly in decline as a result of its tendency to focus on efficiency at the expense of flexibility. In response, Naim and Barlow (2003) combine aspects of the lean and agile paradigms to recommend an 'innovative supply strategy' for customised housing. This demonstrates a further interweaving of SCM discourse with that of other management fashions. Elsewhere, Childerhouse *et al.* (2003) demonstrate a continued allegiance to business process re-engineering (BPR) in their approach to managing material flow in the construction supply chain. It is seemingly impossible to draw discrete boundaries between SCM and other alleged 'new ways of working'.

Perhaps the most influential model of SCM for the construction sector is the 'work cluster' approach derived from the action-research project known as *Building Down Barriers* (Holti *et al.*, 2000; Nicolini *et al.*, 2001). However, it is interesting to note the lack of connectivity between the advocated approach and the generic SCM literature. The espoused ideas owe more to the doctrines of semi-autonomous work groups than they do to SCM as advocated by the likes of Christopher (1998). The *Building Down Barriers* storyline advocates that projects should be broken down into semi-independent clusters to promote collaborative working among designers and suppliers (Nicolini *et al.*, 2001). The adopted perspective is primarily operational; there is little acknowledgement of how broader issues of competitive positioning impinge upon collaborative working on a project level. The associated 'best practice' guidance (for example, Holti *et al.*, 2000) draws extensively from established construction techniques such as value management and through-life costing.

Elsewhere, Nicolini *et al.* (2000) focus on 'target costing' as a means of involving the whole supply chain in achieving the best balance between through-life cost and functionality. However, the reported attempts at pilot target costing were apparently impeded by a lack of trust that forced participants to regress to 'established, non-collaborative practices'. Such tendencies are argued to be 'deep-seated within the industry's culture' (Nicolini *et al.*, 2000). Successful SCM in the construction sector would therefore seem to depend upon culture change, the difficulties of which have already been discussed.

However, a more fundamental problem is that, beyond the notion of managing across boundaries, there is little agreement on what SCM actually *is*. This empirical elusiveness presents a significant obstacle to researchers seeking to demonstrate the efficacy of SCM. If there is no agreed definition of SCM, it becomes impossible to demonstrate any impact on 'improving relations in the supply chain'.

Conclusion

This chapter has questioned the extent to which the diffusion of managerial recipes such as SCM will improve relations in the supply chain. SCM has been seen to display the essential characteristics of a management fashion. The underpinning literature is characterised by a definitional vagueness that maintains maximum scope for 'interpretive flexibility'. In consequence, SCM is commodified by consultants in a variety of ways in order to maximise its commercial appeal in the marketplace. Many management academics are complicit in this process. The broader legitimacy of SCM depends upon the extent to which it resonates with the rhetoric of the 'enterprise culture'. Customer responsiveness, continuous improvement and performance measurement are recurring ideological themes.

The interpretive flexibility of SCM also serves the interests of practising managers, who tend to adopt those scripts that serve their interests within the context of ongoing organisational power games. Improvement ideas such as SCM are debated and contested across a range of organisational arenas. Accommodations are sought with other interest groups to achieve broader political support. Once sufficient support has been achieved among competing managerial groups the initiative will move into the implementation phase, inevitably involving further localised adaptation. The diffusion process is highly socialised and politicised, with the result that the advocated concept frequently takes on different manifestations from those envisaged. Operational managers faced with short-term financial performance measures frequently mobilise the rhetoric of culture change and trust as a means of exerting control over others. This is much more likely to result in instrumental behavioural compliance than in any long-term change in underlying values.

Notwithstanding the above, local variations in the enactment of SCM are inevitably patterned and conditioned by broader structural changes across the sector. In the case of the UK construction industry, extensive restructuring over three decades has involved an increased reliance on subcontracting. The major contractors that characterised the 1970s have transformed into hollowed-out service companies increasingly detached from the physical work of construction. The Holy Grail of 'labour market flexibility' has consistently been given priority over skills development, as dictated by the requirements of the 'enterprise culture'. The failure to invest in the employment relationship sits ill with any espoused commitment to 'relational contracting'. The core capabilities of major construction firms in the twenty-first century have more to with cash flow management and the apportionment of risk than they do with the physical skills of construction. Within this context, SCM is best presented as a means of minimising the risk and loss of control associated with outsourcing strategies.

In conclusion, the current fashion for 'relational contracting' must be understood in the broader context of sectoral change. The discourse of SCM has become attractive to main contractors primarily as a result of their increased reliance on sub-contracting. The competitive advantage of contracting firms increasingly rests on efficiency in managing contracts. The storylines of SCM therefore serve to justify trends that are already well established and institutionalised. SCM is perhaps best understood as a legitimising discourse rather than an instrumental means of improving 'relations in the supply chain'. In many respects, the discourse of SCM provides a comfort blanket to bolster the self-identities of practising managers who feel overwhelmed by change and uncertainty beyond their control.

References

Abrahamson, E. (1996) Management fashion. *Academy of Management Review*, **21** (1), 254–85.

Akintoye, A., McIntosh, G. & Fitzgerald, E. (2000) A survey of supply chain collaboration and management in the UK construction industry. *European Journal of Purchasing and Supply Management*, **6**, 159–68.

Alvarez, J. L. (ed.) (1998) *The Diffusion and Consumption of Business Knowledge*. Macmillan, London.

Antony, P. D. (1994) *Managing Culture*. Open University Press, Milton Keynes.

Benders, J. & van Bijsterveld, M. (2000) Leaning on lean: the reception of a management fashion in Germany. *New Technology, Work and Employment*, **15** (1), 50–64.

Benders, J. & van Veen, K. (2001) What's in a fashion? Interpretative viability and management fashions. *Organization*, **8**, 33–54.

Bennett, J. & Jayes, S. (1998) *The Seven Pillars of Partnering*. Thomas Telford, London.

Bresnen, M. & Marshall, N. (2001) Understanding the diffusion and application of new management ideas in construction. *Engineering, Construction and Architectural Management*, **8** (5/6), 335–45.

Childerhouse, P., Lewis, L., Naim, M. & Towell, D. R. (2003) Re-engineering a construction supply chain: a material flow approach. *Supply Chain Management*, **8** (4), 395–406.

Christopher, M. (1998) *Logistics and Supply Chain Management* (2nd edn). Pitman, London.

Christopher, M. & Towill, D. R. (2000) Supply chain migration from lean and functional to agile and customised. *Supply Chain Management: An International Journal*, **5** (4), 206–13.

Clark, T. & Salaman, G. (1996) The use of metaphor in the client-consultant relationship: A study of management consultancies. In: C. Oswick & D. Grant (eds) *Organization Development: Metaphorical Explorations*. Pitman, London.

Clark, T. & Salaman, G. (1998) Telling tales: management guru's narratives and the construction of managerial identity. *Journal of Management Studies*, **35** (2), 137–61.

Construction Clients' Forum (1998) *Constructing Improvement*. Construction Clients' Forum, London.

Construction Industry Board (1997) *Partnering in the Team*. Report by Working Group 12 of the Construction Industry Board. Thomas Telford, London.

Cooper, M. C., Lambert, M. & Pagh, J. D. (1997) Supply chain management: more than a new name for logistics. *International Journal of Logistics Management*, **8** (1), 1–13.

Cox, A. (1999a) Power, value and supply chain management. *Supply Chain Management*, **4** (4), 167–75.

Cox, A. (1999b) A research agenda for supply chain and business management thinking. *Supply Chain Management*, **4** (4), 209–11.

Cox, A. & Ireland, P. (2002) Managing construction supply chains: the common sense approach. *Engineering, Construction and Architectural Management*, **9** (5/6), 409–18.

de Cock, C. & Hipkin, I. (1997) TQM and BPR: beyond the beyond myth. *Journal of Management Studies*, **34** (5), 659–75.

Department of Environment Transport and the Regions (1998) *Rethinking Construction*. Report of the Construction Task Force. DETR, London.

DiMaggio, P. J. & Powell, W. W. (1983) The iron cage revisited: institutional isomorphism and collective rationality in organizational fields. *American Sociological Review*, **48**, 147–60.

Fincham, R. (1995) Business process re-engineering and the commodification of management knowledge. *Journal of Marketing Management*, **11** (7), 707–20.

Fisher, N. & Morledge, R. (2002) Supply chain management. In: J. Kelly, R. Morledge & S. Wilkinson (eds) *Best Value in Construction*. Blackwell Science, Oxford.

Fuchs, P. (1998) New approach to strategy: dynamic alignment of strategy in execution. In: J. L. Gattorna (ed.) *Strategic Supply Chain Alignment: Best Practice in Supply Chain Management*. Gower, Aldershot.

Gambetta, D. (1988) Can we trust trust? In: D. Gambetta, (ed.) *Trust*. Blackwell, Oxford.

du Gay, P. & Salaman, G. (1992) The cult(ure) of the customer. *Journal of Management Studies*, **29** (5), 615–33.

Green, S. D. (1998) The technocratic totalitarianism of construction process improvement: a critical perspective. *Engineering, Construction and Architectural Management*, **5** (4), 376–86.

Green, S. D. (1999) Partnering: the propaganda of corporatism? *Journal of Construction Procurement*, **5** (2), 177–86.

Hammer, M. & Champy, J. (1993) *Re-engineering the Corporation: A Manifesto for Business Revolution*. Harper Collins, London.

Handfield, R. B. & Nichols, E. L. (1999) *Introduction to Supply Chain Management*. Prentice Hall, Englewood Cliffs, NJ.

Holti, R., Nicolini, D. & Smalley, M. (2000) *The Handbook of Supply Chain Management: The Essentials*. CIRIA, London.

Keiser, A. (1997) Rhetoric and myth in management fashion. *Organization*, **4** (1), 49–74.

Korczynski, M. (2000) The political economy of trust. *Journal of Management Studies*, **37**, 1–21.

Legge, K. (1994) Managing culture: fact or fiction. In: K. Sisson (ed.) *Personnel Management: A Comprehensive Guide to Theory and Practice in Britain*. Blackwell, Oxford.

Legge, K. (1999) Representing people at work. *Organization*, **6** (2), 247–64.

Legge, K. (2002) On knowledge, business consultants and the selling of total quality management. In: T. Clark & R. Fincham (eds) *Critical Consulting: New Perspectives on the Management Advice Industry*. Blackwell, Oxford.

Macneil, I. R. (1980) *The New Social Contract: An Enquiry into Modern Contractual Relations*. Yale University Press, New Haven, CT.

Mangham, I. L. (1990) Managing as performing art. *British Journal of Management*, **1**, 105–15.

Marchington, M. & Vincent, S. (2004) Analysing the influence of institutional, organizational and interpersonal forces in shaping inter-organizational relations. *Journal of Management Studies*, **41** (6), 1029–56.

Naim, M. & Barlow, J. (2003) An innovative supply chain strategy for customized housing. *Construction Management and Economics*, **21**, 593–602.

Nicolini, D., Holti, R. & Smalley, M. (2001) Integrating project activities: the theory and practice of managing the supply chains through clusters. *Construction Management and Economics*, **19**, 37–47.

Nicolini, D., Tomkins, C., Holti, R., Oldman, A. & Smalley, M. (2000) Can target costing and whole life costing be applied in the construction industry? Evidence from two case studies. *British Journal of Management*, **11**, 303–24.

Ogbonna, E. & Harris, L. C. (2002) Organizational culture: a ten year, two-phase study of change in the UK food retailing sector. *Journal of Management Studies*, **39** (5), 673–706.

Palaneeswaran, E., Kumaraswamy, M. M. & Ng, S. T. (2003) Formulating a framework for relationally integrated construction supply chains. *Journal of Construction Research*, **4** (2), 189–205.

Peters, T. J. & Waterman, R. H. (1982) *In Search of Excellence: Lessons from America's Best Run Companies*. Harper & Row, New York.

Rousseau, D. M. (1985) Issues of level in organizational research. In: L. L. Cummings & B. M. Staw (eds) *Research in Organizational Behavior*, Vol. 7. JAI Press, Greenwich, CT.

Sabel, C. (1992) Studied trust: building new forms of co-operation in a volatile economy. In: F. Pyke & W. Sengenberger (eds) *Industrial Districts and Local Economic Regeneration*. International Institute for Labour Studies, Geneva.

Scarbrough, H. (2003) The role of intermediary groups in shaping management fashion. *International Studies of Management and Organisation*, **32** (4), 87–103.

Spender, J-C. (1989) *Industry Recipes*. Blackwell, Oxford.

Strategic Forum (2002) *Accelerating Change*. Rethinking Construction, Strategic Forum, London.

Towill, D. & Christopher, M. (2002) The supply chain strategy conundrum: to be lean *or* agile or to be lean *and* agile? *International Journal of Logistics: Research and Applications*, **5** (3), 299–309.

Vrijhoef, R. & Koskela, L. (2000) The four roles of supply chain management in construction. *European Journal of Purchasing and Supply Management*, **6**, 169–78.

Williamson, O. E. (1985) *The Economic Institutions of Capitalism: Firms, Markets and Relational Contracting*. Free Press, New York.

Willmott, H. (1993) 'Strength is ignorance: slavery is freedom': managing culture in modern organisations. *Journal of Management Studies*, **30** (4), 515–52.

Winch, G. (1989) The construction firm and the construction project: a transactional cost approach. *Construction Management and Economics*, **7**, 331–45.

Zaheer, A., McEvily, B. & Perrone, V. (1998) Does trust matter? Exploring the effects of interorganizational and inter-personal trust on performance. *Organization Science*, **9** (2), 141–59.

11 Relationship management theories and tools in project procurement

Andrew Cox and Paul Ireland

This chapter provides an introduction to the main theoretical approaches to business relationship management, especially those developed in the relationship marketing and partnership sourcing literature. The chapter contends that current practice derived from this literature – in particular in construction – often fails to allow buyers, that is, clients, to manage project relationships effectively. This is because when buyers, the clients, make sourcing decisions they must understand the paradox of value and interests in business relationships when buyers and suppliers interact.

The primary paradox in buyer and supplier exchange is that ideal forms of 'win–win' are only sustainable in a project environment under limited circumstances and that opportunistic forms of 'win–lose' and more restricted forms of collaboration are all that is achievable, despite proponents of relationship marketing and partnership sourcing arguing for and supporting wide-scale adoption. It is also argued that these problems are compounded by the advice contained within government-sponsored industry reports. These reports often unthinkingly promote the adoption of procurement and sourcing approaches that have worked in some industries, without regard to the dissimilar demand and supply and power circumstances operating elsewhere.

The chapter provides practitioners with a framework for understanding effective relationship management choices for buyers or clients and suppliers in industries and supply chains that have a primarily project-based demand and supply structure. It also provides buyers and suppliers with a way of thinking about the appropriateness of different relationship management approaches given a firm's power position within the supply chain. Finally, the chapter explains, from a procurement or sourcing perspective, the limited circumstances under which the proactive and collaborative relationship approaches articulated in the relationship marketing and partnership sourcing literature can be implemented successfully.

Business relationships

Between the 1960s and 1980s 'traditional purchasing practices' were driven by leveraged negotiations in which the buyer used power – a threat of

Table 11.1 The procurement and supply management continuum. Adapted from Moody (1993).

	Late 1940s to 1960s traditional practices	1970s to early 1990s	After mid-1990s partnering for survival, innovation and growth
Nomenclature	Purchasing	Procurement ■ Strategic procurement ■ Strategic purchasing	Supply management ■ Supply chain management
Industry structure	Tiers	Tiers	■ Partnering cluster arrangements ■ Collaborative joint ventures ■ Buyer-supplier alliances
Strategy focus	Cost-driven	Cost-driven	Total value-driven
Timeframe	Short-term	Short-term	Long-term
Tactics	■ Leverage ■ Multiple vendors ■ Legal contracts ■ No information sharing ■ Adversarial negotiation	As in late 1940s to 1960s with: ■ Continuous quality improvement	■ Partnering ■ Few suppliers ■ Blurred organisational boundaries ■ Transparent communications ■ Collaborative problem-solving

withdrawal of contracts, the promise of future business, supplier concern that the buyer would source from alternative sources, or sheer size and scale – to obtain the best price, best service, best delivery terms and acceptable quality from a 'captive' supplier (Forrester, 1958).

However, this approach failed to foresee the movement towards less confrontational approaches initially referred to as 'procurement', encompassing 'strategic purchasing' and 'strategic procurement', and, after the emergence of Japanese sourcing practices, as 'supply management', encompassing 'supply chain management' or 'partnership sourcing' (Macbeth & Ferguson, 1994) (see Table 11.1). These approaches to procurement best practice mirror the parallel and related trends on the supply side in favour of network and relationship marketing (Ford, 2001; Gummesson, 1999; IMP, 1982).

Interestingly, while the subject of relationship management has historically been considered separately from the procurement and supply management activity, the twin developments of outsourcing and supply chain management have seen a blurring of the distinction (cf. Smyth, 2005). Effective business relationship and supply chain management now encompasses a wide range of key commercial and operational issues including:

- The identification of desired value propositions
- The extent of outsourcing
- The nature of supply relationships
- The length of supply relationships
- The nature of contract
- The length of contract
- The extent of ethical sourcing and impact of environmental issues

- The pre-contractual selection of competent and congruent suppliers
- The extent of involvement within the supply chain
- The post-contractual management of suppliers
- The management of commercial risk and uncertainty in supply

Furthermore, the recent emergence of the *supply chain management* concept in buying has represented a significant paradigm shift for procurement strategies and buyer–supplier relationships. This shift has typically focused on the rejection of short-term and highly adversarial approaches in favour of more transparent, equity-based, long-term and collaborative ways of working, such as *partnering*. This practice has been reinforced within the marketing literature. A parallel trend in favour of interactions, networks and relationship marketing approaches also recommends long-term collaborative relationships between suppliers and their customers, based on high levels of operational collaboration, involving non-adversarial transparency, open-book commercial dealings and trust.

However, while the terms supply chain management, partnering and relationship marketing are widely used, the lack of a universally accepted definition of what this commercially non-adversarial and highly operationally collaborative approach to buyer and supplier exchange means in practice, leads to considerable confusion for those attempting to develop a robust approach to business relationship alignment. To understand the appropriateness of these ways of working for project-based industries, it is essential to explain what the concept means, and how it differs from other approaches to business relationship management.

Theories of business relationship management and the paradox of 'win–win' and collaboration

A major issue within the business relationship and supply management literature is the failure to define effectively the nature of commercial and operational exchange (Cox *et al.*, 2004a; Cox, 2004b, 2004c). Buyers and suppliers have their own commercial goals that define their strategic ends, which may exist in state of permanent tension. Buyers are normally concerned operationally with the *functionality* of the goods or services provided and, commercially, with *total costs of ownership*. The buyer is, therefore, always trying to maximise the value for money they receive from the supplier. The supplier normally attempts to increase operationally the *revenue* received and, commercially, the *returns* received from a specific buyer. This inherent tension is shown in Figure 11.1.

This conflict of operational and commercial goals in dyadic exchange needs to be recognised to understand why firms enter into business relationships. However, rather than focusing holistically on the dyadic relationship within exchange transactions between buyers and suppliers and acknowledging the

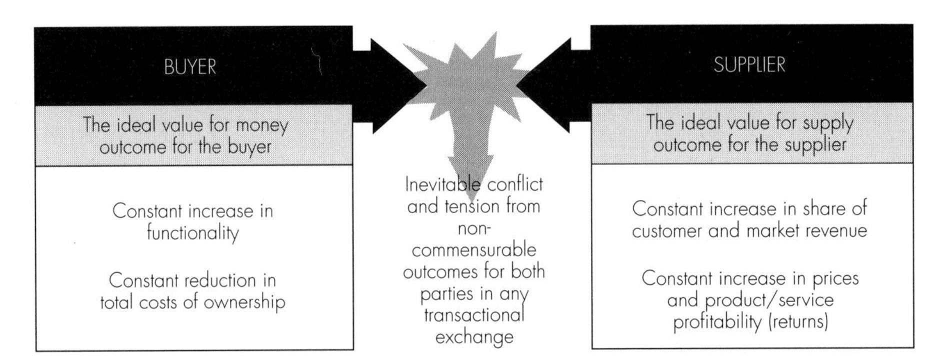

Fig. 11.1 Tension and conflict in buyer–supplier relationships.

conflicts and tensions present, much of current business relationship thinking tends to focus on how one party can achieve their goals without necessarily understanding how this impacts on the other side in the exchange.

In order to outline the strengths and weaknesses of the current thinking within the business relationship literature, the discussion that follows focuses on five broad schools of thought. It is argued that all of these – apart from the power and leverage perspective – have a great deal in common. In particular, there is a tendency to focus on the goals of only one of the parties to the exchange and then, prescriptively, to encourage relationship management styles that favour transparent, collaborative and trusting ways of working, but based on a desire to achieve 'win–win' outcomes for both parties.

One-dimensional buyer-side approaches

Lean supply chain management

In recent years, an approach to procurement and supply improvement has been developed academically that has dominated recent thinking in government and industry reports in a wide variety of industries. Most writers who advocate this approach, known either as partnership sourcing, partnering or *supply chain management*, and encompassing the *lean* and *agile* schools of thought, tend to prescribe the use of a highly transparent, trusting and long-term relationship between the buyer and the supplier to create a physically efficient supply chain through the reduction of waste in processes or responsiveness in product delivery. These approaches tend to focus more on the operational outcomes from exchange than the commercial outcomes.

Although the term *lean* was initially coined by Krafcik (1988), the first comprehensive account of the principles underlying lean production can be traced to Womack *et al.* (1990), who described the pioneering work of Toyota in managing their production processes and supply networks. A key message within the lean literature is the need to add value in high volume customised production to provide end customers with more of what they want at the time they want it. To achieve this end, the lean approach seeks to improve flexibility, narrow sources of supply, reduce waste and improve flow in the supply network through the adoption of long-term and highly collaborative supply relationships.

A major element of the lean school is the need to understand the different types of waste that do not add value to the final product from the point of view of the client. This requires recognition of the seven key wastes (Ohno, 1988):

- Overproduction
- Defects
- Unnecessary inventory
- Inappropriate processing
- Excessive transportation
- Waiting
- Unnecessary motion

Having identified the different types of waste that can exist within production processes and supply chains, the lean literature argues that it is necessary to design a strategy to eliminate it (Womack & Jones, 1996):

- Specify value from the customer perspective, not the perspective of any of the supply chain participants.
- Identify the value stream that delivers the value proposition to the customer.
- Identify those activities within the value stream that add value, are non-value-adding but necessary, or are non-value-adding.
- Make those actions that create value flow without interruptions or impediments.
- Only make what is pulled by the customer. The value stream should be operating a just-in-time pull system, rather than a producer-focused push system.
- Create transparency of strategies and costs within the supply chain. The issue of competitive advantage must be understood at a supply chain level rather than at the individual firm level.
- Constantly pursue perfection – the process of reducing waste never ends.

The lean supply chain management approach has become something of a dominant way of thinking about 'best practice' for buyers. This is because the approach, originally developed in the automotive sector, has provided significant benefits in other sectors such as supermarket retailing, electrical industries, aluminium and computer manufacturing (Cox *et al.*, 2003). However, while it may be highly appropriate for these process-based industries, its universal applicability for practitioners within project-based industries such as construction, oil and gas, and IT systems integration can be questioned (Cox & Townsend, 1998; Cox & Ireland, 2001, 2002).

Research into project-based industries, such as construction (see Table 11.2), has shown that waste is endemic and that there is a need for the development of appropriate waste reduction strategies (Vrijhoef & Koskela, 2000).

Table 11.2 Extent of waste in construction (Koskela, 1997).

Waste	Cost	Country
Quality costs (non-conformance)	12% of total project costs	USA
External quality cost (during facility use)	4% of total project costs	Sweden
Lack of constructability	6–10% of total project cost	USA
Poor materials management	10–12% of labour costs	USA
Excess consumption of materials on site	10% on average	Sweden
Time used for non-value adding activities on site	Approx. $\frac{2}{3}$ of total time	USA
Lack of safety	6% of total project costs	USA

However, as will be shown later, the 'lean thinking' and partnering literature has been applied without a proper understanding of the market, chain demand and supply, and power and leverage structures in project industries in general and in the construction industry specifically.

Despite the recent popularity of the lean approach there is an 'alternative' approach that is now receiving increasing attention. Although having a similar managerial philosophy based on collaboration, an *agile* approach is argued to be more appropriate given the characteristics of project industries: unpredictable, volatile, highly customised and low volume markets.

Agile supply chain management

The 'agile school' argues that while a lean approach makes sense under conditions where demand is predictable and standardised, and volume is high, as in typical process manufacturing, an alternative is needed in unpredictable, volatile, highly customised and low-volume markets, as in construction. A leading proponent of the agile school supports this view:

> *'There are certain conditions where a lean approach makes sense . . . The problems arise when we attempt to implant that philosophy into situations where demand is less predictable [and] requirement for variety is high . . . [As a result, many] firms have been misguided in their attempts to adopt a lean model in conditions where it is not suited.' (Christopher, 2000 p.38)*

In market and supply chain situations, characterised by volatile demand and a high customer requirement for variety, the elimination of waste is a lower priority than responding to the dynamic marketplace. In these circumstances, an innovative and market-responsive strategy, such as agile supply management, is demanded (Christopher, 2000; Harrison *et al.*, 1999; Mason-Jones & Towill, 1999; Stalk & Hout, 1990). Table 11.3 summarises the key attributes of a lean and agile supply management approach.

The agile literature also recognises that 'hybrid' or 'leagile' strategies that combine lean and agile approaches may be appropriate according to the characteristics of demand (Naylor *et al.*, 1999; Christopher, 2000; Christopher & Towill, 2000; Mason-Jones *et al.*, 2000).

Table 11.3 Comparison of lean supply with agile supply. Adapted from Christopher and Towill (2000).

Distinguishing attributes	Lean supply	Agile supply
Marketplace demand	Stable and predictable	Volatile and unpredictable
Nature of product	Standardised and commoditised	Customised
Product variety	Low	High
Product life cycle	Long	Short
Customer drivers	Cost	Availability
Profit margin	Low	High
Dominant costs	Physical costs	Marketability costs
Stockout penalties	Long term contractual	Immediate and volatile
Purchasing policy	Buy goods	Assign capacity
Information enrichment	Highly desirable	Obligatory
Forecasting mechanism	Algorithmic	Consultative

Although the lean and agile supply schools have a difference of focus, there are considerable similarities between the two approaches particularly in the advice offered to buyers for the management of supply networks. The main focus of both approaches is on the development of collaborative relationships between buyers and suppliers based on transparent, coordinated and integrated supply chain activity. The main difference between the lean and agile prescriptions relates to the nature of their operational techniques and how they deal with different demand conditions, rather than this general management philosophy.

Partnering approaches to supply management

The previous discussion has highlighted how integrated supply chain management approaches, whether lean or agile, focus on long-term and closer intra- and inter-organisational relationships, mutual competitive advantage, shared learning, collaboration, greater transparency and trust (Saad *et al.*, 2002; Watson *et al.*, 2003). This approach is frequently termed *partnering*, when the focus is between two parties within a chain – a dyad.

Partnering as a concept has grown out of the development of strategic alliances, which help manage the supply chain in a more coordinated manner. It is based on the premise that entering into a partnering arrangement will allow each firm to meet its own business objectives more effectively, while at the same time achieving the objectives of the supply chain. Its foundation may well be attributed to the Japanese 'Kaizen' management revolution, which focuses upon the principles of total quality management (TQM).

The partnering literature has identified a wide range of criteria that are fundamental to successful collaboration (Barlow, 1997; Bennett & Jayes, 1995, 1998; CII, 1989, 1991, 1994; Godfrey, 1996; Hellard, 1995; Lamming, 1993; McGeorge & Palmer, 1997; Ng *et al.*, 2002; Saunders, 1994). These criteria include commitment, equity, collaboration, trust, mutual 'win–win' goals and objectives, innovation, preparation and training, use of appropriate partnering tools and procedures, inclusion of appropriate parties, risk sharing, continuous joint evaluation and problem solving, effective leadership, robust and open communications, empowerment of stakeholders, customer focus, and timely responsiveness.

Despite the case put forward in favour of partnering, there are several questions that remain unanswered. The most obvious of these is what precisely does partnering entail in practice? While there is broad consensus about the basic philosophy underpinning partnering – a commitment between firms to cooperate – there are contrasting views about a number of features including the precise role and nature of contracts; the duration of partnering arrangements; the role of incentives systems; and the need for formal team-building and facilitation (Barlow, 1997). Thus, partnering can be seen as an imprecise concept capturing within it a wide range of behaviour, attitudes, values, practices, tools and techniques.

As Holti and Standing argue:

> *'Rather than being a separate or definable initiative in its own right, partnering or increasing collaboration is best understood as the result of making progress with one or more of a number of inter-related technical and organisational change initiatives.' (1996 p.5)*

Indeed, more recent accounts of partnering have attempted to inject a greater degree of sophistication by viewing the diversity of partnering practices along a continuum from competition to cooperation, collaboration and coalescence (Thompson & Sanders, 1998). One consequence of this semantic ambiguity (where partnering can also signify an *outcome*) is that it is often difficult to distinguish between partnering as a distinctive practice and partnering as managerial rhetoric (Hinks *et al.*, 1996).

Relationship marketing from the supply side

While there has been much interest in the supposed operational and commercial benefits of partnering or supply chain management among buyers in project- and process-based industries, a similar interest in longer-term more operationally collaborative and 'win–win' ways of working has developed on the supply side, and especially among those in favour of relationship marketing (Christopher *et al.*, 1991; Gummesson, 1987, 1999). In this approach, all types of relationships are considered – indeed Gummesson defines 30 relationship types – placing trust and collaboration between customer and supplier at the heart of long-term exchange so as to counteract the forces of hyper-competition in markets. Such competitive pressure, it is argued, is best counteracted by finding ways to collaborate with customers, for example through loyalty or payment cards, that create trust in the relationship and close markets to competitors by creating forms of long-term repeat business between the two parties, which is predicated on the service provider offering superior reliability and availability to the customer. Furthermore, it is also argued that this collaborative relationship approach can be extended whenever possible into the complex networks and inter-relationships that exist between customers, suppliers and competitors and others throughout the upstream and downstream supply chain.

While relationship marketing offers similar supposed 'win–win' benefits for customers to those espoused in partnering and supply chain management approaches, this way of working has been criticised. This is because, by limiting competition, it may reduce options for buyers where the supplier has high levels of leverage; and while it is obvious that long-term lock-in benefits suppliers, it is not always self-evident that high switching costs for buyers is desirable in all circumstances (Williamson, 1985; Cox, 2004d). One may argue, therefore, that while long-term trusting and collaborative relationships of this type between buyers and suppliers may certainly benefit suppliers who wish to stop competition and create dependent customers, it is not necessarily always in the buyer's best interests.

Relationship marketing can offer benefits to both sides. However, while it is clearly an approach developed in the marketing arena to assist suppliers to use collaboration to close markets to their competitors, it may not be in the objective commercial nor operational interests of buyers under all circumstances. Indeed, it has been argued that the advantages of adopting this approach are increased customer retention and duration, increased marketing productivity and profitability, and increased stability and security for the supplier (Gummesson, 1997). In this sense, it is a one-sided approach that assists suppliers with their problems, but does not always assist buyers with theirs.

Partnering and relationship marketing in the project environment: is it best practice?

The arguments related to the appropriateness of partnering and relationship marketing in the project environment follow the same logic as those related to the adoption of lean and agile supply chain management. While a consensus has developed in the practitioner and academic literature, based on the rejection of adversarial buyer relationships with suppliers in favour of the development of 'value-adding' collaborative relationships, some writers do not agree that this is always appropriate.

Some writers argue that collaborative relationships incur significant costs, either directly through investments, or indirectly through the costs of management resource, and need to be justified by a certain level of return on investment (Watson *et al.*, 2003). It can be argued that a key problem with these approaches is that they do not make sufficiently clear what the mechanisms are that provide the drivers for 'partnering' and collaboration throughout the supply chain.

It is also contended that the 'ideal' sourcing outcome possible for a buyer is not commensurable with the 'ideal' selling outcome possible for a supplier. Therefore, although business relationship alignment, mutuality, is achievable under particular circumstances, a 'win–win' is not feasible commercially in buyer and supplier exchange. As a result, there must also be situations where 'win–lose' or 'lose–lose' in transactional exchange is the only likely outcome, and partnering or collaboration will be extremely difficult to implement successfully (Cox, 2004b, 2004c; Cox *et al.*, 2004a).

Two-dimensional relationship management approaches

While the focus of this chapter is predominantly on the buyer side of business-to-business relationships, there are approaches to business relationship management that focus on both sides of the exchange. These approaches to business relationship management contend that it is necessary to understand the goals and motives of both the buyer and supplier as they interact together so that they can be aligned effectively (Cox, 2004b, 2004c; Cox *et al.*, 2004a, 2004b). However, despite this agreement on the need for a holistic approach, there are significant differences between the two perspectives in their attempts to develop a truly analytical and predictive approach to relationship alignment.

The International Marketing and Purchasing (IMP) Group

The IMP approach has provided an extremely comprehensive descriptive account of the nature of business relationships (Ford, 1980; Håkansson, 1982). The IMP approach contends that the interactions between buyers and suppliers are shaped by 'environmental' factors, which neither party to the exchange can directly control – *limiting factors*. There are some elements of the interaction, it is argued, that buyers and suppliers can influence – *handling factors* – that are referred to as 'atmosphere' (IMP, 1982) (see Figure 11.2).

A major benefit of this approach is its inclusiveness – it is argued that there is a need for buyers *and* suppliers to think about the strategies they should adopt under different types of buyer and supplier interaction and that sometimes these may create a match or mismatch in a relationship (Campbell, 1985). A major weakness of the IMP approach, however, is its

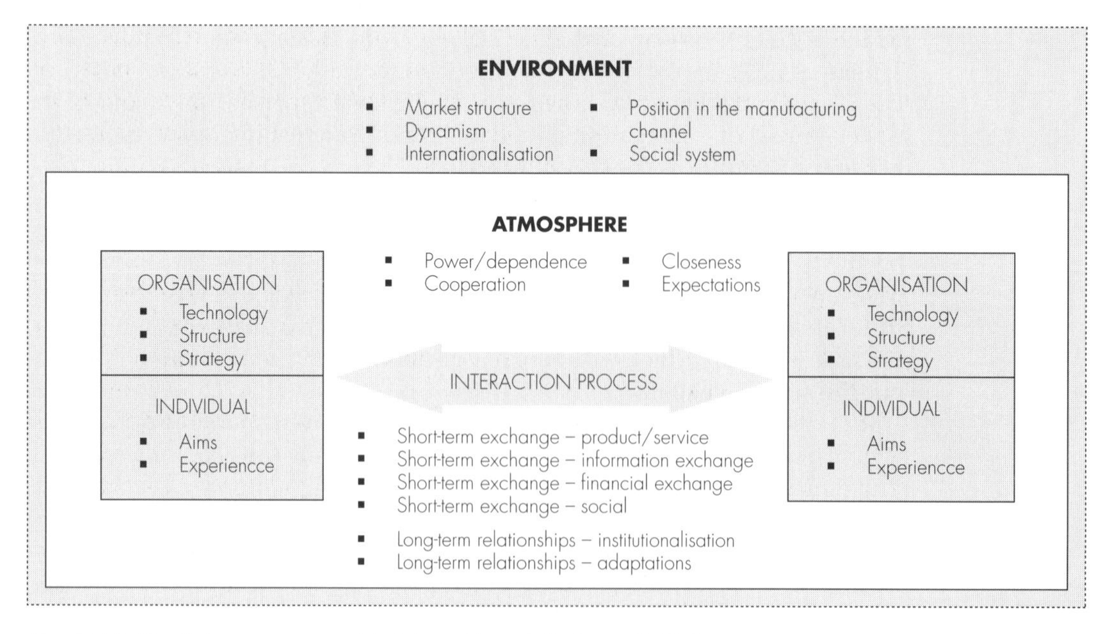

Fig. 11.2 The IMP interaction model. Adapted from IMP Group (1982).

lack of predictive focus on the relative importance of the variables specified in a buyer–supplier interaction. Therefore, while the IMP approach is comprehensive, the fact that it is primarily descriptive provides managers with limited guidance as to which particular relationships are suitable in specific circumstances. Despite these criticisms, the IMP approach, while remaining largely pessimistic about the ability of managers to actually control many of the environmental factors that impinge on relationships (Ford, 1980, 2002; IMP Group, 1982), has provided significant insight into the factors that impinge on buyer and supplier choices.

The power perspective

The power perspective on buyer and supplier relationship management has a long tradition of its own (Emerson, 1962). However, unlike the IMP's approach, the power perspective has always asserted that the relative power of the buyer and supplier is the determining factor in the operational and commercial outcome in any transaction. Writers in this school also contend that the current consensus about what constitutes 'best practice' in business relationship management – namely, the rejection of adversarial buyer–supplier relationships in favour of more long-term collaborative (partnering) approaches based on trust – contradicts the common-sense logic of economic theory. This logic is that the best defence of the buyer's position, and the one that ensures that suppliers innovate and pass value to buyers, is the maintenance of perfectly competitive (or highly contested) supply markets, with low barriers to entry, low switching costs and limited information asymmetries (Cox, 2001).

The fact that buyers and suppliers can pursue symmetrical or asymmetrical relationship management styles with one another to achieve their business objectives also reinforces the argument that there can be no single 'best practice' approach to relationship management. Therefore, competence for buyers *and* suppliers must reside in their respective abilities to make appropriate choices about how they conduct themselves in very different exchange relationships (Cox, 1997). A sourcing or relationship management strategy based on a generic approach may, therefore, be wholly unsuitable (or even impossible) under particular circumstances.

Following this argument, while integrated supply chain management, partnering or relationship marketing can be regarded as a 'best practice' approach for organisations under *some* circumstances, it is highly unlikely to be 'best practice' in *all* external sourcing circumstances. Given this, the first issue to be addressed does not relate to the most effective way to operationalise these collaborative approaches, but to what is the universe of sourcing choices available to practitioners?

To address this issue, practitioners have to first understand the ways in which buyers can work with any supplier (in a reactive arm's length or proactive collaborative manner) and the scope of their activities within a supply chain (at the first tier or in the entire supply chain). Bringing these variables together, it is possible to define four basic sourcing approaches that are always available for buyers to select from when they seek to manage their supply

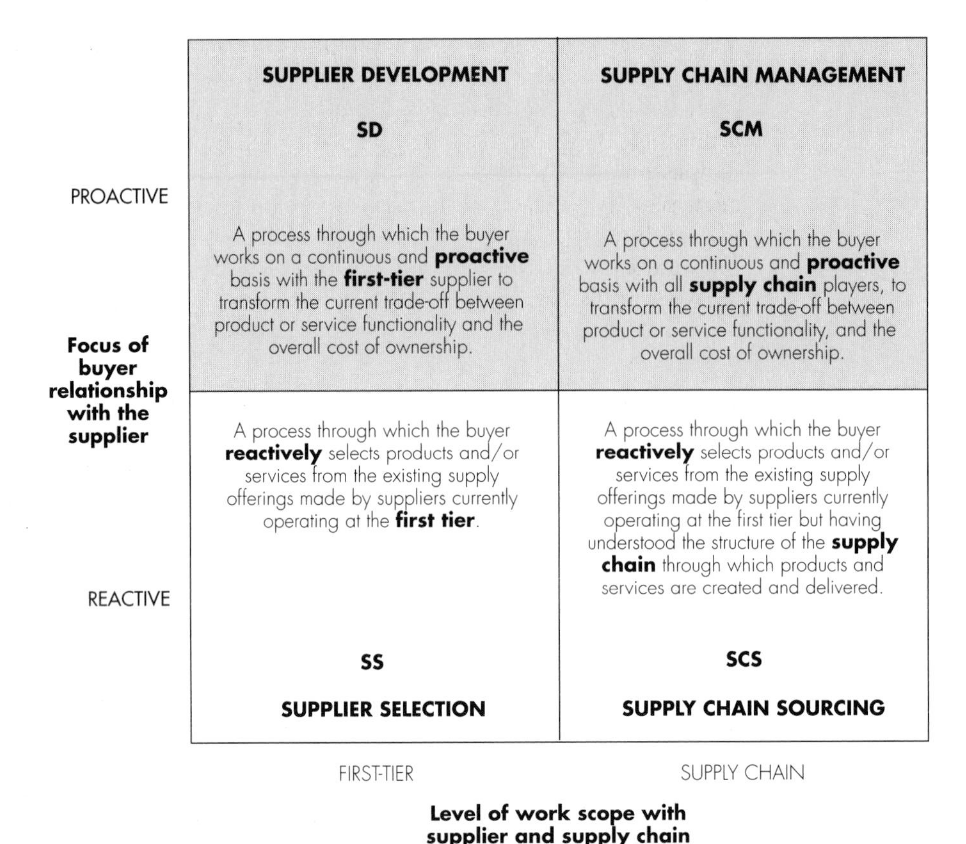

Fig. 11.3 The sourcing options matrix. Adapted from Cox *et al.* (2003).

relationships. The four relationship management options *supplier selection, supply chain sourcing, supplier development* and *supply chain management* are summarised in Figure 11.3.

Supplier selection (SS) is the approach that is most frequently used by buyers in all types of organisation, as it places the lowest demands on scarce internal resources and requires the lowest levels of commitment to long-term collaborative external sourcing relationships. SS implies that the buyer's role is confined to reactive sourcing at the first tier. Therefore, the buyer selects products or services from the supply offerings made by suppliers currently operating in the market and relationships tend to be relatively short term and arm's length in nature. A similar approach may be to demand at arm's length supply inputs throughout the supply chain, *supply chain sourcing* (SCS).

A proactive buyer assesses the scope to undertake *supplier development* (SD) work, which is improving the competence of the immediate supplier operationally and commercially by working in a collaborative manner on improvement initiatives. This is confined to the first tier. Only if this approach is extended throughout the chain, or in significant parts of it, may it be categorised as *supply chain management* (SCM). SCM is arguably the most demanding and challenging of the four options for buyers. It requires extensive coordination of internal business functions and a linking together of all the buyers and suppliers in the chain, so that they are able to focus their procurement and supply strategies on the delivery of improvement in functionality and lower costs of ownership for the ultimate buyer in the chain. Fundamental to achieving this increase in value for money, profitability and long-term performance improvement is longer-term collaboration at a supply chain level. This would, of necessity, require that a supplier is also prepared to adopt the highly collaborative approaches associated with IMP network interaction and relationship marketing thinking.

Given the four alternatives, it is argued that all four sourcing options may be appropriate for buyers and suppliers under particular circumstances and, furthermore, that not all approaches are feasible, or desirable, for buyers or suppliers all of the time. To determine which of these approaches is feasible or desirable, it is argued that buyers need to consider the following key questions (Cox *et al.*, 2003):

(1) What will be the cost to the firm of implementing the different supply management options? To answer this, the buyer and the supplier have to understand the costs of implementing the reactive and proactive approaches, both at the first tier and extended through the supply chain.
(2) What are the likely benefits from adopting the different sourcing options?
(3) What is the probability of the different sourcing options being successfully operationalised and attaining the expected benefits?
(4) What is the nature of the relationship in power terms? How will this power structure impact:
    (a) whether the buyer or supplier is willing to consider a proactive approach, that is collaboration, or is only willing to consider a reactive approach
    (b) whether the approach can be extended beyond the first tier, and
    (c) for each of the four options, the way in which the surplus value is created and distributed.

Having considered these questions practitioners are effectively conducting an investment appraisal of the human and process scope for particular types of relationships, having considered the costs and returns with risk and power as intervening variables. As a result they are able to assess not only which options are likely to be desirable or feasible, but also whether the buyer or supplier will dominate in the exchange relationship. A key criticism of the lean and agile schools is that they have failed to recognise that these types of intra- and inter-organisational power may both facilitate and inhibit efforts to plan, control and manage supply networks (Cox *et al.*, 2003). In other words, this literature that encompasses recent government and industry reports may be guilty of ignoring the objective reality of power and leverage in project industries in their misguided search for a simple answer to a complex problem.

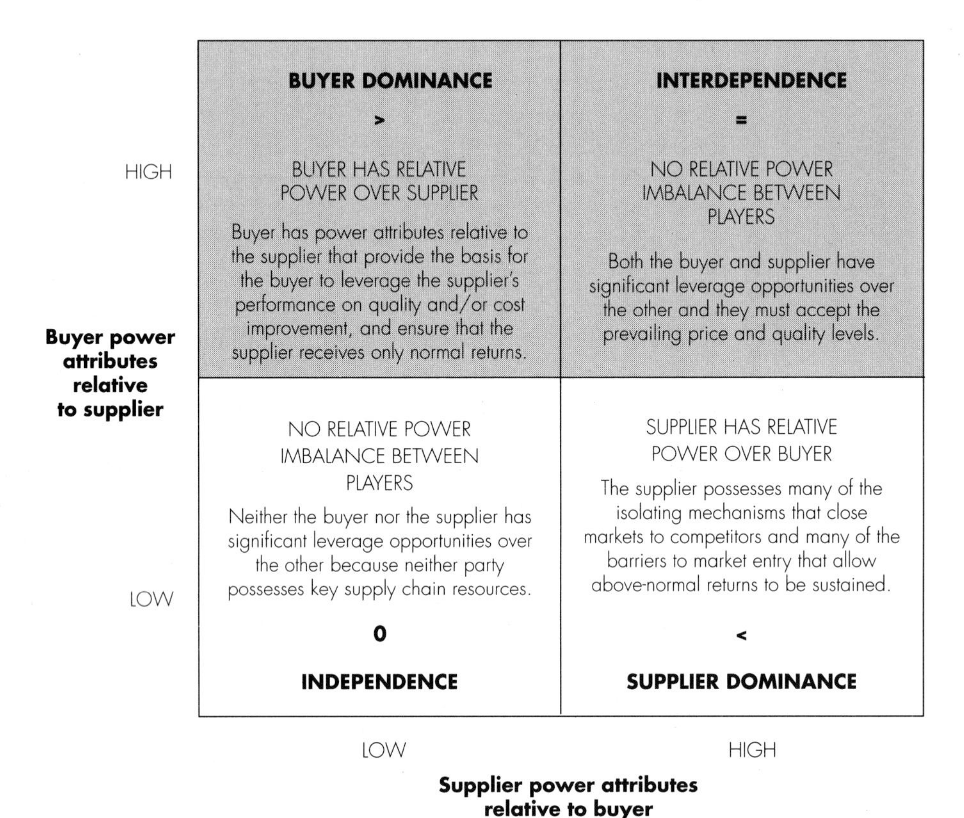

Fig. 11.4 The power matrix. Adapted from Cox *et al.* (2000).

Firms aiming for superior performance in procurement and supply chain management must, therefore, understand the power and leverage situation they are operating in before they decide on which business relationship and sourcing strategies are appropriate. The power matrix, as illustrated in Figure 11.4, provides a framework for the analysis of buyer and supplier power resources. The power circumstances in which buyers and suppliers find themselves can be defined either as *buyer dominance, interdependence, independence* or *supplier dominance*. The matrix is constructed based on the idea that all buyer and supplier relationships are predicated on the relative utility and the relative scarcity of the resources that are exchanged between the two parties (Cox *et al.*, 2002).

To position themselves in the matrix practitioners have to understand the key questions that must be asked in order to understand the power and leverage position of buyers or suppliers. Some of the key questions that need to be addressed before a buyer or supplier can locate their position include:

- The balance between the number of buyers and suppliers
- The salience of the buyer's expenditure to the supplier
- The number of alternative purchasers available to the supplier

- The extent of buyer and supplier switching costs
- The extent to which the product or service is commoditised or standardised
- The level of buyer search cost
- The level of information asymmetry advantage that one party has over the other

The discussion of the power resources in the buyer–supplier relationship is, of course, only a starting point because the power structures within the entire supply chain in which the firm is operating must also be considered. For example, within the construction industry, the dyad between the client and the main contractor is also affected by the relationship that the main contractor has with its subcontractors and their relationship with component, equipment and labour suppliers. Therefore, one has to understand the extended network of dyadic power relationships so that appropriate relationship management strategies can be developed. This is referred to as the *power regime* within which all supply chain relationships must operate. Only by understanding the structure of power within the supply chain or network as a whole is it possible to understand the feasibility or desirability of introducing reactive arm's length or proactive collaborative relationship strategies for the buyer and the supplier (Cox *et al.*, 2000, 2002, 2003).

Furthermore, when considering the impact of power and leverage on relationship management strategies the buyer and the supplier have to recognise that particular operational relationship management approaches will be more conducive to specific relationship outcomes than others. This is indicated in Figure 11.5, where the essential paradox of 'win–win' and collaboration in business relationships between buyers and suppliers is briefly explained. The figure shows that, if there are nine cells that could occur operationally and commercially from exchange relationships, only one of these can be regarded as *ideal collaboration*, that is, a situation in which both parties fully achieve their ideal operational and commercial goals. This is normally referred to as a 'win–win'.

The problem, however, is that – and this is the central paradox of all relationship management approaches between buyer and suppliers – this ideal is not feasible in practice because of the non-commensurability of the objective commercial interests of both parties (Cox, 2004b; Cox *et al.*, 2004a). Put simply, if the buyer is able to achieve its commercial ideal then the ideal commercial goals of the supplier cannot also be achieved, however close the operational collaboration. This occurs because the ideal operational and commercial goals of the buyer are increased operational functionality (a more effective product or service) with continuous reductions in the total costs of ownership (i.e. all unnecessary costs are eradicated and the supplier makes no more than normal returns, typically 0–3% profit margins). Ideally the supplier, on the other hand, would prefer not to have to improve functionality (and especially not if it reduces their profitability), and would prefer to receive all of the customer's business revenue, while also making above normal returns (typically 10%+ profit margins). This means that the

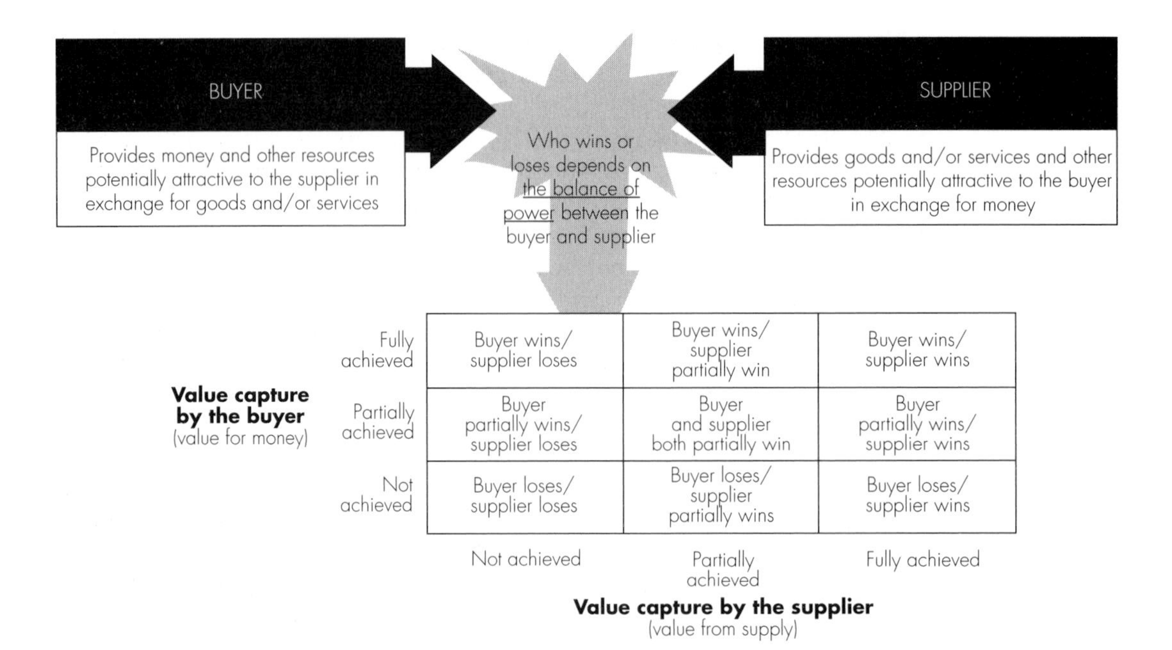

Fig. 11.5 Relationship outcomes in buyer and supplier exchange. Adapted from Cox (2004c).

commercial, if not necessarily the operational, goals of both parties are not fully commensurable.

Nevertheless, while a 'win–win' in which both parties fully achieve their ideal commercial goals is only feasible under certain conditions, as Figure 11.5 reveals, it is still possible for relationship outcomes to occur that do, and also that do not, involve some form of mutuality. Thus, it is perfectly possible for buyers and suppliers to achieve operationally and commercially all or something of what they value ideally, without it being possible for both parties simultaneously to achieve everything that they value operationally and commercially. This is what collaboration in buyer and supplier exchange means (Cox, 2004c).

Thus, in Figure 11.5 there are three mutuality outcomes in which long-term collaborative relationships between buyers and suppliers are feasible and sustainable. These are *win for the buyer/partial win for the supplier*; *partial win for both the buyer and the supplier*; and *win for the supplier/partial win for the buyer.* In each of these three situations it is feasible for both parties to envisage the maintenance of long-term collaborative relationships operationally. This is because both parties are achieving either all or something of what they desire operationally and commercially from the exchange. It must be recognised, however, that the operational and commercial outcome in these three cells is not the same for both parties. This means that, while it is one thing to favour longer-term collaborative relationship approaches, this does not obviate the need to consider also the specific operational and, especially, the commercial consequences of collaboration for both parties.

This is because collaboration does not eradicate the use of power and leverage operationally and commercially between buyers and suppliers, and this has a major significance for those interested in understanding the full range of relationship management choices available to buyers and suppliers. What this means is that, while operational forms of collaboration are always theoretically possible ways of working for buyers and suppliers to consider, this does not mean that the three mutuality cells in Figure 11.5 are necessarily the ideal relationship outcomes for any buyer or supplier.

While neither party would wish to operate in the 'lose–lose' cell, if business relationships do not have to be sustainable because a buyer or a supplier can quite easily find alternative business partners, then a 'lose' outcome for one party is not necessarily a problem for the winner. If buyers or suppliers do not have to consider repeat games with the same partners – and can countenance one-off games – then a 'win–lose' or a 'partial win–lose' outcome (where they are not the loser) can be perfectly acceptable for them individually. The significance of this point is that, while proponents of relationship marketing, partnering and supply chain management all tend to espouse the virtues of extensive collaboration, this may not always be sensible. If there is no need for continuous operational linkages to deliver what is required, and the business relationship can be played as a one-off game, then buyer or supplier opportunism may be a far preferable outcome for one party than collaboration.

It is this scenario that some proponents of a more transparent and trusting approach to business relationship management simply do not seem to understand in their recommendations about best practice for relationship management in the project environment. Non-adversarial forms of collaboration may be seen as one way of managing buyer and supplier exchange. However, when the demand profile in the project environment is unique and irregular, rather than standardised and regular or episodic, then alternative forms of exchange – where opportunism replaces collaboration as the most sensible approach – become desirable. This conclusion is reinforced in the following discussion of the limited circumstances in which relationship marketing and partnering forms of relationship management are feasible in construction.

Project sourcing in practice: on the appropriate management of business relationships in construction

It should by now be self-evident that proactive approaches to business relationship management based on long-term collaboration are not always possible in project industries in general, and in many areas of the UK construction industry specifically. This does not mean that collaborative approaches, such as partnering and relationship marketing or SD and SCM are impossible to implement; rather it means that they are likely to be of benefit only to certain clients, contractors and suppliers who have power resources of *buyer dominance* or *interdependence*, because they are able to manage a regular high-volume and highly standardised spend with the same suppliers over a long period of time. Given the structure of demand and supply in project industries like construction, this is rarely the case.

Supply chain characteristics in a typical construction project environment

Within the UK construction industry, like many project-based industries, practitioners are confronted with considerable difficulties when attempting to manage their supply chains. Construction clients are often faced with a highly competitive and adversarial supply market, from which they attempt to obtain value for money when sourcing their construction requirements. The difficulties are compounded by the fact that they do not always fully understand their demand profile and the nature of the supply market. As a result, they are often open to opportunistic behaviour from construction firms operating at the first tier.

Construction companies, as suppliers and buyers, are faced with the bifurcated challenge of obtaining a regular workload from clients that is sufficiently profitable for long-term survival, plus the difficulty of managing subcontractors and materials suppliers who are attempting to be successful in their own fight for survival. Adversarial behaviour and opportunism are rife, as low barriers to entry encourage low levels of profitability and investment within these markets.

There is a significant debate within the academic and practitioner communities as to how to address and overcome these inherent problems. In the UK construction industry there have been a number of government and industry-sponsored reports aimed at raising competitiveness by focusing on improvements in supplier performance and customer satisfaction (DETR, 1998; Latham, 1994). However, despite the advice provided, these initiatives appear, by and large, to have failed most actors in the industry. This is because construction supply chains have largely remained contested, fragmented and highly adversarial with low margins.

Given this, many of those who called for the adoption of collaborative supply chain management tools and techniques that have proven highly successful in the automotive industry would probably argue that the problem is in poor understanding and implementation by practitioners, and that things would be better for everyone if only practitioners implemented these approaches as prescribed. It may well be, however, that it is not the implementation that is at fault but the approach itself. As Figure 11.6 indicates, in the project environment there are a myriad of supply chains for materials, labour, equipment and professional services that have to be integrated by construction firms in the delivery of completed projects to the end customer. Each of these supply chains has different demand and supply properties, as well as operational and commercial properties that practitioners need to understand before they can develop appropriate relationship management and sourcing strategies for the specific products and services involved.

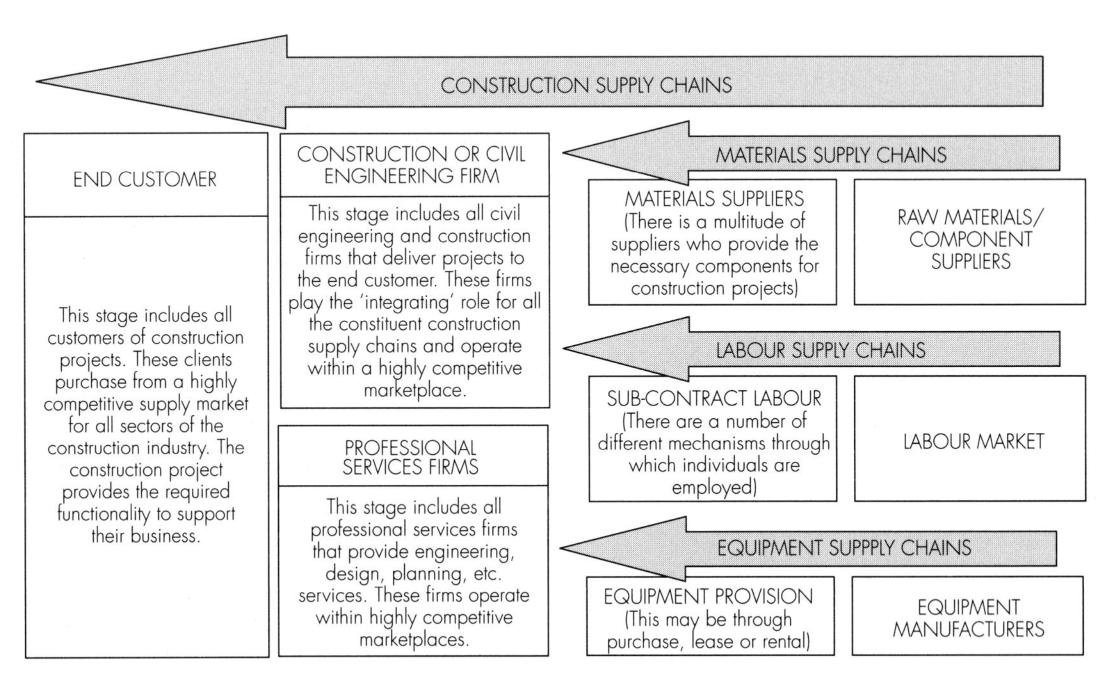

Fig. 11.6 The myriad of construction supply chains. Adapted from Cox and Ireland (2002).

Figure 11.6 suggests that the generic supply chains that are required for a 'typical' construction solution are rather simple and linear but the reality is quite different. A 'typical' project does not exist, as the ultimate level of complexity involved in the management of a construction project (and the exact number of constituent materials, equipment and labour supply chains that have to be integrated) will be determined by the specific *one-off* requirements of the client as defined in the original design and specification. The client typically appoints professional services firms to assist in the development of this design and specification and the selection of a construction firm. The construction firm normally fulfils the major 'integrating' role for all upstream supply chains and coordinates the appointment of third-party subcontractors to deliver 'packages' that can be integrated within the solution.

Surprisingly, throughout the project procurement process little control or management of the entire supply chain is taken up by the focal organisation, the client (London & Kenley, 2000). As a result, each tier of the supply chain is able to manage their supply relationships in such a way that they can effectively act as a 'gatekeeper' that controls the sourcing of products and services from their upstream supply market. The first-tier construction firm typically acts as a gatekeeper to the second-tier specialist subcontractors, who in turn subsequently act as a gatekeeper to the materials suppliers operating at the third tier. Therefore, regardless of whether it is the client or the construction firm that is attempting to control the procurement activity, this restricted access to upstream supply chain stages creates considerable challenges for those attempting to undertake effective procurement, relationship marketing and supply chain management.

However, there are a number of other characteristics of construction supply chains that may create further significant difficulties for practitioners responsible for construction procurement and relationship management. These characteristics are also evident within other project-based industries with similar demand and supply characteristics.

'Manufacture-to-order' one-off supply chain

It is widely stated that the construction industry is a relatively unique industry in the way that it establishes projects to deliver one-off products (Cox & Ireland, 2001, 2002; Cox & Townsend, 1998; Ireland, 2004). However, it may be contended that other industries display very similar characteristics. For example, within the IT industry systems integrators develop complex one-off IT infrastructures for clients, shipbuilders construct large unique naval destroyers for government organisations within the defence industry, and pipeline contractors within the oil and gas industry also develop one-off solutions.

In construction, like these other project-based industries, it is the client who identifies a need for a new (or refurbished) asset and takes the initiative to start a construction project. Therefore, construction is often considered to be a client-driven one-off production process. This leads to the frequent conceptualisation of the construction supply chain as a process explicitly starting and ending with the end user. A common representation of a construction process is shown in Figure 11.7.

Temporary supply chain

Construction projects tend to be considered as unique and temporary (Morris, 1997; Turner, 1993; Turner & Müller, 2003). In contrast to manufacturing, this implies a temporary organisation of production for each project characterised

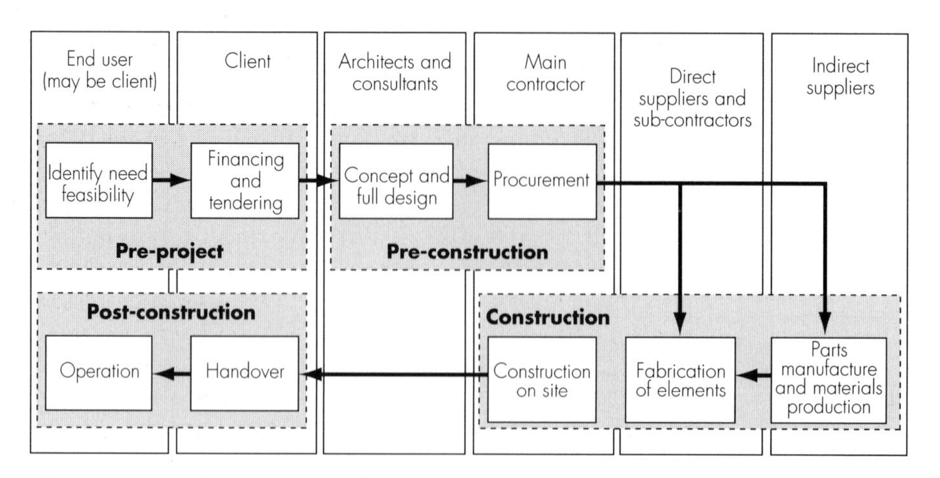

Fig. 11.7 Typical representation of the construction process.

by a short-term coalition of participants with frequent changes of membership, often termed 'temporary multi-organisation' (Cherns & Bryant, 1984; Koskela, 1997; Luck & Newcombe, 1996). The participants are totally interdependent, but operate through a variety of contractual arrangements and specific procedures with considerable fluctuations in productivity (Hellard, 1995).

While this literature argues that a construction project is unique and temporary, there is a literature that argues that this is not the case for all construction projects (Cox & Ireland, 2002; Cox & Thompson, 1998; Ireland, 2004). Cox and Thompson (1998) contend that clients possessing regular process spends are unlikely to constitute more than 25% of the total UK market, while Blismas (2001) contends that multi-projects accounted for 10% of entire industry output and as much as 30% of contractor output in 1999. This literature highlights that while unique, temporary and one-off projects constitute a large percentage of the market, there are also a number of major construction clients, such as those on Egan's Construction Task Force, whose construction programmes may be classified as regular, with a regular and fairly standardised process-orientated demand structure (Cox & Ireland, 2002; Cox & Townsend, 1998).

The regular nature of their construction means that it is possible to standardise the procurement process, design and specification of particular elements of their construction requirements and manage them in a similar manner to a manufacturing process. This would offer more opportunity for suppliers to adopt relationship marketing approaches, since the buyer in this case would welcome longer-term collaboration and both parties would benefit operationally and commercially from joint learning through standardisation and waste reduction exercises. Despite this, these demand and supply circumstances are likely to be the exception rather than the rule in the construction industry.

In summary, the construction industry's temporary and one-off nature, together with the high level of fragmentation, the existence of adversarial relationships and the tendency for self-interest rather than common objective, affects the efficient and effective satisfaction of customer demands (Cutting-Decelle, 1997; Nam & Tatum, 1988). In particular, it creates power situations that do not very often support the development of long-term collaborative *partnering* or relationship marketing approaches based on trust. Despite these unique characteristics, it is still widely believed that the key to the solution of many of the inherent problems of the construction industry is to be found in promoting commitment, trust and openness among project participants, and adopting such relationship management approaches as partnering and integrated supply chain management to achieve an integrated collaborative project organisation (DETR, 1998; Latham, 1994).

The implementation of long-term collaborative relationships in the project environment

Given the predominantly project-specific demand and supply structures outlined above, the majority of practitioners responsible for construction procurement decisions have typically adopted reactive and arm's length *supplier selection* (SS) approaches to sourcing in the past. This is largely attributable to the power structures that make more proactive, long-term and collaborative relationship management approaches unfeasible or undesirable both operationally and commercially. This is why construction relationships tend to be short term and often adversarial. It can be argued that, rather than this being, as perceived, a mistake on the part of construction buyers, it may in fact be simply a common-sense relationship approach to adopt in most demand and supply environments. Indeed, while the highly competitive nature of the construction industry has historically 'guarded' contractors against pre-contractual opportunism, buyers have to be aware that suppliers may regard a post-contractual opportunistic approach as highly appropriate and critical to revenue generation, cash-flow and profitability in one-off rather than repeat games.

However, there are buyers who require construction on a more regular basis and are in a position to standardise their processes and requirements across multiple projects. With a high frequency and volume of demand, these buyers have the opportunity to consider alternative relationship and supply management strategies, and the feasibility or desirability of a more proactive sourcing approach with a smaller number of preferred suppliers is likely to be much higher. It is important to understand that whether or not a buyer can develop such a proactive and collaborative approach is not just a matter of choice. Proactive supply management requires that the buyer has both the internal resources and capabilities and also a conducive external power structure to ensure partners are willing and able to be transparent and act in a collaborative manner, and are open to operational and commercial leverage. In short, the internal and external risk and incentives structures must support being proactive and facilitate collaboration.

For this to occur, it is contended that the external power structures at the first tier and throughout the supply chain in the power regime have to be buyer dominant or interdependent, and they must persist over time (Cox *et al.*, 2003). This is because when buyer dominance occurs (since all options are open to them) the buyer is able to select whether or not to use proactive (collaborative) or reactive (arm's length) sourcing approaches. It is also because they can use their power resources to leverage the maximum share of commercial value from the supplier whether they collaborate or not. In interdependence, if both the buyer and supplier choose to collaborate, given their equivalence of power resources it is likely that the sharing of value commercially in the relationship will be relatively equitable. Clearly the latter approach supports relationship marketing rather more than buyer dominant-led collaboration does. This is because the relationship outcome in interdependence is likely to be 'partial win–partial win' rather than 'win–partial win' under buyer dominant-led collaboration.

In the absence of these power structures there is little point in buyers attempting proactive sourcing initiatives as the required investment is unlikely to yield cost or value benefits. In other words, unless the demand and supply structures can be constructed in such a way that a repeat game rather than a one-off game is being played, proactive sourcing and relationship marketing on the part of suppliers carries a net cost and will lead potentially to a 'lose' outcome. This, in part, explains why attempts at project partnering and collaboration on a one-off basis are likely to fail in practice (Bennett & Jayes, 1995, 1998). This is because it is difficult, but not impossible, to incentivise potential suppliers as partners to make the necessary relationship-specific investments without a guaranteed return, and the project-specific learning is difficult to replicate or transfer across projects with dissimilar partners.

A key buyer power resource in project procurement: frequency of demand

While it is evident that the possession of conducive positions of relative power are of high importance for construction buyers, one can argue that it is the possession of specific resources that are critical within a project-based procurement environment. Within the construction industry, the majority of clients are not in a position of dominance over their supply base because of the nature of their ad hoc construction programmes. Those clients who possess a high *frequency* of construction demand and who can standardise their requirements are clearly in a better position to leverage the supply chain effectively and implement proactive and collaborative supply chain management approaches successfully. It is interesting to note that the client members of the Construction Task Force for the Egan Report all had demand profiles characterised by high frequency, which made proactive sourcing possible for them. It is ironic that these practitioners did not consider the fact that just because such a relationship and sourcing approach was possible for them it might be neither feasible nor desirable for everyone else in the construction industry.

The belief that power relationships in supply chains are highly transaction-specific, and that there are variations between firms and transactions in relation to the dimensions of utility and scarcity, is supported by the following two statements (Cox *et al.*, 2002):

- The resources that give a buyer power over a specific supplier in the context of a particular transaction will not necessarily give it power over other suppliers in the context of an equivalent transaction.
- The resources that give a buyer power over a specific supplier in the context of a particular transaction will not necessarily give it power over the same supplier in the context of other transactions.

Missing from these two statements is the variable of *time* and how this may impact on the power resources in buyer–supplier relationships. This variable is of particular importance when developing appropriate relationship management and sourcing strategies in the project environment – especially in construction supply chains because of the project-specific and one-off nature of construction procurement for the majority of clients and construction firms. Therefore, before considering the issue of time, in the context of frequency, it is necessary to clarify the definition of a project and the distinction between this and a regular or frequent construction process spend.

In a definitive work on project management, Turner defined a project as:

'an endeavour in which human, material and financial resources are organized in a novel way, to undertake a unique scope of work, of given specification, within constraints of cost and time, so as to achieve beneficial change defined by quantitative and qualitative objectives.' (1993 p.5)

This definition is reinforced by Turner and Müller, who argue that a project should be viewed as:

'a temporary organisation to which resources are assigned to undertake a unique, novel and transient endeavour managing the inherent uncertainty and need for integration in order to deliver beneficial objectives of change.' (2003 p.7)

This second definition emphasises that projects are unique, require novel processes and are transient with a clear start and end. However, as stated previously, although the majority of clients procure discrete one-off projects, there are those with an ongoing demand that source construction in a relatively regular manner. This regularity of process spend may lead to a construction programme that involves:

- *Either* projects that are similar because of a standardised design and specification and are undertaken in a consecutive or overlapping manner
- *Or* projects that have distinct design and specifications but still form part of an ongoing construction programme for a client or client type

However, irrespective of whether the design and specification is standardised or not, it is evident that the client with a high frequency of demand also has a choice with regard to the nature of the supply relationship to adopt. With a standardised specification the client may wish to select a small number of preferred suppliers and develop a long-term collaborative framework through which it can leverage these suppliers. Obviously the sharing of value here will vary depending on whether the buyer is dominant or there is interdependence where the commercial value has to be shared more equitably by the buyer with the supplier. In the absence of a standardised specification, whether the client is able to develop a preferred supplier strategy and use long-term relationships will depend on the procurement competence of the client and the capabilities of the supplier to deliver the range of requirements demanded by the client.

From this, it is clear that the nature of regularity or frequency involves two separate variables: *regularity of workload* and *regularity of relationship*. These two variables related to regularity are shown in Figure 11.8.

Given this distinction, it is clear that the appropriate management of construction spend, hence buyer and supplier relationships, may vary according to the nature of the frequency of the demand and supply variables that have to be managed:

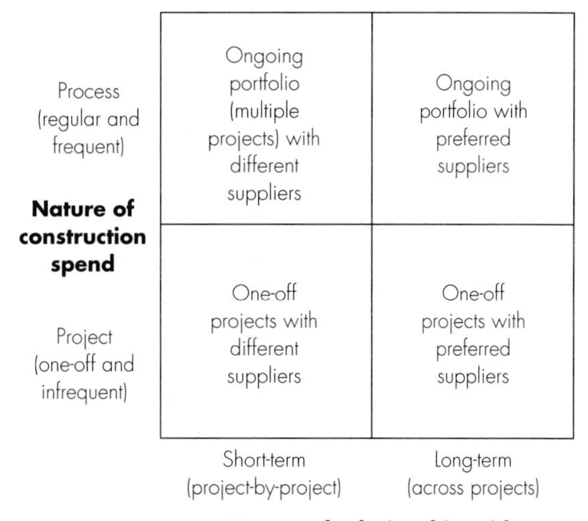

Fig. 11.8 The nature of regularity in construction procurement. Adapted from Ireland (2004).

(1) For *one-off projects with different suppliers* it is likely that arm's length *reactive sourcing* approaches, such as SS, based on adversarial and opportunistic short-term relationships, are likely to persist, with the buyer and supplier equally as interested in these 'win–lose' and 'partial win–lose' as in any of the three collaborative outcomes in Figure 11.3 that tend to favour them in the short-term.
(2) When there are *one-off projects with preferred suppliers* it is possible for opportunistic and arm's length adversarial *reactive sourcing* approaches to be adopted that could include SS and SCS, with the buyer and the supplier similarly interested in the opportunistic one-off game outcomes that favour them more than repeat game outcomes.
(3) For those with an *ongoing programme with different suppliers* it is possible to be both *reactive and proactive*, but the proactive and collaborative sourcing approach would probably only extend to relationship marketing and SD activities with the first-tier supplier, with the buyer preferring collaborative outcomes and the supplier those that favour them when proactive approaches are adopted.
(4) When there is an *ongoing programme with preferred suppliers* it is also possible to pursue both reactive and proactive approaches. There may be scope for both long-term collaborative relationship marketing, and the comprehensive and continuous SD and SCM or network sourcing approaches advocated by proponents of lean and agile thinking. Buyers and suppliers may adopt the same tactics over outcomes, as described previously, when the particular reactive or proactive approach is adopted.

Table 11.4 The implementation of arm's length reactive and proactive collaborative sourcing options in different conditions of regularity.

Nature of construction spend		Nature of relationship with construction suppliers	Supply management option: Reactive arm's length relationships – Supplier selection	Supply management option: Reactive arm's length relationships – Supply chain sourcing	Supply management option: Proactive collaborative relationships – Supplier development	Supply management option: Proactive collaborative relationships – Supply chain management
Nature of construction spend	Process spend high frequency of demand (regular and ongoing)	Long term (ongoing)	Possible	Possible	Possible	Possible
		Short term (by project)	Possible	Possible	Possible	Not possible
	Project spend low frequency of demand (one-off)	Long term (ongoing)	Possible	Possible	Not possible	Not possible
		Short term (by project)	Possible	Not possible	Not possible	Not possible

Obviously, since very few construction clients have the type of portfolio of spend that occurs in categories 3 or 4, it is highly unlikely that these types of proactive and highly collaborative relationship practices will be very prevalent in construction. This is because the demand and supply structures in construction are rarely conducive to relationship marketing or partnership sourcing and supply chain management. Despite this, for some clients, for example those with regular or serial and standardised demand, it is feasible to adopt these approaches. This would include the oil and gas companies, the retail and financial institutions, the utilities, the rail industry, and major manufacturing and retail companies, as well as the Highways and Environmental Agencies and the Ministry of Defence (MOD), among others.

For those – the majority of client demand in construction – where demand is ad hoc, irregular and non-standard, there is little scope for relationship marketing or partnering approaches to be successful. For these buyers, and the suppliers who service them, a more short-term, adversarial and opportunistic relationship management approach will continue.

Table 11.4 summarises the limited and unique circumstances, in relation to frequency, under which collaborative relationship marketing, supply chain management and partnering approaches are feasible.

Conclusions

It follows from this discussion that, while government-sponsored and industry reports may encourage those in project industries in general, and in construction specifically, to implement collaborative and proactive approaches whenever possible, such recommendations are in reality poor thinking. This is because it is impossible to generalise about the way that particular buyers and suppliers should operationalise collaboration under particular power and leverage circumstances. In reality, when a construction project is designed and specified it results in the creation of a particular and often unique power regime that has to be managed. Sometimes it is impossible to create power regimes that will allow the effective development of collaborative proactive sourcing approaches that will support relationship marketing or partnering. On the contrary, one can conclude, given the ad hoc and one-off nature of demand combined with the fragmented and contested supply industry, that conducive power regimes for proactive sourcing are unlikely to exist for the majority of players within project industries.

Despite this, there is scope for the adoption of collaborative relationship marketing and partnering types of approaches in a project environment. In construction, for example, we have seen that for those clients (and also those contractors and suppliers in the chain) who have a high frequency of demand, there may well be opportunities for the creation of highly collaborative relationship marketing approaches in the form either of first-tier supplier development or network sourcing approaches. Therefore, the key is not to try to implement these approaches unthinkingly in any circumstances, but to understand the underlying demand and supply characteristics within supply chains and the power regime structure that these create in order to understand when there is a conducive environment for trust, transparency and collaboration to work together to create improved operational and commercial value for both parties.

Unfortunately, two things follow from this conclusion. First, even if some players in project industries can find opportunities to create collaboration outcomes for themselves, the conflict and tension over who appropriates the lion's share of value from the relationship does not disappear even when relationship marketing and partnering approaches are introduced. This struggle over value is endemic and continues because, paradoxically, operational collaboration does not mean that a commercial 'win–win' is necessarily achievable (Cox, 2004c). Second, while forms of collaboration are feasible in relationships in the project environment, this is not the case for most participants, who struggle to make acceptable returns from participating in a highly fragmented, adversarial and opportunistic industry. This is because neither suppliers nor their clients have any incentives to make the necessary investments of time, money or people to make relations collaborative. As a result, reactive sourcing will continue to be, as it has been in the past, the primary tool of construction procurement and supply management in the project environment.

References

Barlow, J. (1997) Institutional economics and partnering in the British construction industry. *AEA Conference on Construction Econometrics*, February, Neuchatel, Brazil.

Bennett, J. & Jayes, S. (1995) *Trusting the Team: The Best Practice Guide to Partnering in Construction*. Reading Construction Forum, UK.

Bennett, J. & Jayes, S. (1998) *The Seven Pillars of Partnering*. Thomas Telford Publishing, London.

Blismas, N. G. (2001) *Multi-project environments of construction clients*. Unpublished PhD Thesis, Loughborough University, Loughborough.

Campbell, N. C. G. (1985) An interaction approach to organisational buying behaviour. *Journal of Business Research*, **13**, 35–48.

Cherns, A. B. & Bryant, D. T. (1984) Studying the client's role in construction management. *Construction Management and Economics*, **2** (2), 177–84.

Christopher, M. (2000) The agile supply chain. *Industrial Marketing Management*, **29** (1), 37–44.

Christopher, M. & Towill, D. R. (2000) Supply chain migration from lean and functional to agile and customised. *Supply Chain Management: An International Journal*, **5** (4), 206–15.

Christopher, M., Payne, A. & Ballantyne, D. (1991) *Relationship Marketing*. Heinemann, London.

Construction Industry Institute (1989) *Partnering: Meeting the Challenges of the Future*. CII Special Publication, Construction Industry Institute, Austin, Texas.

Construction Industry Institute (1991) *In Search of Partnering Excellence*. CII Special Publication, Construction Industry Institute, Austin, Texas.

Construction Industry Institute (1994) *Benchmarking Implementation Results, Teambuilding and Project Partnering*. Construction Industry Institute, Austin, Texas.

Cox, A. (1997) *Business Success*. Earlsgate Press, Stratford-upon-Avon.

Cox, A. (2001) The power perspective in procurement and supply management. *Journal of Supply Chain Management*, **37** (2), Spring, 4–7.

Cox, A. (2004a) The art of the possible: relationship management in power regimes and supply chains. *Supply Chain Management: An International Journal*, **9** (5), 346–57.

Cox, A. (2004b) Business relationship alignment: on the commensurability of value capture and mutuality in business exchange. *Supply Chain Management: An International Journal*, **9** (5), 410–20.

Cox, A. (2004c) *Win–Win? The Paradox of Value And Interests in Business Relationships*. Earlsgate Press, Stratford-upon-Avon.

Cox, A. (2004d) Strategic outsourcing: avoiding the loss of critical assets and the problems of adverse selection and moral hazard. *Global Purchasing and Supply Chain Strategies*, Business Briefings Ltd, London.

Cox, A. & Ireland, P. (2001) Managing construction supply chains: the common-sense approach for project-based procurement. *Proceedings of the 10th International Annual IPSERA Conference*, 9–11 April, Jönkoping, Sweden.

Cox, A. & Ireland, P. (2002) Managing construction supply chains: a common-sense approach. *Engineering, Construction and Architectural Management*, **9** (5/6), 409–18.

Cox, A. & Thompson, I. (1998) *Contracting for Business Success*. Thomas Telford Publishing, London.

Cox, A. & Townsend, M. (1998) *Strategic Procurement in Construction*. Thomas Telford Publishing, London.

Cox, A., Ireland, P., Lonsdale, C., Sanderson, J. & Watson, G. (2002) *Supply Chains, Markets and Power: Mapping Buyer and Supplier Power Regimes*. Routledge, London.

Cox, A., Ireland, P., Lonsdale, C., Sanderson, J. & Watson, G. (2003) *Supply Chain Management: A Guide to Best Practice*. Financial Times/Prentice Hall, London.

Cox, A., Lonsdale, C., Sanderson, J. & Watson, G. (2004a) *Business Relationships for Competitive Advantage: Managing Alignment and Misalignment in Buyer and Supplier Transactions*. Palgrave Macmillan, London.

Cox, A., Sanderson, J. & Watson, G. (2000) *Power Regimes: Mapping the DNA of Business and Supply Chain Relationships*. Earlsgate Press, Stratford-upon-Avon.

Cox, A., Sanderson, J. & Watson, G. (2001) Supply chains and power regimes: toward an analytical framework for managing extended networks of buyer and supplier relationships. *Journal of Supply Chain Management*, **37** (2), 28–35.

Cox, A., Watson, G., Lonsdale, C. & Sanderson, J. (2004b) Managing appropriately in power regimes: relationship and performance management in 12 supply chain cases. *Supply Chain Management: An International Journal*, **9** (5), October, 357–71.

Cutting-Decelle, A. F. (1997) The use of industrial management methods and tools in the construction industry: application to the construction process. *Concurrent Engineering in Construction: Papers Presented at the First International Conference* (eds C. J. Annumba & N. F. O. Evbuomwan), Institution of Structural Engineers, London.

Department for Environment, Transport and the Regions (1998) *Rethinking Construction*. DETR, London.

Emerson, R. E. (1962) Power-dependence relations. *American Sociological Review*, **27** (1), 31–41.

Ford, D. (1980) The development of buyer-seller relationships in industrial markets. *European Journal of Marketing*, **14** (5/6), 339–54.

Ford, D. (ed.) (2001) *Understanding Business Marketing and Purchasing*. Thomson Learning, London.

Forrester, J. (1958) Industrial dynamics: a major breakthrough for decision makers. *Harvard Business Review*, **36** (4), 37–66.

Godfrey, K. A., Jnr (ed.) (1996) *Partnering in Design and Construction*. McGraw-Hill, New York.

Gummesson, E. (1987) The new marketing: developing long-term interactive relationships. *Long Range Planning*, **20** (4), 10–20.

Gummesson, E. (1997) Relationship marketing as a paradigm shift: some conclusions from the 30R approach. *Management Decision*, **53** (4), 267–72.

Gummesson, E. (1999) *Total Relationship Marketing*. Butterworth-Heinemann, Oxford.

Håkansson, H. (ed.) (1982) *International Marketing and Purchasing of Industrial Goods: An Interaction Approach*. John Wiley, Chichester.

Harrison, A., Christopher, M. & van Hoek, R. (1999) *Creating the agile supply chain*. School of Management Working Paper, Cranfield University, Cranfield.

Hellard, R. B. (1995) *Project Partnering: Principles and Practice*. Thomas Telford Publishing, London.

Hinks, A. J., Allen, S. & Cooper, R. D. (1996) Adversaries or partners? developing best practice for construction industry relationships. In: D. A. Langford and A. Retik (eds) *The Organisation and Management of Construction: Shaping Theory and Practice*. E & FN Spon, London.

Holti, R. & Standing, H. (1996) *Partnering as Inter-related Technical and Organisational Change*. Tavistock, London.

IMP Group (1982) An interaction approach. In: H. Håkansson (ed.) *International Marketing and Purchasing of Industrial Goods: An Interaction Approach*. John Wiley and Sons, Chichester.

Ireland, P. (2004) Managing appropriately in construction power regimes: understanding the impact of regularity in the project environment. *Supply Chain Management: An International Journal*, **9** (5), 372–82.

Koskela, L. (1997) Lean production in construction. In: L. F. Alarcón (ed.) *Lean Construction*. A.A. Balkema, Rotterdam.

Krafcik, J. (1988) Triumph of the lean production system. *Sloan Management Review*, **30** (1), 41–52.

Lamming, R. (1993) *Beyond Partnership*. Prentice Hall, New York.

Latham, M. (1994) *Constructing the Team*. The Stationery Office, London.

London, K. & Kenley, R. (2000) Mapping construction supply chains: widening the traditional perspective of the industry. *Proceedings of the 7th Annual European Association of Research in Industrial Economics*, EARIE Conference, Lausanne, Switzerland.

Luck, R. & Newcombe, R. (1996) The case for integration of the project participants' activities within a construction project environment. In: D. A. Langford & A. Retik (eds) *The Organisation and Management of Construction: Shaping Theory and Practice*. E & FN Spon, Glasgow.

Macbeth, D. K. & Ferguson, N. (1994) *Partnership Sourcing*. Financial Times/Pitman, London.

Mason-Jones, R. & Towill, D. R. (1999) Total cycle time compression and the agile supply chain. *International Journal of Production Economics*, **62** (1/2), 61–73.

Mason-Jones, R., Naylor, B. & Towill, D. R. (2000) Lean, agile or leagile? Matching your supply chain to the marketplace. *International Journal of Production Research*, **38** (17), 4061–70.

McGeorge, W. D. & Palmer, A. (1997) *Construction Management: New Directions*. Blackwell Science, London.

Moody, P. E. (1993) *Break-through Partnering*. Oliver Wright Publications, Essex Junction, VT.

Morris, P. W. G. (1997) *The Management of Projects*. Thomas Telford Publishing, London.

Nam, C. H. & Tatum, C. B. (1988) Major characteristic of constructed products and resulting limitations of construction technology. *Construction Management and Economics*, **6** (2), 133–48.

Naylor, J., Naim, M. M. & Berry, D. (1999) Leagility: integrating the lean and agile supply chain. *International Journal of Production Economics*, **62** (1/2), 107–18.

Ng, S. T., Rose, T. M., Mak, M. & Chen, S. E. (2002) Problematic issues associated with project partnering – the contractor perspective. *International Journal of Project Management*, **20** (6), 437–49.

Ohno, T. (1988) *Toyota Production System: Beyond Large-Scale Production*. Productivity Press, Cambridge, MA.

Saad, M., Jones, M. & James, P. (2002) A review of the progress towards the adoption of supply chain management (SCM) in construction. *European Journal of Purchasing and Supply Management*, **8** (3), 173–83.

Saunders, M. J. (1994) *Strategic Purchasing and Supply Chain Management*. Pitman Publishing, London.

Smyth, H. J. (2005) Procurement push and marketing pull in supply chain management: the conceptual contribution of relationship marketing as a driver in project financial performance. Special issue on Commercial Management of Complex Projects. *Journal of Financial Management of Property and Construction*, **10** (1), 33–44.

Stalk, G. & Hout, T. M. (1990) *Competing Against Time: How Time Based Strategies Deliver Superior Performance*. The Free Press, New York.

Thompson, P. J. & Sanders, S. R. (1998) Partnering continuum. *Journal of Management in Engineering*, **14** (5), 73–8.

Turner, J. R. (1993) *The Handbook of Project Based Management*. McGraw-Hill, London.

Turner, J. R. & Müller, R. (2003) On the nature of the project as a temporary organisation. *International Journal of Project Management*, **21** (1), 1–8.

Vrijhoef, R. & Koskela, L. (2000) The four roles of supply chain management in construction. *European Journal of Purchasing and Supply Management*, **6** (3/4), 169–78.

Watson, G., Cox, A., Lonsdale, C. & Sanderson, J. (2003) Thinking strategically about supply chain relationship management: the issue of incentives. In: D. Waters (ed.) *Global Logistics and Distribution Planning*. Kogan Page, London.

Williamson, O. E. (1985) *The Economic Institutions of Capitalism*. Free Press, New York.

Womack, J. P. & Jones, D. T. (1996) *Lean Thinking: Banish Waste and Create Wealth in Your Organisation*. Simon & Schuster, New York.

Womack, J. P., Jones, D. T. & Roos, D. (1990) *The Machine that Changed the World*. Rawson Associates, New York.

12 The economics of relationships

Graham Ive and Kai Rintala

In 1992, the government of the United Kingdom introduced the Private Finance Initiative (PFI) as a policy to allow and regulate privately financed public projects. However, from the perspective of organisational economics, PFI can be better understood as an innovative procurement method for public services. A PFI project involves a private sector entity, hereafter referred to as the *ProjectCo*, taking the responsibility to design, build, finance and operate (DBFO) an asset or a facility as part of the provision of a public service for a contract period of up to four decades.

The ProjectCo typically consists of at least a number of designers, a design-and-build contractor, an operator and a financier. In the project development process, some of these private sector actors organise themselves initially into a bid vehicle (BV) and subsequently, once the financial close of the project becomes imminent, into a special purpose vehicle (SPV). The relationships of the remaining actors to the BV/SPV are through a complex set of contracts and agreements. Thus, the term ProjectCo refers to a wider set of private sector actors that are involved in the project and not just those that are parties to the contracts and agreements that govern the BV/SPV.

Conventional procurement of public services acquires design and construction separately from the inputs required to operate an asset. PFI is innovative because it procures the design, construction and operation of an asset together. Conventional procurement may involve any form of construction procurement plus, for operating services, either use of a public sector organisation or separate purchasing from private providers. Thus one DBFO contract replaces several separate contracts or agreements for D, B, F and O.

PFI procurement is also innovative because, uniquely, it divides ownership rights in the project-asset between the public sector client and the private sector ProjectCo. Typically, in the UK, PFI projects that involve construction are built on land for which the freehold owner is the client. This land is then leased, for a period usually exceeding the term of the DBFO contract, to the private sector. The ProjectCo then finances the construction of the project-asset, and receives quarterly payments from the client for provision of the asset-based services specified in the contract. During the contract period, the ProjectCo has residual control rights over the use of the project-asset. That is, in matters on which the contract is silent, the power of decision and authority over project-asset use is with the ProjectCo.

However, it would be incorrect simply to describe the ProjectCo as the 'owner', for two reasons: not only does the public client reserve extensive rights to itself as both freeholder and purchaser, but also it has acquired the option (but not the obligation) to take over the ownership of the project-asset without further payment on expiry of the contract, and the contract is designed to protect the long-term ('residual') value of this option. Nonetheless, when compared with conventional procurement, the feature of PFI is the extent to which the ownership of the project-asset has been transferred to the private sector. On why it matters whether one party or another is the owner of the asset required to provide a service, this chapter follows the approach developed by Hart (1995). The ownership of the project-asset matters when contracts are incomplete, because in such circumstances it gives power.

This chapter analyses the key relationship in a PFI project between the client and the BV/SPV (the nexus of the ProjectCo) using the framework of transaction cost economics, as developed by Williamson (1975, 1985, 1996) from original insights by Coase (1937, 1960), as its theoretical framework. The analysis compares and contrasts the relationships in PFI projects with those in traditionally procured projects.

The basic premise of transaction cost economics is that the cost of acquiring a product or a service is a combination of the cost of its production and the cost, to buyer and seller, of transacting for it. The gap between the value of a service to its buyer and its production-plus-transaction cost represents the economic surplus created by the activity. This surplus will be appropriated as 'quasi-rent' by the transactors. Insofar as the seller obtains a price in excess of its cost, the 'cost to the purchaser' will differ from the economic cost.

The total quasi-rent of a project is the excess of its value to the buyer over the opportunity cost of the resources the supplier requires in order to produce it. This total rent then gets divided up between the buyer (whose rent-share is the excess of project value to them over the price they pay to the seller) and the seller (whose rent-share is the excess of the price they receive from the buyer over the resource cost). These two shares always make up 100% of the total quasi-rent. Thus, as price is altered, so respective rent-shares of buyer and seller change by equal but opposite amounts.

If contracts were complete and easily enforceable, economic efficiency would be able to dominate struggles over rent-appropriation, in the sense that the more efficient of two uses of scarce and net present value-yielding resources would prevail. Efficient investments would then be made. Put at its simplest, a bigger cake (bigger economic surplus) would permit agreement on bigger slices all round. However, where contracts are incomplete and it is expensive to transact (exchange) ownership-rights, the result may be that an inefficient solution prevails.

The Coase Theorem (Coase, 1960) posits that, if the parties bargain to an efficient agreement (for themselves), then the value-creating activities that they will agree upon do not depend on the bargaining power of the parties or on what assets each owned when the bargaining began. Efficiency alone determines the activity choice. Even if the main gainer from the greater efficiency is not the project owner (and therefore initially lacks the authority to implement the more efficient use of the assets), it will be worth their while to buy out the owner, at a price that owner will find attractive. The larger total 'cake' means that there will always be some such price, that leaves both parties better off than they would be receiving shares of a smaller cake. The other (non-efficiency) factors only affect decisions about how the costs and benefits are to be shared (see also Milgrom & Roberts, 1992, chapters 2 and 9). However, if contracts are incomplete or hard to enforce, post-contract there may be attempts to renegotiate the agreed shares of the economic surplus by exploiting the sunk costs (and thus vulnerability to 'hold-up' demands) of the other party.

Sunk costs are expenditures made as part of a transaction that cannot be recovered or salvaged if the resources have to be redeployed to an alternative use. The *hold-up* problem arises because of the existence of sunk costs. Consider an example. Suppose a ProjectCo has agreed to provide prison services to a government department. It expects its total building costs (including its opportunity cost of capital) to be C and its revenue (net of operating costs) to be R (thus its expected quasi-rent is R – C). However, in order to obtain this quasi-rent, it had to invest in building a prison, which is an asset whose value (net revenue) in its best use other than as a prison is much lower than R (and the government department has the monopoly over use as a prison, thus in this case finding an alternative *user* means finding an alternative *use* as well). Suppose, for simplicity of illustration, that the value in alternative use happens to equal two-thirds of C. Then, we can usefully break the ProjectCo's total costs, C, into a part (two-thirds) that is salvageable and a part (one-third) that is irrevocably 'sunk'.

Now, once the ProjectCo has irrevocably incurred this sunk cost, in the absence of a binding long-term contract the department could come back to it with a revised offer to pay it only, say, C. What is the least-bad option now available to the ProjectCo? It could refuse to concede to this 'hold-up' demand, and redeploy its asset. In that case, it ends up with revenue of 2/3rds of C and costs of C (a loss equal to one-third of C). Or, it could concede the hold-up demand, in which case it ends with revenue just equal to its costs – clearly, this is the least-bad option. In the out-turn, after changing the terms of the bargain, the buyer in this example instead of sharing the economic surplus with the seller has appropriated all of it. Fear of such hold-up may deter parties from making investments in sunk costs whose value is transaction specific. This in turn means a loss of potential economic surplus and efficiency (Klein *et al.*, 1978; Williamson, 1979, 1996).

A large part of the argument for the use of PFI procurement proceeds from the assumption that, under conventional procurement, efficiency losses of this kind are substantial, because the hold-up problem remains unresolved. In particular, it is argued that potential efficiency gains from whole life value/cost optimisation solutions are being lost, because under conventional procurement it is not possible to set up enforceable agreements that give parties the incentives to invest in such solutions. It is then further argued that PFI overcomes this problem and gives the ProjectCo both incentives and opportunities to optimise whole life benefit/cost. This argument assumes that PFI contracts can be made to be binding and enforceable. By offering a long-term contract with strong penalties for cancellation, the government department is held to have offered the ProjectCo a valuable 'counter-hostage' sufficient to give it confidence that it will not have to concede to hold-up demands.

This chapter explores the procurement of *accommodation service* PFI projects. A public sector client consumes an accommodation service as part of its provision of a core service. Education and healthcare are examples of core public services. Thus, an accommodation service is a support service as opposed to a core service. The chapter focuses on projects in education and healthcare as those sectors are the two largest PFI market segments if measured by the number of projects that have reached financial close (HM Treasury, 2004). Consequently, the project-asset that the ProjectCo designs, builds, finances and operates as part of its service provision is a building such as a school or a hospital. It must be noted that a client can also use PFI to procure core public services, such as custodial services where the private sector is responsible for the full service provided for the benefit of the end users, or to procure support services such as the provision of serviced heavy transport equipment for the Ministry of Defence where the project-asset is not a building. A much fuller description of PFI as a mode of procurement and of the consequences of using PFI for the management of projects is offered in Ive (2004); see also HM Treasury (1997).

This chapter uses the term *traditional procurement* to refer to a specific variant of conventional procurement, involving a combination of traditional design-bid-build construction procurement and outsourcing of operational services – that is, hard and soft facilities management. It uses this particular setting as a point of reference as it is widely used in the procurement of accommodation services and, consequently, allows the reader to relate easily to the differences in the two procurement settings.

In addition, this chapter examines how and why mutually beneficial relationships may evolve in PFI projects and explains the reasons why such relationships would have been unlikely to develop if the projects had been procured traditionally. It also shows how PFI greatly increases, for each party, the amount at stake in the form of sunk investments with transaction specific value – investments that both make each party vulnerable to 'hold-up' by the other and at the same time create efficiency gains and allow pay-off gains for each party, but that would result in significant pay-off losses if the relationship should in fact break down, and that would not be made if fear of such loss could not be overcome. The chapter illustrates this by way of anecdotal evidence from real life projects. The two case study projects analysed in this chapter are the King's College London (KCL) Site Rationalisation project and the University College London Hospitals NHS Trust (UCLH) Gower Street Redevelopment project. Hereafter, these case studies are simply referred to as the KCL project and the UCLH project respectively. The procurement and development of these two projects are described in explicit detail in Rintala (2004).

KCL is one of the premier universities in the UK. In October 1995, it initiated the formal PFI procurement of a site rationalisation project in order to consolidate its academic accommodation into the proximity of its Strand Campus in the London Borough of Westminster. As part of a wider rationalisation of facilities, the project included the design, construction and 25-year operation of New Hunt's House for teaching and research in basic medical sciences on Guy's Campus in the London Borough of Southwark. New Hunt's House was the only building within the overall site rationalisation project that was to be newly constructed. In December 1997, KCL entered into an accommodation service provision contract with the private sector and, in August 1999, the 22 000 m^2 teaching and research facility opened for the beginning of the new term as scheduled. The capital cost of the entire site rationalisation was around £100 million of which approximately £40 million can be attributed to the capital cost of New Hunt's House. It must be noted that the KCL project cannot be considered a 'typical' accommodation service PFI project. This is because KCL awarded the project using a design, build and operate (DBO), as opposed to DBFO, contract as the project included the disposal of a number of properties that became surplus to its requirements, enabling it to finance the capital cost of the project from the resulting receipts.

UCLH is one of the largest and best performing NHS trusts in the UK. In November 1995, it advertised the procurement of an accommodation service PFI project in order to overcome the problems it was facing as a result of having to work in facilities that were scattered in Central London and unsuitable for the provision of modern healthcare services because of their age. In July 2000, the project reached financial close based on a DBFO contract, which had duration of 40 years. The private sector provider completed the first phase of construction slightly ahead of schedule in 2005 and the second phase is due in 2008. The capital cost of the 72 500 m^2 new single site hospital is in the region of £225 million.

Key theoretical concepts

Transaction costs are influenced by asset transaction specificity, uncertainty and transaction frequency. In combination, high asset transaction specificity, high uncertainty (which makes contract incompleteness unavoidable) and low transaction frequency (which makes design of transaction specific forms of governance uneconomic, while also reducing the sanction of loss of pay-offs from future projects) give rise to the threat of opportunistic behaviour from one of the contracting actors (Williamson, 1975, 1985, 1996). *Opportunism* is a phenomenon where one of the contracting actors seeks to take advantage of a post-contract shift in relative bargaining power to appropriate pay-offs originally promised to the other by reneging on promises and demanding renegotiation of contract. This is enabled by asymmetric information between the contracting actors. By definition, *asymmetric information* exists when the actors possess different amounts of information on the actual and hypothetical 'efficient' production cost or on the quality of the product or the service being transacted. In the case of an accommodation service PFI project, the production cost is the cost of designing, building, financing and operating the project-facility.

Asset transaction specificity refers to the degree that the investment by party X (or Y) in an asset A (or B) intended for use in (that is, to improve the efficiency of) the transaction with party Y (or X) has less value in transactions with alternative parties, should the original transaction break down. In an accommodation service PFI project, asset transaction specificity is high for the ProjectCo as the project-facility it owns is designed and built to meet the needs of the client procuring the project. As a result, the ProjectCo is unlikely to be able to recover the full value that the building has in the context of the specific project if the service provision contract is terminated prematurely and it has to sell the facility or use it to provide services to an alternative customer.

The asset transaction specificity is also determined by the asset itself. It is higher in PFI projects where the project-facility is, for example, a prison than in ones where it is, say, an office building, as the pool of alternative clients is smaller. In fact, it is sometimes useful to divide asset transaction specificity into the part of asset value that would be lost if another *user* had to be found, but keeping the same *type of use* (we can call this transaction specificity); and the part of asset value that would be lost if another type of use had to be found for the asset (we can call this asset specificity). On the other side, asset transaction specificity is also high for the client. In this case, it is best thought of as the cost of switching to an alternative provider should the transaction with the original provider break down. This comprises the difference between the sum of the net amounts it will then have to pay to both original and replacement providers, and the amount it originally contracted to pay to the initial provider (together with consequent loss in the value of the service provided) (Chang, 2002; Chang & Ive, 2000; Chang & Ive, 2004).

Investments in relationships between actors are transaction specific investments, as they cannot be utilised in transactions with other actors. *Uncertainty* is inherent in the future and arises from the interplay of bounded rationality and complexity. *Bounded rationality* refers to the computational limitations of the human mind. The future is uncertain by definition and, therefore, it is impossible for the contracting actors to anticipate all the possible states of the world and cater for them by writing a complete contract, covering every possible contingency, to govern the transaction. In addition, the product or the service being transacted for may be complex. *Complexity* may make it impossible for the contracting actors to specify all the desired attributes of the transaction in all states of the world (on one side, the payments that will then apply, and on the other the attributes of the product or the service that will then be provided) in a contract, which consequently remains incomplete. The absence of complete contracts enables opportunism.

Transaction *frequency* refers to how often similar transactions between these parties are likely to occur. If the frequency is high, the future benefits from those future transactions deter the contracting actors from behaving opportunistically. In other words, the short-term gains that either of the actors can obtain by behaving opportunistically are smaller than the benefits from future transactions with the other actor that they stand to lose as a result of the breakdown of the relationship. If transaction frequency is low, the actors have little incentive to refrain from behaving opportunistically as there are no future benefits to be had by not doing so (ignoring, as we shall for simplicity throughout this chapter, wider effects on their reputation with third parties). On the one hand, a university is an example of a non-repeat client that is unlikely to procure further PFI projects in the foreseeable future. As a result, it is unable to exercise bargaining power arising from the prospect of future projects and, thus, it is more likely to encounter opportunistic behaviour from the private sector. On the other hand, HM Prison Service is an example of a repeat client with a high frequency of transactions as it is responsible for procuring all custodial service PFI projects in the UK. Consequently, it has more bargaining power when operating in the PFI market than a university.

There is, however, another side to transaction frequency. If your transaction requirements are highly specific or highly uncertain, your transaction is more likely to require a specialised *governance structure* (a set of arrangements for dealing with disputes or changes). However, the costs of designing a bespoke governance structure are high and are fixed costs. It only pays to incur these fixed costs if you can spread them over a high frequency (actually, high total value) of transactions of this specialised kind. The PFI client with only one small project to place is therefore caught between two unattractive alternatives. Either they pay the relatively large fixed costs (legal and financial fees, for example) of designing and enforcing a bespoke contract, or they accept the hazard of entering into a transaction for a specialised and uncertain requirement using a standard form of contract ill-adapted to their specific requirements and problems. To deal with this problem, the government departments with overall responsibility for areas such as health have themselves undertaken the fixed costs of developing forms of PFI contract that are both specialised in the sense that they are designed specifically to cope with the specialised requirements of a hospital trust, as opposed to other kinds of client, but at the same time standardised, in the sense that they can be used by any hospital trust.

In accommodation service PFI projects, *transaction duration* becomes important. This is because such projects have contract periods of several decades. This will deter the contracting actors from opportunism in the project as long as the value of potential loss of future benefits from its unexpired duration outweighs the short-term gains to be had. In a complete contract, duration gives comfort and assurance to each party contemplating a vulnerable transaction specific investment. In an incomplete contract, however, one long-duration contract may increase total vulnerability compared to a series of short-term contracts. Without such a long-term contract, each party would be less willing to invest in an asset with a degree of transaction specificity. On the other hand, the very length of the contract increases the transaction specificity of some assets, and increases the cost, at any moment, of switching.

It is the combination of high asset transaction specificity, high uncertainty and low transaction frequency that gives rise to the threat of opportunism.

First, if asset transaction specificity did not exist, neither actor would stand to lose from the premature termination of the transaction. This is because they would have not incurred sunk costs in the transaction and would be able to switch to an alternative client or provider at no cost. Second, if the future were certain (a limited set of possibilities, each with known probability), it would be possible for the actors to write complete contracts that would cater for all the contingencies of the future and specify all the attributes of the product or the service being transacted for, in order to protect them from opportunism. Third, if the frequency of the transactions (potential 'repetition') was high, the loss of future benefits to the actors from transactions yet to take place would prevent them from taking advantage of each other. Moreover, if single-transaction value is taken as a given, the value of the whole set of transactions is a direct function of frequency. With high overall value, it would then be economical to invest heavily in tailor-made arrangements designed to monitor and enforce that series of contracts. In the absence of such investments in specialised governance structures, the contracting parties are more vulnerable to opportunism.

A fuller explanation of all of the terms discussed in this section of the paper can be found in appropriate textbooks on the economics of organisation (Douma & Schreuder, 2002; Milgrom & Roberts, 1992).

Transaction cost economics, because of its assumption that an unknown proportion of actors will behave opportunistically if they think it will pay them to do so, can be fairly characterised as being cynical about trust. On the other hand, TCE does argue that the giving of reciprocal hostages can be an effective way to build functional long-term relationships, because it makes it in the self-interest of each party to prevent relationship-breakdown.

In effect, TCE reasoning is analogous to the argument that, in personal relationships, institutions that have the effect of making divorce painful to obtain for each party will have the effect of encouraging parties to realise that their best options are either to 'make their relationships work' or, if reflection suggestions the odds on that being achievable are poor, to avoid entering into marriages altogether. If your situation is highly unusual, it may be worth supplementing the standard marriage contract by a special pre-nuptial agreement. Making the right choices about whether to marry will be more likely, the more complete is the information that each party has about the other prior to the decision to marry or not. TCE argues for the importance of symmetry – equal vulnerability and equal information on each side.

Hostage arrangements recommended by TCE are various. In addition to the more direct kinds of hostage (investments whose value will be lost, financial penalties that will be payable in the event of breakdown), if when parties transact they let their reputation in the wider world depend to a measurable extent upon the publicised opinion of their co-transactor(s), then they too are offering a kind of hostage, where the moral of the story is to make sure that you only relate to parties that value their reputations (and remember that symmetry requires that the reputation of those who make false allegations must also suffer).

A model of the comparative structure of transaction costs in PFI and traditional procurement

One way of thinking about the idea of PFI is that benefits are meant to flow from reducing the total of such costs as the consequence of transforming the composition or structure of transaction costs. In traditional procurement the main transaction costs are invisible – the lost potential efficiency gains from investments in asset-specificity that are not made because the hold-up problem remains unsolved. The idea is to reduce these lost potentials, by creating a framework within which investments in asset-specificity will be made. This will come at the cost of some increase in 'visible' transaction costs – the administrative and legal costs of doing business. This distinction between the visible, administrative costs of transacting and the partially invisible, efficiency-loss costs of unsolved hold-up problems is derived from Chang (2002) and Chang and Ive (2000).

The administrative costs of running a procurement method (traditional or PFI) can be thought of as a kind of 'tax', assessed so that for a given total level of direct spending on procurement, P, total 'tax' bill is in part a function of the number of transactions, in part a function of the total value of expenditure, and in part a function of the average level of complexity of transactions.

Let total administrative transaction costs in traditional procurement ATC_{trad} be expressed:

$$ATC_{trad} = (P.t) = a.n + b.P + n.S^c$$

where a, b and c are coefficients; n is the number of transactions into which P is divided; S is the average size of a transaction; c measures the impact of the average level of complexity of transactions on administrative cost; and t is total administrative costs of traditional procurement expressed as a proportion of total value of direct procurement expenditure.

Thus, by definition: $n.S = P$ and is constant across the procurement methods being compared.

Let total administrative transaction costs in PFI procurement, ATC_{PFI} be expressed:

$$ATC_{PFI} = (P.f) = a\{n/r\} + b.P + \{(S.r)^c.(n/r)\}$$

Where r = ratio between the number of transactions comprising P in traditional and in PFI procurement. (If we simplify, we might imagine typically three contracts, one each for D, B and O, are replaced by one contract for DBO. In that case, r = 3.) The average size of contract in PFI will then be r.S.

By definition, $a\{n/r\} < a.n$. Let us call the amount of the difference x.

Wherever $c > 1$, it is clear that $n.S^c < \{(S.r)^c.(n/r)\}$. Let us call the amount of difference y.

Thus $ATC_{PFI} > ATC_{trad}$ whenever $y > x$.

Suppose, for example, that c has a value of 2. Then, comparing PFI and traditional, y increases in proportion to the square of the difference in transaction size, while x reduces in proportion to the difference in transaction size.

The argument that the complexity element of ATC rises as 'size' of contract rises needs to be clarified. We are not necessarily arguing (though see below, on 'bundling') that if, say, three similar building contracts were combined into one, the bigger contract would be more complex than the average complexity of the smaller ones. The argument rather is that when we combine contracts with different content (D, B, F and O) into one integrated contract, by increasing the heterogeneity of the work-content, and by attempting to capture and internalise interdependencies, we do increase complexity. The impact of increased complexity of the 'typical' contract on ATC for the whole expenditure programme, P, is partially, but not fully, offset by the fact that there are fewer contracts in total.

We believe this result will arise whenever the only change between PFI and traditional is one of scope of contract. In practice, the picture is complicated by the fact that 'other things' are not left 'equal'. In particular, PFI is likely to involve (as we argue below) a greater element of output specification, which may (as experience is gained) eventually make DBO contracts less complex.

This increase in administrative transaction costs is 'visible' as the increased fee costs for legal, financial and technical advice and as the increased amount of contract documentation incurred under PFI as compared to traditional procurement.

Now, it should be clear that transaction costs fall into two kinds: the costs of acquiring information (search costs, decision costs, monitoring costs) such as would arise even in the absence of a hold-up problem, on the one hand (which we have discussed above, as 'administrative' transaction costs); and the costs of dealing with the hold-up problem and post-contract renegotiation on the other. The former are in part 'per transaction' costs because while in part they vary with the size and complexity of the transaction, in part they are fixed in amount per transaction. Thus, in total they can be diminished by transforming a large number of small transactions into a small number of larger ones, if this can be done without an offsetting increase in complexity of the transactions. The hold-up problem, on the other hand, creates three kinds of cost:

(1) The cost of attempting to write complete contracts.
(2) The largest cost, though 'invisible' to accounting systems, comprises the potential efficiency gains lost by not undertaking investments that are vulnerable to hold-up.
(3) The cost of proceeding with a transaction-specific investment without the protection of a complete contract. This cost is the probable risk of a hold-up attempt occurring, multiplied by the costs created by that attempt – in an efficiency perspective, these are the costs of the resources used up in resolving or litigating the dispute, or in switching; in the perspective of a single party, they also include the amount of 'ransom' paid, even though this is only a transfer of quasi-rent and not a resource-cost.

Which if any of these three types of hold-up costs have an element that is fixed 'per transaction', and does not vary with the size or complexity of the transaction?

(1) The cost of contract writing, no doubt, has a fixed ('per transaction') as well as a variable (in proportion to the size of the transaction) element.
(2) The potential efficiency gains lost by avoiding transaction specific investment will be reduced by 'integrating' D, B and O, if that practice encourages some investments to be made that would not otherwise have been made. Therefore in effect they too have a 'fixed per transaction' element.
(3) The cost of proceeding and accepting hold-up risk is a function (*inter alia*) of the total level of investment in asset-specificity across the whole set of contracts of a client, and as such is increased by integrating smaller projects up into fewer, larger ones, insofar as this practice induces such investment – but this increase must be expected to be less than the decrease in the second case, otherwise the extra investments would not have been made.

In a traditional setting, the client would, for example, procure each of a series of school new builds and their operation independently and sequentially, whereas PFI, beyond simple integration of D, B and O contracts for each school, encourages further 'integration' of these into a single contract. Bundled contracts will present greater opportunities for 'hold-up' and contract renegotiation, but also greater opportunities for investment in transaction specific assets.

Opportunity cost, delay and asset transaction specificity

One key investment each party makes is to hold back their resources from alternative possible uses while bidding and negotiation are in progress. By the end of that process, the best value alternative use of these resources may be substantially less than it was before the process started. Let us call the best alternative option for the resources O. Like any project, O has a positive NPV (net present value) only if it is timely. At time $t = 0$ the expected value of O is Y, and $Y < X$ (X being the expected pay-off from investing those resources in the PFI project). Let us consider the client's perspective, and let O be a traditionally procured version of the same project (also known as the public sector comparator [PSC]) for which they are developing PFI procurement. At time $t = n$, if the client then has to switch to O, all the flows of costs and benefits associated with O will be put back in time by n years, as compared to the streams yielding Y at $t = 0$. Let us call the NPV of O at n, Z. What does putting back both benefits and costs do to the NPV of O? It is likely to reduce it. If it does not, then Y cannot be more than Z. But in that case, at time $t = 0$, the better choice would have been to defer decision on O and hold the option to procure it later on. But if the value of O is improved by deferral, surely it is plausible that the same would be true for the value of the PFI version – in which case, its procurement should not have been initiated at t = 0.

Another way of looking at this question is to consider the effects of uncertainty. A project is conceived at time t = 0 intended to deliver benefits from, say, t = 3 to t = 53. After 50 years in use, the asset wears out. It is understood that unforeseen social and technological changes are more likely to remove expected benefits the further they lie in the future. Cost, on the other hand, has a large component of construction cost, and is front-loaded, that is, a smaller proportion of future cost, if any, is likely to be saved by any effects of unforeseen change. Delay and switching (at, say, t = 2) means that we move to a project conceived at t = 0 but now intended to deliver benefits from t = 5 to t = 55. It is thus exposed to two more years' worth of the negative effects of unforeseen obsolescence on each of its expected annual flows of benefit. Thus it is likely that Z < Y, and thus Z < X, even if Y = X.

Even if initially (at time t = 0) there was no difference between the expected value-for-money of the PFI bid (X) and the PSC (Y), if, after attempting to negotiate a PFI contract the client has to revert to the PSC, by that time (say, t = 2), the expected value-for-money from the PSC will be less than it was at t = 0. The amount by which it has fallen measures the loss of value-for-money to the client resulting from delay. It is a measure of the cost to the client of having to switch mode of supply because of failure to agree terms with the PFI supplier. If this failure to agree arises because of opportunistic attempts by the ProjectCo to renegotiate terms in their favour between the submission of tender and financial and commercial close, then it is an instance of loss of transaction-specific sunk (opportunity) costs. If, on the other hand, agreement is finally reached with the ProjectCo at a price higher than originally tendered, offering only Z value-for-money to the client, we have a case of vulnerability arising from sunk costs leading to appropriation of X-Z from the client by the ProjectCo. Initially, had the ProjectCo only offered Z (or any amount less than Y) its bid would have been rejected and the client would have chosen the superior option of the PSC. However, by the time negotiations reach their 'crunch point', at t = 2, the client will have to accept an offer of Z, because it is equal to the value of the best alternative then still available (the PSC, delayed 2 years).

Two case studies of PFI procurement

The procurement of an accommodation service PFI project has four principal stages. The first is *procurement preparation*, where the client compiles a case for procuring a specific facility-based service. The stage ends when the project is advertised, simultaneously initiating the formal procurement process. The second is the *bidding* stage, where the client invites ProjectCos to bid for the service provision contract being advertised. The stage concludes as the client appoints a private sector provider offering the most economically advantageous service as the preferred bidder. The third stage is the *preferred bidder* stage, where the client and the ProjectCo negotiate and, subsequently, agree the detailed service provision solution and contractual terms for the project. The fourth is the *implementation* stage that begins after the financial close of the project. This stage consists of construction and operational phases. The transition between the two phases takes place as the project-facility is commissioned. The implementation stage ends with the expiry of the accommodation service provision contract (Treasury Taskforce, 1999a).

Asset and transaction specificity

In the procurement preparation stages of the case study projects, as we would expect, opportunism arising from asset or transaction specificity was not a problem. This was because the clients and ProjectCos had yet to build relationships and, thus, to make transaction specific investments. KCL and UCLH had developed most of the invitation-to-negotiate documentation, but they would have been able to re-use the documentation if they had had to recommence the bidding process at a later stage. The ProjectCos, in turn, had yet to become aware of the project and, thus, make any investments.

The threat of opportunism began to arise in the bidding stage as KCL and UCLH prequalified six and four ProjectCos respectively to bid for the projects. This was because they made transaction specific investments in the selection of the ProjectCos and incurred sunk costs that could not be redeployed in an alternative procurement process. The transaction specific investments each of the clients made were the costs arising from evaluating all the potential bidders and drafting the shortlist. If the procurement processes had collapsed, the clients would have had to repeat the prequalification and, thus, incur the vast majority of the costs again. The ProjectCos too had made transaction specific investments in both preparing their prequalification submissions and building relationships with the clients. The vast majority of that work would have been lost if the procurement of the project had not proceeded.

Once invited to bid for the projects, the ProjectCos began to develop service provision solutions for the projects. In other words, they started to make transaction-specific investments as they developed the solutions to meet the specific needs of KCL and UCLH. If the procurement processes had not gone forward, those investments would have become irrecoverable sunk costs. The clients incurred further transaction specific investments as they interacted with the ProjectCos and answered their queries, evaluated their proposals and prepared and issued additional procurement documentation, and as they allowed time to pass (see above, 'Opportunity cost, delay and transaction specificity'). The bidding stage culminated in the clients nominating the ProjectCos offering the most economically advantageous service provision solutions as the preferred bidders. As a result of the competitive nature of PFI procurement, the transaction specific investments became irrecoverable sunk costs for the ProjectCos that were not nominated as the preferred bidders.

It must be noted that the UCLH project experienced a considerable delay in its bidding stage because of the political uncertainty over the future of PFI as a public procurement method. However, at this stage of the procurement process, UCLH only had a single ProjectCo bidding for the project – because of that ProjectCo's superior service provision solution that included the acquisition of additional land for the project. During policy reviews that went on for over a year, the ProjectCo suspended the future development of its solution, but kept the project active. In other words, it made additional transaction specific investments in order to protect the future benefits that it might receive if the project were implemented. UCLH appreciated the ProjectCo's commitment. This was because if the ProjectCo had abandoned the project, UCLH would have needed to recommence its procurement. The relationship of the contracting actors deepened considerably during this period of uncertainty.

HM Treasury Taskforce (1999b) recommends that the client should agree the vast majority of the details of the project prior to nominating the preferred bidder. Thus, for example, once KCL and UCLH had nominated the preferred bidders the ProjectCos were no longer subject to competitive pressure. Changes introduced by the client at this stage, whether intended to incorporate desiderata omitted from the original specification or, conversely, intended to make the project affordable, will therefore potentially widen the ratio of price to cost. It is at this stage of the PFI procurement process that the threat of ProjectCo opportunism is at its greatest. The clients had as yet no contract offering them even incomplete protection, yet they could not easily switch to alternative providers. Potentially this enabled the ProjectCos to extort marginal prices (the change in total price corresponding to a change in total specification or risk transfer) from them that were not reflective of competitive market prices. The main deterrent, at this point, was the danger that 'price creep', if pushed too far, would make the PFI project unaffordable, or make it fail the value-for-money test – either of which would lead to no contract, and no positive pay-off, for the ProjectCo. There was no obvious evidence of this distortion in the case study projects. It is likely that the ProjectCos assessed that if price/cost ratio increased there would be a real increase in the probability that the projects might not reach financial close.

In both case study projects, the clients nominated the preferred bidders at a stage where the scheme design solutions had yet to be completed and, thus, the final price had yet to be fully agreed. This eventually resulted in a situation where, as designs were completed, the projects proved to be unaffordable to the clients, which is not an uncommon occurrence on PFI projects. At this stage the ProjectCos had incurred considerable costs in developing the service provision solutions for the projects. If the projects remained unaffordable to the client and, consequently, could not reach financial close, the development costs would become irrecoverable sunk costs. If the clients had had to cancel the project in its preferred bidder stage, they would have incurred three types of losses. First, the transaction specific investments they had incurred in procuring the project would have been lost. Second, the client would have had to incur substantial additional costs from organising a new procurement competition to select another preferred bidder. Third, the clients would have lost the benefits that the projects would have generated in the period of delay arising from the cancellation.

As a result, cancelling the PFI version of the projects was not an attractive alternative for either the clients or the ProjectCos. This is an example of resolution of a potential hold-up problem. The client could have 'held up' (to ransom) the ProjectCo by demanding that it lower its asking price, without corresponding reduction in risk accepted or in specification of outputs. The ProjectCo, in turn, could have held up the client by demanding an increased asking price per unit or item in the specification. Because one way to reduce total price is to reduce either the risks allocated to the ProjectCo or the specification, it is possible for a ProjectCo at this point to reduce its overall price while simultaneously demanding an increase in its profit margin or in its ratio of return to risk.

The fact that the ProjectCos were no longer in a competitive situation made their position stronger. Prior to preferred bidder stage, the client would not have agreed a revised asking price higher than the price obtainable from the second-lowest bidder. After appointing the preferred bidder, the client may have to accept a revised asking price so long as it is lower than the cost (price and new transaction cost) of switching provider by re-procuring the project. However, the joint objective of the two actors had to be to reach financial close on the project in order to enable the benefits from the project for both parties to be materialised.

It was evident in both case study projects that the client and the ProjectCo strived in close cooperation to value engineer the service provision solution to an affordable price, enabling the projects to progress to the next stage. This included both the re-evaluation of the client's requirements and the ProjectCo's profit expectations. In the KCL and UCLH projects the ProjectCos were £10 million and £8 million at risk respectively just prior to financial close. The fact that they stood to lose that amount of money gave them a strong incentive to make their relationships with the clients work and to enable them to agree a mutually beneficial service provision solution that could reach financial close.

After financial close, the projects entered into their construction phases. At this point, the focus of the potential problem of opportunism shifts from 'hold-up' re-bargaining over price, to issues of the impact of design and construction quality and costs on the operational service. If quality can only be monitored and enforced imperfectly, or expensively, and if the ProjectCo can reduce its construction costs by measures that will reduce quality, then the potential exists for 'shaving' of contracted quality by the ProjectCo. However, two parties will be adversely affected by any such actions – the client and the operator. The latter has a strong interest in preventing design and construction decisions that will adversely affect its costs of providing the operational service, or that will increase the risk of its suffering deductions in its service payments. If opportunistic behaviour occurs in the construction phase, it cannot be easily detected at that point. This is because the client only begins to evaluate and pay for the accommodation service in the operational phase.

It is primarily the potential threat to the future revenues from the project that protects the client from the ProjectCo behaving opportunistically in the construction phase. Opportunism is more of an internal problem for the ProjectCo in the construction phase (in relations between the design-and-build contractor, the operator and the SPV) and, thus, outside the scope of the analysis presented in this chapter. However, it must be noted that the client does have contingency contracts in place to protect itself during the construction phase in the event that the ProjectCo runs into financial difficulties from which it can only extricate itself by opportunistic behaviour towards the client. The cost of drawing up these contingency contracts is a transaction cost.

In the operational phase of an accommodation service PFI project the asset specificity is high as the service provision is tied into the project-facility. However, the client has contingency contracts in place to ensure the continuity of service if its relationship with the ProjectCo breaks down. These contracts mitigate the threat of opportunism arising from asset specificity. However, it is likely that the benefits tied into the implementation of the project are more crucial in alleviating the problem.

In traditional contracting there is both the problem of actual hold-up demands arising from process specificity of expenditure on the incomplete transaction (see below) and the problem that potential efficiency gains are lost precisely because insufficient related investment is made.

In traditional procurement, asset specificity is still an issue especially for the client during construction. The contractor can hold the client to ransom by (in effect, though not explicitly) threatening to delay completion of the building (for example, by saying it cannot proceed until the client provides more information) if the client does not meet its various demands, such as claims it puts forward. The cost of switching to another supplier, or of resolving claims in court, is high, and this makes the money paid to the contractor something of a sunk cost, representing a form of asset specificity called process specificity. In the operational phase the asset specificity is considerably reduced as the operational service providers mostly use assets, such as cleaning equipment, that they can use to provide services to alternative clients, or assets owned by the client.

Uncertainty

In the KCL and UCLH projects, uncertainty decreased as the projects progressed through their development. In the procurement preparation stage, uncertainty was at its highest as the clients had yet to specify the type of accommodation services they wanted to obtain. For example, initially KCL invited DBFO bids that would include both New Hunt's House and the conversion of Cornwall House (a larger, existing building on another campus) and disposal of surplus sites. The latter was then split off and the financing from site sales used to change the contract from DBFO to DBO; though they bid together, separate design and construction teams were used by the ProjectCo for Cornwall House and New Hunt's House. After bids were received the amount of accommodation required at New Hunt's House was revised upwards by 10%. Uncertainty gradually reduced as the ProjectCos developed service provision solutions to meet the needs of the clients, which eventually materialised as buildings in the construction phase. KCL sought bids on the basis of an outline 'reference design' prepared by its design consultants. However, after bidding on this basis, the ProjectCo later developed a completely separate alternative design, which was eventually built. However, in the operational phase of the projects considerable uncertainty remains. This is because the clients cannot know whether they will require changes in the service provided; or the price they will have to pay to obtain consent to a change in the service requirement.

The operational uncertainty associated with the KCL and UCLH projects is considerable as they both have contracts of over 25 years in duration. The permutations of the possible events that may occur in that space of time are almost infinite. UCLH has the right to introduce new medical equipment at any point during the contract, and is responsible for the direct costs of that equipment. However, if there are indirect consequential costs of adapting the building or extra consequential operational costs, the Trust and the ProjectCo will have to agree the amount that will be paid for this. Thus, it was impossible for the contracting actors to draw up complete contracts to protect them under all possible future conditions. In traditionally procured projects, the uncertainty associated with construction and operational service contracts is considerably smaller as such contracts are considerably shorter in duration. Thus, they are able to be more complete as the near future is easier to predict.

Monitoring quality

Grout (1997) has suggested that PFI contracts are easier to write and monitor and are more complete than construction contracts. For Grout (1997 pp. 63–64), this is because the clients procure a service using an output-based specification instead of an input-based specification:

'In practice, building contracts are notoriously problematic and there are large information problems . . . resolutions of uncertainty, unexpected events . . . all lead to disputes as to whether this is a fresh specification, hence falling on the purchaser, or a response to a reasonably foreseeable event that the contractor should have built into the price . . . In addition, monitoring can be difficult, so there are incentives for the builders to cut on quality where the consequences are not immediately obvious. The result is that build contracts need to include service specifications to ensure that they have been properly conducted. The argument for PFI is that many of these problems disappear if the contract is written over the flow of services rather than the build process.'

However, because of limitations to the type of service contract that can be written:

'PFI projects may still include monitoring of the build process.'

And:

> *'Probably the biggest problem of the service-based approach is that the service needs specification in advance, which may not always be optimal. For example, it is difficult to get a good idea of what reasonable standards of service quality will be like in 20 or 30 years' time.'*

And in conclusion:

> *'[W]here build contracts are easy to specify but service contracts are not, then . . . the incentives for it [the asset] to remain in the private sector are less obvious. At the other extreme, where service contracts are easy to write and build contracts are difficult, the PFI approach may be particularly sensible.'*

Grout tends to assume that the specification in PFI procurement is normally less complex than in traditional procurement, thus reducing the writing and monitoring costs for the party bearing the risk of fitness-for-purpose, inherent in the project. In the KCL and UCLH projects, the clients specified, for example, the condition requirements for various areas of the facilities instead of the technical solutions to be used. It may indeed be easier to make and to monitor a definition of requirements than to specify and monitor the quality of a technical solution. However, the relative inexperience of all parties with output-specification may work against this when PFI is still novel. Moreover, attempts to foresee what may be required over a long period, and to write such contingent provision into the output specification, will greatly increase the complexity and cost of writing and monitoring.

Insofar as an output-based specification does indeed reduce the cost of monitoring a category of transaction costs, it will be because it involves a smaller and more 'visible' number of attributes being measured and evaluated. The contract in the KCL project was one of the very first contracts drawn up for an accommodation service PFI project. As a result, it was somewhat ambiguous. The contract clearly outlined the accumulation of penalty points for underperformance and how they would lead to deductions in the service payment. For example, the contract specified the temperature tolerances for each of the areas of the building. However, it did not specify the proportion of those areas that would have to be outside the tolerances before the ProjectCo incurs a penalty point. The question became: does a temperature irregularity in a single office trigger a penalty when the total floor area of office space is several thousand square metres? The client and the ProjectCo were able to agree a solution without having to trigger the dispute resolution procedure. The fact that this was the case can be seen as an illustration of the willingness of the transactors to work together to achieve mutually beneficial solutions.

KCL mistakenly specified the opening hours and thus the heating and ventilation requirement for part of the facility, New Hunt's House, to only include standard office hours. This resulted in a number of complaints from academic staff, as they tend to work during the evenings and weekends. The

ProjectCo understood the situation and agreed to heat the building outside the hours specified in the contract. This was not a cost issue directly as the client was responsible for the cost of energy consumption in the building. However, the ProjectCo will need to utilise the heating and ventilation plant more than it had anticipated, in order to maintain acceptable environmental conditions. This will require the ProjectCo to maintain the plant more frequently than it had anticipated and, eventually, replace it earlier than it had predicted. However, the ProjectCo was willing to do this as it appreciated that KCL had made an unintentional mistake and by doing so it could greatly enhance its relationship with the client. The additional costs that the ProjectCo will incur as a result can be seen as a transaction-specific investment into its relationship with the client.

New Hunt's House contains some naturally ventilated offices on the periphery of the building. On a limited number of midsummer days the ProjectCo had been unable to keep the temperature in that office space within the agreed tolerances. KCL has noted this, but chose not to impose penalty points for underperformance. Instead, it allowed the ProjectCo some time to look into the technical solution that it is using in order to prevent the problem from occurring in the future. In this illustration the potential gains for the client of not having to pay the full service payment for the naturally ventilated offices would have been smaller than the cost of the damage done to its relationship with the ProjectCo from imposing a payment deduction for the inconvenience it had been caused. Not imposing the deduction in the service payment can be seen as a transaction-specific investment by the client into its relationship with the ProjectCo.

In traditionally procured projects, the cost of monitoring for the client can be considerably greater than in PFI projects. This is because what is being monitored in traditional projects is compliance with a detailed construction specification and (usually) with an input-based operating specification. In PFI projects the compliance is with an output-based service specification. Thus, the complexity of the object of monitoring is greater in traditional projects. However, it must be noted that an output-based specification for operating services can also be used in traditional procurement. Moreover, it is possible that for some services, such as cleaning, it is easier to monitor input- than output-based specification (Ive, 2004).

Both KCL and UCLH projects have dispute resolution procedures for when disagreements in the operational phase of the project arise. The costs associated with running these systems can be seen as enforcement costs, which are a category of transaction costs. The dispute resolution procedures allow the client and the ProjectCos to continue working together without having to resort to litigation that might be detrimental to their relationships. It must be noted that in the KCL project, which is the one of the case studies that is already in its operational phase, the dispute resolution procedure had not been activated by April 2002, several years after commencement of service delivery. This can be credited to the good relationships of the client and the ProjectCo.

In traditional procurement similar dispute resolutions procedures have become increasingly common (Alternative Dispute Resolution approaches).

However, it must be noted that both the complexity of the object of the dispute and the number of parties to whom blame might be attributed contributes to the complexity of the dispute and, thus, to the contract enforcement costs.

Conclusions

The introduction of the PFI is said to have improved the relationships of the public and private sector actors. NAO (2001) found that 72% of clients and 80% of ProjectCo actors in PFI projects perceive their relationship to be either good or very good. Similar evidence on traditionally procured projects is not widely available. However, experience would suggest that satisfaction for client and ProjectCo in their relationships with each other may frequently be low under traditional procurement, although the chapter has not reviewed the evidence for this.

Good working relationships between the clients and ProjectCos were evident in both KCL and UCLH projects. The relationships can be best described as partnerships. The public and private sector actors were aware that they were about to enter or had entered into mutually beneficial long-term contractual relationships. In both case study projects, this was evident in the behaviour of the contracting actors. This experience may suggest that the satisfaction levels are generally higher under PFI procurement (see also Chapter 4).

The analysis of accommodation service PFI projects presented in this chapter reveals that good relationships develop because they make economic sense to the contracting actors. Large transaction-specific investments by each party, because they are in some degree balanced or symmetrical, seem so far to have tended, rather than encouraging opportunistic renegotiation, to encourage each party to make considerable efforts to maintain the transaction in being. They avoid actions that may lead to termination and take actions to maintain a good 'atmosphere' to the transaction and, thus, enable them to economise on transaction costs. However, somewhat surprisingly, it emerged that it was not the scope to reduce 'administrative' transaction costs, but the desire of the contracting actors to protect their future benefits from the project, that drove them to foster their relationships. The reason for such relationships not developing in traditional procurement is evident in that the transaction duration is short, except where strategic partnering is categorised as 'traditional'. Hence, benefits are retained under PFI procurement, which are partially or totally lost by the breakdown of relationships under traditional procurement.

A PFI project has a long contract period, which in effect increases transaction duration. This has a positive effect, creating incentives for the private sector actors to build relationships. It is not easily conceivable on a traditional short-term contract basis how contractors would make equivalent transaction-specific investments into their relationship with the client. This is the contribution that PFI makes to relationship development within the range of procurement options available.

References

Chang, C-Y. (2002) *Economic interpretation of construction procurement behaviour for commercial and industrial buildings*. PhD thesis, UCL, London.

Chang, C-Y. & Ive, G. (2000) Comparison of two ways of applying a transaction cost approach: the case of construction procurement routes. *Bartlett Research Paper*, **13**, UCL, London.

Chang, C-Y. & Ive, G. (2004) New frontiers of construction organisation. *CIB 2004 World Building Congress*, Toronto, Canada.

Coase, R. (1937) The nature of the firm. *Economica*, **4**, November, 386–405.

Coase, R. (1960) The problem of social cost. *Journal of Law and Economics*, **3**, 1–44.

Douma, S. & Schreuder, H. (2002) *Economic Approaches to Organizations*. Pearson Education, Harlow.

Grout, P. (1997) The economics of the Private Finance Initiative. *Oxford Review of Economic Policy*, **13** (4), 53–66.

Hart, O. (1995) *Firms, Contracts and Financial Structure*. Clarendon Press, Oxford.

Ive, G. (2004) The Private Finance Initiative and the management of projects. In: P. W. G. Morris & J. K. Pinto (eds) *The Wiley Guide to Managing Projects*. Wiley, Englewood Cliffs, NJ.

Klein, B., Crawford, R. & Alchian, A. (1978) Vertical integration, appropriable rents and the competitive contracting process. *Journal of Law and Economics*, **21**, 297–326.

HM Treasury (2004) *PFI List of Signed Projects – April 2004*. HM Treasury, London.

HM Treasury Taskforce (1997) *Partnerships for Prosperity: The Private Finance Initiative*. Series 1 – Generic guidance. HM Treasury, London.

HM Treasury Taskforce (1999a) *A Step-by-Step Guide to the PFI Procurement Process*. HM Treasury, London.

HM Treasury Taskforce (1999b) *How to Appoint and Work with a Preferred Bidder*. HM Treasury, London.

Milgrom, P. & Roberts, J. (1992) *Economics, Organization and Management*. Prentice Hall, Englewood Cliffs, NJ.

National Audit Office (2001) *Managing the Relationship to Secure a Successful Partnership in PFI Projects*. HM Stationery Office, London.

Rintala, K. (2004) The economic efficiency of accommodation service PFI projects. *VTT Publications*, **555**, VTT, Espoo, Finland.

Williamson, O. (1975) *Markets and Hierarchies: Analysis and Antitrust Implications*. The Free Press, New York.

Williamson, O. (1979) Transaction cost economics: the governance of contractual relations. *Journal of Law and Economics*, **22**, 233–61.

Williamson, O. (1985) *The Economic Institutions of Capitalism: Firms, Markets and Relational Contracting*. The Free Press, London.

Williamson, O. (1996) *The Mechanisms of Governance*. Oxford University Press, New York.

Section V

Conclusion

13 A relationship approach to the management of complex projects

Stephen Pryke and Hedley Smyth

We want to draw out the implications of this book for the future of industry and academic practice. We will consider:

- Where did we start from when setting out to put together this collection?
- What have we learnt?
- What does this mean for industry and academic endeavours?
- What are the next steps?

Where did we start?

There has been some neglect of the human and social dimension of managing projects. People and the relationships between them have often hidden behind more technically orientated ideas, concepts, activities and tasks. There was a need to address and bring together some of those that have recognised this need.

In the **Introduction** we outlined three main approaches to managing projects, the relationship approach being a fourth:

(1) Traditional project management approach
(2) Functional management approach
(3) Information processing approach
(4) Relationship approach

It was stated that the primary *aim* was to provide frameworks for understanding relationships, exploring a range of theoretical approaches from a number of disciplines. The *objectives* of the book were to set out the parameters of a relationship approach, conceptually and practically in a framework, and then to develop detailed analysis of the parameters through the contributions from other authors. The 'framework' was not therefore conceived to provide a unified body of theory. However, it was in mind to develop a framework of reference for a discourse to be developed and continue, specifically covering:

(1) Interpersonal and inter-firm relationships at the project interface and hence attendant networks of relationships
(2) Different concepts for understanding and for the inception, development and management of relationships
(3) How relations form and develop to aid project delivery for contractors and project success for clients
(4) Some of the key issues that require development, theoretically, through applied research and in practice

What have we learnt?

The **Introduction** and **Chapter 1** provide the context and scope for developing a relationship approach. While there may be theoretical and conceptual reference points for development in the future, the primary aim was to provide a point of departure for a discourse rather than necessarily provide a definitive or a tentative way forward. However, we believe there are some learning points that arise from the overview, namely, that a more integrated approach is needed to place relationships in a more central position and stimulate understanding of their contribution to adding service value in projects; understanding how relationships and other technical and management tools interface and interact; and developing a more integrated approach to how we know about managing projects and the methods we use for understanding them in both research and practice.

Morris in **Chapter 2** envisaged the networks of human interaction in the project environment being used to import knowledge from a wider context beyond the immediate confines of the project itself. He emphasised the importance of creating and managing knowledge within the project and identified the problems associated with transferring the project knowledge into the enterprise. Kumaraswamy and Rahman in **Chapter 7** were also concerned about sources of knowledge and dissemination; their focus was upon the role of the team within the project and the importance of personality profiles in the context of the relational contracting model. Loosemore's **Chapter 8** contrasts the homeostatic, or traditional, perception of risk with what he refers to as an irrational viewpoint. If we accept the importance of irrational fears relating to projects and their outcomes, then the management of such irrational fears becomes of fundamental importance in achieving project success, particularly where projects involve particularly high levels of risk or are contentious in some way. The case study dealt with the manner in which stakeholders must be engaged in order to deal with hazard mitigation.

In terms of understanding, developing and managing relationships, a number of contributors to this book had widely diverging views about how these activities occur and what might be achieved through such understanding and management. Smyth in **Chapter 4** outlined the stages in the development of trust and the importance of trust as a form of human capital

– moving through and away from self-interested trust in pursuit of less adversarial and opportunistic behaviour in project teams. Cova and Salle in **Chapter 5** discussed the ritualisation of relationships in the *milieu* – the creation of friendship and mutual debt as a means of maintaining relationships through which a project can be carried out. The French case study related by Cova and Salle highlights the importance of identifying and effectively managing project stakeholders, a theme that is developed by Wilkinson in **Chapter 6** regarding service community. Finally, in terms of conceptualisation of relationships, Ive and Rintala offer in **Chapter 12** the transaction cost economists' view of relationships in major projects. They investigate transaction specific investments in relationships within two private finance initiative projects and explain how mutually beneficial relationships may evolve in the major project environment.

Wilkinson looks at client handling models for continuity of service. **Chapter 6** wrestles with the age-old problems of fragmentation and complexity in projects and draws upon the account handler or key account manager model to propose a solution to the problem of managing client relationships within and through project teams. Druskat and Druskat in **Chapter 3** bring together an awareness of the social sciences and construction industry experience through their discussion of emotional intelligence. They describe how self-awareness, social awareness, relationship management and self-management – four emotional intelligence competencies – help project teams to deal with the unexpected challenges of the temporary, unique, culturally diverse and complex, interdisciplinary project environment. Individuals with high levels of emotional intelligence exhibit optimism, initiative and organisational awareness and contribute very positively to project success.

Green identifies some key issues for further development through applied research and in practice, but from a quite different position. In **Chapter 10** he dismisses supply chain management as the most recent in a long line of management fashions – products designed by consultants to sell their services in apparently new ways and to new markets. He suggests that the professional service providers have a relatively small range of talents and expertise and rely upon standard *industry recipes* (Spender, 1989) rather than an innovation-based approach to service provision. Cox and Ireland, **Chapter 11**, have a more positive use for the concept of supply chain management but are, however, more critical of the extent to which relationships can make substantive differences against market forces generally and power positions between parties exchanging goods and services. Thus, they deal with the tensions and conflict in buyer–supplier relationships. They highlight the importance of leverage within the supply chain and explain how this is a position that relatively few exploit for continuous improvement, especially those that are small or relatively inexperienced. The issue of leverage is also dealt with by Pryke in **Chapter 9** on social network analysis in project coalitions. Pryke's chapter draws upon an ongoing programme of international research utilising social network analysis in the study of relationships in construction projects. The study of information exchanges, contractual relationships and financial incentives, all within the framework of social network analysis, enables the governance of projects to be explored in a structured and quantitative manner.

Taking the body of this work together, there are a number of strands that can be drawn out. There is the issue that relationships are established before a project is won and commenced upon site and relationships have the capacity to be enduring and sustained beyond the life cycle of project tasks *per se*. Managing such relationships, therefore, is a valid focus that is both conceptually distinct from projects, yet is also central to successful projects at a practical level. The book has focused upon aspects of this, including the forming of relationships as a means to form projects (see Cova & Salle in **Chapter 5**), the configuration and continuity of relationships throughout the project life cycle (see Pryke in **Chapter 9** and Wilkinson in **Chapter 6**), the quality of those relationships (see especially Druskat & Druskat in **Chapter 3**; Smyth in **Chapter 4**, although other chapters also feed into this aspect). The book has also drawn attention to the need to manage relationships to fulfil project tasks and achieve particular outcomes, for example learning to improve performance (see Morris in **Chapter 2**), stakeholder management (see Loosemore in **Chapter 8**). The importance of relationships in managing the market has also been addressed, for example the power of market forces upon relationships (see Cox & Ireland in **Chapter 11**), in managing PFI procurement (see Ive & Rintala in **Chapter 12**), and across company boundaries where teams are working together (see Kumaraswamy & Rahman in **Chapter 7**), and supply chains are in operation (Green in **Chapter 10**).

We are beginning to improve learning in these areas and across a range of areas concerned with relationships. Therefore this book has made a contribution towards developing greater understanding of the role of relationships in managing projects. Our learning must take two forms. There is learning concerning *what is*, that is, to say describing and analysing the state of play in managing projects. Then there is also learning considering what *ought to be*, that is, normative thinking. The range of theories and concepts drawn upon here, and elsewhere, approaches including the framework provided in this book, are not simply tools for understanding, but are also tools for practice. There is learning on managing projects in order to improve project performance.

What does this mean?

What are the implications of the learning from this book? The book has tried to recognise that social processes and arrangements are both complex and yet important to *understanding* projects (Bresnen *et al.*, 2005a; Morris, 1994). In order to understand projects in greater depth there has been an increasing focus upon the social sciences; however, this has generally taken the form of picking and mixing aspects of the social sciences in ways that seem to fit. In other words, social science tools have become additions to the 'kit bag' (Green

& May, 2005; Bresnen *et al.*, 2005a), rather than another 'kit bag', that provide explanations to issues. Research into managing projects has primarily seen the 'project' as the focus in the sense of it being a series of technical and managerial tasks to be completed. The human dimension has frequently been absent and thus the social nature of both technical work and management is relegated to a secondary position.

Effort has already begun to correct this. For example, Green (1996) has addressed relations of power and organisational politics at the briefing stage, drawing upon the work of Morgan (1986), culture and leadership have been addressed by Liu *et al.* (2003) and Phua and Rowlinson (2004), social networks by Pryke (2001, 2005, 2006, see also **Chapter 9**) and trust by Edkins and Smyth (2006) and Smyth (2005; see also **Chapter 4**). All these examples provide deeper understanding as to what is happening either generally or specifically in complex settings. However, they tend towards two characteristics. First, most of those cited and many other contributions from the social science arena tend to be stronger on describing and analysing what is happening and thus weaker on what could, might or should happen in future. Second, the disparate nature of methodologies used provides difficulties in addressing where the connections are between different studies, and between the studies and the realm of application, and hence practice (Smyth *et al.*, 2006). This hampers progress in understanding projects and therefore in turn hampers developing practice improvements.

As we were coming to the end of the editing of this book we were encouraged by the special issue of *Building Research & Information* covering managing projects as complex social settings (Bresnen *et al.*, 2005a), which embraces a range of important issues: Cicmil and Marshall (2005) utilise interpretative approaches in the analysis of collaboration in procurement processes, Koch and Bendixen (2005) and Rooke and Clark (2005) all utilise ethnographic approaches, while Bresnen mobilises a sociological approach using structure-agency notions drawn from Giddens' (1984, 1990) structuration theory (Bresnan *et al.*, 2004, 2005b). The problem with such approaches is that they tend to *explain* where the tensions, conflicts and contradictions are located, and hence promote *understanding*, yet the studies have less to say on what ought to happen. In other words, practitioner guidance tends to be of the type that points out what *causes* the tensions, conflicts and contradictions, and thus what pitfalls to avoid where they are avoidable, rather than saying that as a consequence of the analysis the following norms, behaviours, actions, systems and controls should be put into practice (cf. Smyth *et al.*, 2006).

We agree with those that contend that there is no theory of projects *per se*, and thus seeking to develop laws or generalisations that are always applicable is fruitless because management and project management provide two important contexts (cf. Sayer, 1992). Each or every tendency will depend upon context. The project context is inherently unstable. Winch (2002) has clearly drawn attention to issues concerning uncertainty, lack of information and risk. However, we do not agree with those that go on to say that because projects are inherently unstable trying to create stability is a misguided objective (see for example Cicmil & Marshall 2005). It may be difficult to improve matters, but complexity and instability are challenges, not total barriers, as Winch (2002) clearly addresses.

Academic endeavour does not always help. There is a lack of definition concerning many key terms – for example, the definition of the project was raised in **Chapter 1**, supply chain management definitions have been contested (see Green in **Chapter 10**), therefore the aims and objectives of putting such notions and concepts to work is problematic. It can be argued that different definitions and applications are sources of service differentiation and competitive advantage in practice; however, some applications are clearly better that others both in terms of project success and in terms of business ethics (cf. Green in **Chapter 10**). There is also a lack of understanding as to how different management and project management practices and technological applications interface and how these can be managed effectively. While this concerns systems, procedures and other protocols, human relationships are central, even though there have been tendencies to conflate IT and intranet solutions with knowledge management, or IT and telecommunication solutions with customer relationship management, or to conflate project programming with continuity of service, and hence to ignore the human dimensions.

What are the next steps?

Above all else, we want to see a discourse on developing the management of projects in a way that recognises and addresses relationships. We wish to see that conducted as a research task of description and analysis, and a research task of normative concepts and applications. Therefore we recommend:

(1) The academic community needs to place social science approaches on an equal footing to the natural sciences in research and teaching, and this means placing relationships at the heart of the arena, rather than subsuming them under issues that are technical tasks or that have been artificially isolated.
(2) The project community needs to recognise the need to manage relationships, not just tasks or individuals undertaking tasks.
(3) The research community should use methodologies that recognise context and ensure that research is articulated in contextual terms in order for academics and practitioners to make connections between other areas of research and areas of technical and management practice.
(4) The academic community needs to more deliberately acknowledge relationships as the source of social capital, and hence improved performance on projects.
(5) There needs to be greater integration of traditional project management with the functional management approach, and the information processing approach with the relationship approach.

(6) Further research is needed on:
 (a) interpersonal and inter-firm relationships at the project interface and networks
 (b) a deeper level of understanding and integration of the concepts for understanding and for the inception, development and management of relationships on projects
 (c) how relations form and develop to aid project delivery for contractors and project success for clients specifically requires further attention.

A general recognition in practice is also needed, and one function of the academic community it to facilitate that process. Drawing upon the learning from the book, there are some current lessons for practice too:

(7) The industry bodies of knowledge need to take on board the broader spectrum, and thus the currently competing bodies will converge.
(8) There needs to be greater recognition of the importance of managing in a holistic way the relationships from courtship onwards, that is, beyond the confines of the project life cycle *per se*, thus the role of the forming of relationships as a means to deliver projects, the configuration and continuity of relationships throughout the project life cycle, and the quality of those relationships.
(9) The need to manage relationships to fulfil functional project tasks and achieve particular outcomes, in other words, the relationship–function and relationship–technology interfaces are important in addition to the person–person relationship.
(10) The importance of relationships in managing the market also needs addressing, particularly issues concerning market forces and the leverage of key actors, plus the development of social capital in market contexts.
(11) Individual organisations need to develop tactics and strategies for managing relationships internally and at the interface with other key stakeholders, particularly the client, that takes on board many of the issues raised in this book.
(12) Managing relationships is a vast area in its own right. However, organisations may wish to focus upon particular relationships in order to build strength in particular areas of service provision.

References

Bresnan, M., Goussevskaia, A. & Swan, J. (2004) Embedding new management knowledge in project-based organizations. *Organization Studies*, **25** (9), 1535–55.
Bresnan, M., Goussevskaia, A. & Swan, J. (2005a) Managing projects as complex social settings. *Building Research & Information*, **33** (6), 487–93.

Bresnan, M., Goussevskaia, A. & Swan, J. (2005b) Implementing change in construction project organizations: exploring the interplay between structure and agency. *Building Research & Information*, **33** (6), 547–60.

Cicmil, S. & Marshall, D. (2005) Insights into collaboration at the project level: complexity, social interaction and procurement mechanisms. *Building Research & Information*, **33** (6), 523–35.

Edkins, A. J. & Smyth, H. J. (2006) Contractual management in PPP projects: evaluation of legal versus relational contracting for service delivery. Special issue on Legal Aspects of Relational Contracting. *ASCE Journal of Professional Issues in Engineering Education and Practice*, **132** (1).

Giddens, A. (1984) *The Constitution of Society: Outline of the Theory of Structuration*. Polity, Cambridge.

Giddens, A. (1990) *The Consequences of Modernity*. Stanford University Press, Stanford, CA.

Green, S. D. (1996) A metaphorical analysis of client organizations and the briefing process. *Construction Management and Economics*, **14**, 155–64.

Green, S. D. & May, S. C. (2005) Lean construction: areas of enactment, models of diffusion and the meaning of 'leanness'. *Building Research & Information*, **33** (6), 498–511.

Koch, C. & Bendixen, M. (2005) Multiple perspectives on organizing: projects between tyranny and perforation. *Building Research & Information*, **33** (6), 536–46.

Liu, A., Fellows, R. & Fang, Z. (2003) The power paradigm of project leadership. *Construction Management and Economics*, **21** (8), 819–29.

Morgan, G. (1986) *Images of Organization*. Sage, London.

Morris, P. W. G. (1994) *The Management of Projects*. Thomas Telford, London.

Phua, F. T. T. & Rowlinson, S. (2004) The antecedents of co-operative behaviour among project team members: an alternative perspective on an old issue. *Construction Management and Economics*, **22** (9), 913–25.

Pryke, S. D. (2001) *UK construction in transition: developing a social network approach to the evaluation of new procurement and management strategies*. PhD thesis, The Bartlett School, University College London.

Pryke, S. D. (2005) Analytical methods in construction procurement and management: a critical review. *Journal of Construction Procurement*, **10** (1), 49–67.

Pryke, S. D. (2006) Legal issues associated with emergent actor roles in innovative UK procurement: a prime contractor case study. *ASCE Journal of Professional Issues in Engineering Education and Practice*, **132** (1), 67–76.

Rooke, J. & Clark, L. (2005) Learning, knowledge and authority on site: a case of safety practice. *Building Research & Information*, **33** (6), 561–70.

Sayer, R. A. (1992) *Method in Social Science: A Realist Approach* (2nd edn). Routledge, London.

Smyth, H. J. (2005) Trust in the design team. *Architectural Engineering and Design Management*, **1**, 193–205.

Smyth, H. J., Morris, P. W. G. & Cooke-Davies, T. (2006) Understanding project management: philosophical and methodological issues. *Proceedings of EURAM 2006*, May 17–20, Oslo.

Spender, J-C. (1989) *Industry Recipes: An Enquiry into the Nature and Sources of Managerial Judgement*. Basil Blackwell, Oxford.

Winch, G. M. (2002) *Managing the Construction Project*. Blackwell, Oxford.

Index